Recent Titles in This Series

158 **A. K. Kelmans, Editor,** Selected Topics in Discrete Mathematics: Proceedings of the Moscow Discrete Mathematics Seminar, 1972–1990
157 **M. Sh. Birman, Editor,** Wave Propagation. Scattering Theory
156 **V. N. Gerasimov, N. G. Nesterenko, and A. I. Valitskas,** Three Papers on Algebras and Their Representations
155 **O. A. Ladyzhenskaya and A. M. Vershik, Editors,** Proceedings of the St. Petersburg Mathematical Society, Volume I
154 **V. A. Artamonov, et al.,** Selected Papers in K-Theory
153 **S. G. Gindikin, Editor,** Singularity Theory and Some Problems of Functional Analysis
152 **H. Draškovičová, et al.,** Ordered Sets and Lattices II
151 **I. A. Aleksandrov, L. A. Bokut', and Yu. G. Reshetnyak, Editors,** Second Siberian Winter School "Algebra and Analysis"
150 **S. G. Gindikin, Editor,** Spectral Theory of Operators
149 **V. S. Afraĭmovich, et al.,** Thirteen Papers in Algebra, Functional Analysis, Topology, and Probability, Translated from the Russian
148 **A. D. Aleksandrov, O. V. Belegradek, L. A. Bokut', and Yu. L. Ershov, Editors,** First Siberian Winter School in Algebra and Analysis
147 **I. G. Bashmakova, et al.,** Nine Papers from the International Congress of Mathematicians 1986
146 **L. A. Aĭzenberg, et al.,** Fifteen Papers in Complex Analysis
145 **S. G. Dalalyan, et al.,** Eight Papers Translated from the Russian
144 **S. D. Berman, et al.,** Thirteen Papers Translated from the Russian
143 **V. A. Belonogov, et al.,** Eight Papers Translated from the Russian
142 **M. B. Abalovich, et al.,** Ten Papers Translated from the Russian
141 **Kh. Drashkovicheva, et al.,** Ordered Sets and Lattices
140 **V. I. Bernik, et al.,** Eleven Papers Translated from the Russian
139 **A. Ya. Aĭzenshtat, et al.,** Nineteen Papers on Algebraic Semigroups
138 **I. V. Kovalishina and V. P. Potapov,** Seven Papers Translated from the Russian
137 **V. I. Arnol'd, et al.,** Fourteen Papers Translated from the Russian
136 **L. A. Aksent'ev, et al.,** Fourteen Papers Translated from the Russian
135 **S. N. Artemov, et al.,** Six Papers in Logic
134 **A. Ya. Aĭzenshtat, et al.,** Fourteen Papers Translated from the Russian
133 **R. R. Suncheleev, et al.,** Thirteen Papers in Analysis
132 **I. G. Dmitriev, et al.,** Thirteen Papers in Algebra
131 **V. A. Zmorovich, et al.,** Ten Papers in Analysis
130 **M. M. Lavrent'ev, et al.,** One-dimensional Inverse Problems of Mathematical Physics
129 **S. Ya. Khavinson; translated by D. Khavinson,** Two Papers on Extremal Problems in Complex Analysis
128 **I. K. Zhuk, et al.,** Thirteen Papers in Algebra and Number Theory
127 **P. L. Shabalin, et al.,** Eleven Papers in Analysis
126 **S. A. Akhmedov, et al.,** Eleven Papers on Differential Equations
125 **D. V. Anosov, et al.,** Seven Papers in Applied Mathematics
124 **B. P. Allakhverdiev, et al.,** Fifteen Papers on Functional Analysis
123 **V. G. Maz'ya, et al.,** Elliptic Boundary Value Problems
122 **N. U. Arakelyan, et al.,** Ten Papers on Complex Analysis
121 **D. L. Johnson,** The Kourovka Notebook: Unsolved Problems in Group Theory
120 **M. G. Kreĭn and V. A. Jakubovič,** Four Papers on Ordinary Differential Equations

(Continued in the back of this publication)

Selected Topics in Discrete Mathematics

American Mathematical Society

TRANSLATIONS

Series 2 • Volume 158

Selected Topics in Discrete Mathematics

Proceedings of the Moscow
Discrete Mathematics Seminar
1972–1990

A. K. Kelmans
Editor

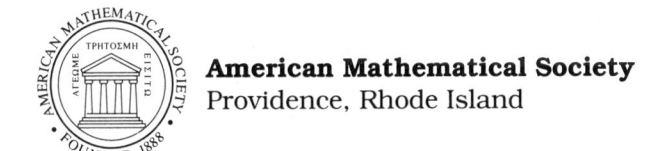

American Mathematical Society
Providence, Rhode Island

Translated by A. D. VAĬNSHTEĬN
Translation edited by SIMEON IVANOV

1991 *Mathematics Subject Classification.* Primary 68R10, 90B10, 05C38, 05B40, 68R05; Secondary 05C10, 05B35, 68P10.

Library of Congress Cataloging-in-Publication Data
Moscow Discrete Mathematics Seminar.
  Selected topics in discrete mathematics: proceedings of the Moscow Discrete Mathematics Seminar, 1972–1990/A. K. Kelmans, editor.
    p. cm. — (American Mathematical Society translations; ser. 2, v. 158)
  "A collection of translations of. . .papers"—Pref.
  Includes bibliographical references.
  ISBN 0-8218-7509-4 (acid-free)
  1. Mathematics—Congresses. 2. Computer science—Mathematics—Congresses. I. Kelmans, A. K. (Alexander K.) II. Title. III. Series.
QA3.A572 ser. 2, vol. 158
[QA1]
510 s—dc20   93-48534
[511'.5]   CIP

Copying and reprinting. Individual readers of this publication, and nonprofit libraries acting for them, are permitted to make fair use of the material, such as to copy an article for use in teaching or research. Permission is granted to quote brief passages from this publication in reviews, provided the customary acknowledgment of the source is given.

Republication, systematic copying, or multiple reproduction of any material in this publication (including abstracts) is permitted only under license from the American Mathematical Society. Requests for such permission should be addressed to the Manager of Editorial Services, American Mathematical Society, P.O. Box 6248, Providence, Rhode Island 02940-6248. Requests can also be made by e-mail to reprint-permission@math.ams.org.

The owner consents to copying beyond that permitted by Sections 107 or 108 of the U.S. Copyright Law, provided that a fee of $1.00 plus $.25 per page for each copy be paid directly to the Copyright Clearance Center, Inc., 222 Rosewood Drive, Danvers, Massachusetts 01923. When paying this fee please use the code 0065-9290/94 to refer to this publication. This consent does not extend to other kinds of copying, such as copying for general distribution, for advertising or promotional purposes, for creating new collective works, or for resale.

© Copyright 1994 by the American Mathematical Society. All rights reserved.
Translation authorized by the
All-Union Agency for Author's Rights, Moscow.
Printed in the United States of America.
The American Mathematical Society retains all rights
except those granted to the United States Government.
∞ The paper used in this book is acid-free and falls within the guidelines
established to ensure permanence and durability.
♻ Printed on recycled paper.
This publication was typeset using $\mathcal{A}_{\mathcal{M}}\mathcal{S}$-TEX,
the American Mathematical Society's TEX macro system.
10 9 8 7 6 5 4 3 2 1   98 97 96 95 94

Contents

Preface	xi
L. G. Babat, Approximate evaluation of a linear function at the vertices of the unit n-dimensional cube	1
L. G. Babat, On the growth of coefficients in an integral linear aggregation	11
B. V. Cherkasskiĭ, A fast algorithm for constructing a maximum flow through a network	23
V. P. Grishukhin, On the extremality of the rank function of a connected semimodular lattice	31
V. P. Grishukhin, On polynomial solvability conditions for the simplest plant location problem	37
A. V. Karzanov, Minimal mean weight cuts and cycles in directed graphs	47
A. V. Karzanov, An algorithm for determining a maximum packing of odd-terminus cuts, and its applications	57
A. V. Karzanov, Maximum- and minimum-cost multicommodity flow problems having unbounded fractionality	71
A. V. Karzanov, On a class of maximum multicommodity flow problems with integer optimal solutions	81
A. K. Kelmans, On edge mappings of graphs preserving subgraphs of a given type	101
A. K. Kelmans, On edge semi-isomorphisms of graphs induced by their isomorphisms	113
A. K. Kelmans, Constructions of cubic bipartite 3-connected graphs without Hamiltonian cycles	127
A. K. Kelmans, Nonseparating circuits and the planarity of graph-cells	141
A. K. Kelmans and V. P. Polesskiĭ, Extremal sets and covering and packing problems in matroids	149
E. V. Kendys, V. M. Makarov, A. R. Rubinov, and E. M. Tishkin, Optimal distribution sorting	175

P. A. Pevzner, Branching packing in weighted graphs 185
P. A. Pevzner, Non-3-crossing families and multicommodity flows 201
A. D. Vaĭnshteĭn, The vector shortest path problem in the l_∞-norm 207
A. D. Vaĭnshteĭn, Lower performance bounds for on-line algorithms in the simple two-dimensional rectangle packing problems 217

Russian Contents*

Л. Г. Бабат, Приближенное вычисление линейной функции на вершинах единичного n-мерного куба, «Исследования по дискретной оптимизации», Москва, 1976, 156–169

Л. Г. Бабат, О росте коэффициентов при целочисленном линейном программировании, «Математические методы решения экономических задач», Сб. 8, Москва, 1979, 34–43

Б. В. Чершасский, Быстрый алгоритм построения максимального потока в сети, «Комбинаторные методы в потоковых задачах», Вып. 3, Москва, 1979, 90-96

В. П. Гришухин, Экстремальность ранговой функции связной полумодулярной решетки, «Задачи дискретной оптимизации и методы их решения», Москва, 1987, 218–223

В. П. Гришухин, Об условиях полиномиальной разрешимости простейшей задачи размещения, «Экономико-математическое моделирование и анализ дискретных систем», Москва, 1988, 52–65

А. В. Карзанов, О минимальных по среднему весу разрезах и циклах ориентированного графа, «Качественные и приближенные методы исследования операторных уравнений», Ярославль, 1985, 72–83

А. В. Карзанов, Алгоритм максимальной упаковки нечетнополюсных разрезов и его приложения, «Исследования по прикладной теории графов», Новосибирск, 1986, 126–140

А. В. Карзанов, Максимальные и минимальные стоимостные многопродуктовые потоки неограниченной дробности, «Задачи дискретной оптимизации и методы их решения», Москва, 1987, 123–135

---

*The American Mathematical Society scheme for transliteration of Cyrillic may be found at the end of index issues of *Mathematical Reviews*.

А. В. Карзанов, Один класс задач о максимальных многопродуктовых потоках с целочисленными оптимальными решениями, «Моделирование и оптимизация систем сложной структуры», Омск, 1987, 103–121

А. К. Кельманс, Об отображениях ребер графов, сохраняющих подграфы заданного вида, «Модели и алгоритмы исследования операций и их применения к организаци работы в вычислительных системах», Ярославль, 1984, 19–30

А. К. Кельманс, О реберных полуизоморфизмах графов, индуцируемых их изоморфизмами, «Модели исследования операций в вычислительных системах», Ярославль, 1985, 80–95

А. К. Кельманс, Конструкции кубических двудольных 3-связых графов без гамильтоновых циклов, Сб. Трудов Всесоюз. Научно-Исслед. Инст. Систем Исслед. 1986, No. 10, 64–72

А. К. Кельманс, Неразделяющие циклы и планарность графов-клеток, «Задачи дискретной оптимизации и методы их решения», Москва, 1987, 224–232

А. К. Кельманс и В. П. Полесский, Экстремальные множества и задачи покрытия и упаковки в матроидак, «Исследования по прикладной теории графов», Новосибирск, 1986, 140–168

Е. М. Тишкин, В. М. Макаров, А. Р. Рубинов и Е. В. Кендыс, Математические основы метода комбинаторной сортировки вагонов, Вестник Всесоюз. Научно-Исслед. Инст. Железнород. Транспорта, 1989, No. 8, 1–8

П. А. Певзнер, Эффективный алгоритм упаковки ветвлении во взвешенном графе, «Комбинаторные методы в потоковых задачах», Вып. 3, Москва, 1979, 113–127

П. А. Певзнер, Линейность мощности 3-незацепленных семейств множеств, «Задачи дискретной оптимизации и методы их решения», Москва, 1987, 136–142

А. Д. Вайнштейн, Векторная задача о кратчайшем пути в равномерной норме, Эконом. и Мат. Методы, Том 21 (1985), 1132–1137

А. Д. Вайнштейн, Нижние оценки для задачи упаковки прямоугольников в полосу в режиме реального времени, «Теория и методы автоматизации проектирования сложных систем и автоматизации научного исследования», Москва, 1985, 22–25

Preface

This volume is a collection of translations of some papers on discrete mathematics, published earlier in the USSR.

Many interesting results in discrete mathematics, found in the USSR, have remained unknown in the West for many years, some to the present day. And the language barrier was hardly the only reason for that. In the USSR, there were virtually no journals where one could publish results on discrete mathematics, and sending papers to Western journals was for many a practical impossibility. Thus, many interesting results have remained to this day either unpublished, or published (sometimes with great delay) in small, badly prepared collections, barely accessible not only in the West, but also in the USSR. The present collection should fill that communication gap to some extent. I believe that it will help not only Western, but also ex-Soviet mathematicians to get acquainted with such inaccessible publications.

Why are these papers put together in the same collection? Not only because they are united by a common theme and are dedicated to various aspects of discrete mathematics. The authors of these papers have been for many years members of a common seminar, the Moscow Seminar on Discrete Mathematics. The ideas and results presented in this collection have not only been debated in detail, but on occasion also been born in the process of stormy discussions in the seminar.

I was lucky enough (and it was a great privilege for me) to be the leader of that seminar for many years, from its inception in the spring of 1972 to the fall of 1990, and, I dare say, its story is well worth recounting. The sessions of the seminar were, as a rule, weekly and lasted about three hours with a small break. Sometimes additional more specialized subseminars for discussing special themes (for example, network multiflow, matroids, etc.) were organized, and then the results of these subseminars were brought to the main seminar. I never had a strict schedule of reports. The subject of a report could be changed at the last moment (some interesting mathematician would wander into town, or somebody has thought of something interesting or unexpected, such as a proof of a previously considered conjecture, or for some other reason). And that never hampered the work of the seminar, since everybody used to come to the seminar, not to a report. Everyone strived to take active participation in the discussion regardless of whether the subject was related to his current interests or not. There was no restrictions on the subject matter either. We were ready to consider any mathematical picture if we thought it was pretty and nontrivial and we were able to comprehend it: hence the wide range of themes.

Here are some of them:

Various topics in Graph Theory, Network Flow and Multicommodity Flow, Linear Programming and Combinatorial Optimization, Matroid Theory and Submodular Systems, Matrix Theory and Submodular System, Matrix Theory and Combinatorics, Parallel Computing, Algorithm Complexity, Random Graphs and Statistical Mechanics, Coding Theory, Algebraic Combinatorics and Group Theory.

The atmosphere at the seminar was always one of wild joviality, sparkling humor, creativity and benevolence (as always, when different talented people are united by a common interest and want to work together). And that made each session a moving event, a celebration. Maybe, that was the pledge of the longevity of this (I would say unique in those conditions) seminar. Every session was extremely productive and helpful to all of us. The "enormous" contribution of every participant, original and inimitable, could hardly be exaggerated. And I use the occasion to express my great gratitude to all of them.

We learned a lot, came to know a lot, to comprehend, to discover for overselves, to find ourselves in mathematics, thanks to that seminar. Our friendship extended to the family level; common evenings were organized devoted to various happenings and holidays; we entertained our friends who were to emigrate to the West (and whom we did not expect to see again); we spent our vacations together on summer marches (in the woods, on water, etc.). I would say that it was not merely a mathematical seminar; it was a friendly mathematical family, which played an important role in our lives. For many of us this was a "ray of light in a kingdom of darkness". Possibly the reader senses traces of nostalgia in my story. Yes, I would like to restore that seminar at least in part, or to organize something similar. Would I succeed in this whenever, wherever? I do not know.

Here is the list of mathematicians that took participation in the seminar at various times: Some of the regular participants were G. Adel'son-Vel'skiĭ (now at the Weizmann Institute, Israel), V. Grinberg (now at Carnegie Mellon University), E. Dinits (now at the Technion, Israel), B. Cherkasskiĭ, A. Karzanov, A. Kelmans (now at Rutgers University), A. Leontovich, M. Lomonosov (now at Ben-Gurion University, Israel), V. Grishukhin, the late B. Papernov, V. Polesskiĭ, P. Pevzner (now at Pennsylvania State University), A. Rubinov, A. Vainstein. Some of the other lecturers at the seminar were G. Margulis (now at Yale University), J. Bernstein (now at Harvard University), M. Burstein (now in the USA), M. Goldberg (now at Rensselaer Polytechnic Institute), L. Levin (now at Boston University), L. Vaserstein (now at Pennsylvania State University), B. Weisfeiler (later at the Institute of Advanced

Study, Princeton, and at Pennsylvania State University), L. Babai (Loránd Eötvös State University, Budapest), I. Bárány, G. Katona, J. Pach (all three at the Mathematical Institute of the Hungarian Academy of Sciences, Budapest), the late Yu. Burtin, L. Khachiyan (now at Rutgers University), A. Ivanov (now at the University of Michigan), M. Klin (now at Ben-Gurion University, Israel), S. Lavrenchenko, V. Liskovets, A. Razborov, S. Shlosman (now at the University of California, Irvine), L. Sholomov. This (incomplete) list shows that many of the participants are now abroad. "Some are no more, others are distant." And this is one of the reasons why the seminar practically ceased its existence after 1990.

Some Western mathematicians made efforts to help us keep at least a minimal contact with the Western mathematical world through the Iron Curtain. Great help in this respect was given to us by Hungarian mathematicians who had the occasion to spend some time in Moscow: L. Babai, I. Bárány, P. Gács (now at Boston University), G. Katona, A. Krámli, P. Major, J. Pach, and many others, and also P. Graham and P. Hammer (USA), B. Korte (Germany) and T. Walsh (Canada). Many thanks to all of them.

We are extremely grateful to Professor W. T. Tutte, who enthusiastically accepted my suggestion to give his seminar at the University of Waterloo a series of lectures on the results of our seminar, and also to T. Walsh, who took upon himself to deliver these lectures.

Finally, in the name of the entire seminar I wish to express our deep appreciation to the American Mathematical Society for their initiative and the organization of this collection of translations.

Deliberately, I say no word about the results in this collection. Let them speak for themselves.

ALEXANDER KELMANS
Rutgers University
New Brunswick, NJ

Approximate Evaluation of a Linear Function at the Vertices of the Unit n-dimensional Cube

L. G. BABAT

Introduction

When we search for approximate solutions of the knapsack problem, as a rule, we fix $\varepsilon > 0$ and consider equivalent values h_1 and h_2 of the objective functional h that satisfy the relation $|h_1 - h_2|/\min(|h_1|, |h_2|) \leqslant \varepsilon$. The following question arises: What is the minimal number of groups into which the values of the functional h at the vertices of the unit n-dimensional cube can be split, provided the values in each group are equivalent in the above sense? The appropriateness of posing this problem has become evident on finding out that the minimal number of such groups grows linearly with n; this fact allows one to construct an algorithm for calculating the maximal and minimal values of h in each group with the number of steps increasing as n^2, and to construct an algorithm for finding approximate solutions of the knapsack problem with the number of steps increasing as n^3.

We introduce some basic notation. Denote by Q the unit n-dimensional cube, and by Q^0 the set of its vertices. Denote by $Q(i_1, \ldots, i_k)$ the subset of the vertices of Q that have at least one nonzero coordinate in positions $i_1, \ldots, i_k$. We use Q_k to denote the set of vertices equal to

$$Q(1, \ldots, k) \setminus Q(k+1, \ldots, n),$$

and Q_{k+1}^* to denote $Q_{k+1} \setminus Q_k$. The vertices of the cube themselves are denoted by the sets of the indices of nonzero coordinates. By a δ-*comparison of numbers* a *and* b we mean the comparison of the number δ with the difference $a - b$. By a *relative* δ-*comparison of* a *and* b we mean the comparison of δ with the ratio a/b. The symbol log is used for binary logarithms.

1991 *Mathematics Subject Classification.* Primary 68R10.
Translation of Studies in Discrete Optimization (A. A. Fridman, editor), "Nauka", Moscow, 1976, pp. 156–169; MR **56** #4154.

1. Estimating the number of ε-groups

Let $a, b > 0$. We say that a, b belong to the same ε-group ($\varepsilon > 0$), if

$$\max(a, b)/\min(a, b) \leqslant 1 + \varepsilon,$$

i.e. $|a - b|/\min(a, b) \leqslant \varepsilon$.

Let h be a function defined at the vertices of the cube Q as follows: if $x_1, \ldots, x_n$ are the coordinates of a vertex $a \in Q^0$, then

$$h(a) = \sum_i c_i x_i.$$

LEMMA 1. *Let the coefficients $c_1, \ldots, c_n$ of the function h be positive and admit a partition into two disjoint sets, $M = \{c_{i_1}, \ldots, c_{i_k}\}$ and $N = \{c_{j_1}, \ldots, c_{j_t}\}$, such that $M \neq \varnothing$ and*

$$1 = c_{i_1} \leqslant \cdots \leqslant c_{i_k}, \tag{1}$$

$$c_{j_1} + \cdots + c_{j_t} \leqslant 1. \tag{2}$$

Then for any $\varepsilon > 0$ one can split the values of h at the vertices from $Q(i_1, \ldots, i_k)$ into ε-groups such that the number of the groups does not exceed $k + k/\log(1 + \varepsilon)$.

PROOF. Denote by R the number of the above ε-groups. According to (1) and (2), we have

$$0 \leqslant \log h(a) \leqslant \log(c_{i_1} + \cdots + c_{i_k} + 1)$$

for $a \in Q(i_1, \ldots, i_k)$. Moreover, the relation $h(a_1)/h(a_2) > 1 + \varepsilon$ is equivalent to the relation $\log h(a_1) - \log h(a_2) > \log(1 + \varepsilon)$, $a_1, a_2 \in Q^0$. Therefore,

$$R \leqslant \lceil \log(c_{i_1} + \cdots + c_{i_k} + 1)/\log(1 + \varepsilon) \rceil, \tag{3}$$

where $\lceil m \rceil$ denotes the minimal integer $\geqslant m$.

We prove the lemma by induction on the cardinality of M, denoted by μ. Inequality (3) and the condition $c_{i_1} = 1$ yield the proof for $\mu = 1$. Let us prove it for $\mu = k$, assuming that it holds for $\mu \leqslant k - 1$. If

$$c_{i_m} \leqslant 2^{m-1} \tag{4}$$

for $m = 1, \ldots, k$, then $\log(c_{i_1} + \cdots + c_{i_k} + 1) \leqslant \log 2^k = k$ and the assertion of the lemma follows from (3). Let c_{i_l} be the minimal number from M that does not satisfy (4). Then

1) if $a \in Q(i_1, \ldots, i_{l-1}) \setminus Q(i_1, \ldots, i_k)$, we have

$$0 \leqslant \log h(a) \leqslant \log(1 + 2 + \cdots + 2^{l-2} + 1) = l - 1;$$

2) if $a \in Q(i_l, i_{l+1}, \ldots, i_k)$, we have

$$(l - 1) \leqslant \log c_{i_l} \leqslant \log h(a),$$

and therefore
$$R \leqslant \lceil (l-1)/\log(1+\varepsilon) \rceil + s, \qquad (5)$$
where s is the number of ε-groups into which the values of h at the vertices from $Q(i_l, i_{l+1}, \ldots, i_k)$ are split. To estimate s, let us introduce the following notation:
$$c_1' = c_1/c_{i_l}, \ldots, c_n' = c_n/c_{i_l},$$
$$M' = \{c_{i_l}', c_{i_{l+1}}', \ldots, c_{i_k}'\}, \qquad N' = \{c_{j_1}', \ldots, c_{j_t}'\} \cup \{c_{i_1}', \ldots, c_{i_{l-1}}'\},$$
$$h' = \sum c_i' x_i.$$

Since $c_{i_l} \geqslant 2^{l-1}$, $c_{j_1} + \cdots + c_{j_t} \leqslant 1$, and $c_{i_m} \leqslant 2^{m-1}$ for $m < l$, the sets M' and N' satisfy all the conditions on the coefficients of h' implied by the lemma. By the relation $c_{i_l}' = 1 = 2^0$, the set M', consisting of numbers not less than 1, has cardinality $\leqslant k-1$. Comparing this with the fact that $h'(a_1)/h'(a_2) = h(a_1)/h(a_2)$ for all a_1, a_2, we infer, by the induction hypothesis, that
$$s \leqslant \lceil k - (l-1)/\log(1+\varepsilon) \rceil + \lceil k - (l-1) \rceil,$$
which, along with (5), proves the lemma.

THEOREM 1. *If $c_1, \ldots, c_n > 0$, then for any $\varepsilon > 0$ the values of h at the vertices from $Q(1, \ldots, n)$ can be split into ε-groups in such a way that the number of the groups does not exceed $n(1 + 1/\log(1+\varepsilon))$.*

PROOF. Without loss of generality, we can assume that $c_1 \leqslant c_2 \leqslant \cdots \leqslant c_n$. Consider $c_1' = c_1/c_1, c_2' = c_2/c_1, \ldots, c_n' = c_n/c_1$. We have $1 = c_1' \leqslant \cdots \leqslant c_n'$. Hence, we can apply Lemma 1 to the function $h' = \sum c_i' x_i$. Comparing the conclusion of this lemma with the fact that $h'(a_1)/h'(a_2) = h(a_1)/h(a_2)$ for all a_1, a_2, we derive the theorem.

2. Approximate evaluation of a linear function

Given a real function f on the vertices of the cube Q, consider a sequence of vertices of the cube. The sequence is called *nondecreasing* if f does not decrease as the index of the vertex grows. The empty sequence is assumed to be nondecreasing by definition.

Fix a number $\delta > 0$. If a third vertex is found between two vertices a and b of a nondecreasing sequence such that $|f(a) - f(b)| \leqslant \delta$, then this vertex can be deleted. Such an operation is called *δ-reduction*, while the original sequence is said to be *δ-reducible*.

The following two lemmas are obvious.

LEMMA 2. *A δ-reduction transforms a nondecreasing sequence into another nondecreasing sequence.*

LEMMA 3. *If a sequence of vertices is δ-irreducible, then at most two vertices can be mapped by f into a segment of length $\leq \delta$.*

LEMMA 4. *A number x belonging to a segment of length $\leq \delta$ whose endpoints are the images of two vertices of a nondecreasing sequence retains this property under a δ-reduction of this sequence.*

PROOF. Let a and b be the vertices such that $f(a)$ and $f(b)$ are the endpoints of the above segment. If neither a nor b is deleted under the reduction, then the lemma is evident.

Let a be deleted. Then the sequence includes vertices c and d such that a lies between them and $0 \leq f(d) - f(c) \leq \delta$. The following two options are distinguished: $x \geq f(d)$ and $x < f(d)$.

Let $x \geq f(d)$. Since the initial sequence is nondecreasing, we have $f(c) \leq f(a) \leq f(d) \leq x \leq f(b)$. Comparing this with the relation $0 \leq f(b) - f(a) \leq \delta$, we find that the vertices satisfying the lemma with $x \geq f(d)$ are d and b.

Let $x < f(d)$. Then it is easy to see that $f(c) \leq f(a) \leq x \leq f(d) \leq f(b)$. Comparing this with the relation $0 \leq f(d) - f(c) \leq \delta$, we conclude that the vertices c and d satisfy the lemma.

The case when b is deleted is considered similarly.

By induction on the number of reductions, one can derive from Lemma 4 the following two lemmas.

LEMMA 5. *If a sequence N is obtained from a sequence M by a number of δ-reductions and there is a vertex in M that does not occur in N, then its image under the mapping f lies in a segment of length $\leq \delta$ whose endpoints are the images of two vertices from N.*

LEMMA 6. *If a number x lies in a segment of length $\leq \delta$ whose endpoints are the images of two vertices from M, then the same is true with M replaced by N.*

LEMMA 7. *Let a nondecreasing sequence N consist of k vertices. Then, after at most k δ-comparisons, one can indicate a sequence of δ-reductions taking N to a δ-irreducible sequence.*

PROOF. We prove the lemma by induction on k. If $k \leq 2$, then N is already δ-irreducible, and the lemma is obvious.

Let the lemma holds for $k \leq m$ ($m \geq 2$); we prove it for $k = m + 1$. Let $N = \langle a_1, \ldots, a_{m+1} \rangle$. Let us compare $f(a_3) - f(a_1)$ with δ. If $f(a_3) - f(a_1) \leq \delta$, then a_2 can be deleted and the lemma follows from the induction hypothesis. Let $f(a_3) - f(a_1) > \delta$. Since the sequence N is nondecreasing, we have $f(a_i) - f(a_1) \geq f(a_3) - f(a_1)$ for $i > 3$, and therefore the vertex a_2 can never be deleted. Hence, all δ-reductions of N are at the same time δ-reductions of the sequence $N' = \langle a_2, \ldots, a_{m+1} \rangle$. Applying the induction hypothesis and taking into account that only one δ-comparison was performed, we derive the assertion of the lemma.

A nondecreasing sequence N of vertices from some set M is called δ-*enclosing* for the set M, if for any vertex $a \in M \setminus N$ there exist vertices $b, d \in N$ such that $f(b) \leq f(a) \leq f(d)$ and $f(d) - f(b) \leq \delta$.

Let $N = \langle a_1, \ldots, a_t \rangle$ be a sequence of vertices. The sequence of vertices $\langle \{k\}, a_1 \cup \{k\}, \ldots, a_t \cup \{k\} \rangle$ is said to be the k-*shift* of N. Recall that a vertex of the cube Q is denoted by the set of indices of nonzero coordinates at this vertex.

Consider now a particular case when $f(a) = \log h(a)$, where h is the function defined in §1 with the coefficients $c_1, \ldots, c_n > 0$. Assume the number δ is given in the form $\delta = \log(1 + \varepsilon)$ $(\varepsilon > 0)$.

In this case, it follows from the proof of Lemma 1 that the values of f at vertices from Q_k belong to at most $k(1 + 1/\log(1 + \varepsilon))$ nonintersecting segments of length $\log(1 + \varepsilon)$. Comparing this with Lemma 3, we get

LEMMA 8. *If a sequence of vertices is δ-enclosing for the set Q_k and δ-irreducible with $\delta = \log(1 + \varepsilon)$, then it contains at most $2k(1 + 1/\log(1 + \varepsilon))$ vertices.*

For any real positive numbers z and y, the relations $0 \leq \log z - \log y$ and $w \geq 0$ yield the relation $0 \leq \log(z + w) - \log(y + w) \leq \log z - \log y$. Therefore, in our particular case, the following lemma holds for the function f.

LEMMA 9. *If $a, b \in Q_k$ and $0 \leq f(a) - f(b)$, then $0 \leq f(a \cup \{k+1\}) - f(b \cup \{k+1\}) \leq f(a) - f(b)$.*

We deduce from Lemma 9 the following fact:

LEMMA 10. *If N is a $\log(1 + \varepsilon)$-enclosing sequence of vertices for the set Q_k, then the $(k+1)$-shift of N is a $\log(1 + \varepsilon)$-enclosing sequence of vertices for Q_{k+1}^*.*

LEMMA 11. *Let $v_k = 2k(k+1)(1 + 1/\log(1+\varepsilon))$. Then one can construct a $\log(1+\varepsilon)$-enclosing and $\log(1+\varepsilon)$-irreducible sequence for the set Q_k, using the evaluation of the function f at at most $v_k/2 + k$ vertices of the cube Q, and performing at most $v_k + k$ comparisons of the values of f and at most $v_k + k$ $\log(1+\varepsilon)$-comparisons of the values of f.*

PROOF. Let us prove the lemma by induction on k. For $k = 1$ it is obvious, since Q_1 contains only one vertex $\{1\}$.

Let the lemma be true for $k \leq m$; we prove it for $k = m + 1$. Let N_m be a δ-enclosing and δ-irreducible sequence constructed for Q_m with $\delta = \log(1 + \varepsilon)$. Denote by N_m^* the $(m+1)$-shift of N_m. Let us evaluate f at the vertices from N_m^*. According to Lemma 8, $1 + 2m(1 + 1/\log(1 + \varepsilon))$ evaluations are required. It is known that the number of comparisons of f allowing one to construct the nondecreasing sequence N_{m+1}^* that consists of all the vertices from both N_m and N_m^* does not exceed the total number of points in N_m and N_m^*, that is, does not exceed $1 + 4m(1 + 1/\log(1+\varepsilon))$. One can use, for example, the merge sorting algorithm. Then, applying Lemma 7, we can obtain an irreducible subsequence N_{m+1} of the sequence N^*_{m+1} by at most $1 + 4m(1 + 1/\log(1 + \varepsilon))$ $\log(1 + \varepsilon)$-comparisons of f. Using the induction hypothesis, Lemmas 5, 6, and 10, and the relation $Q_{m+1} = Q_m \cup Q^*_{m+1}$, we infer that the sequence N_{m+1} is the one required. Together with the above estimates of the number of comparisons and evaluations of f performed while passing from N_m to N_{m+1}, this proves the lemma.

For any numbers $z, y \geqslant 0$, the relation $\log z \leqslant \log y$ is equivalent to the relation $z \leqslant y$, while the relation $\log y - \log z \leqslant \log(1 + \varepsilon)$ is equivalent to the relation $y/z \leqslant 1 + \varepsilon$. Therefore, comparisons of values of f can be replaced by comparisons of values of h. Then Lemma 11 yields

THEOREM 2. *Let* $v = 2n(n+1)(1+1/\log(1+\varepsilon)))$. *If the function* $h = \sum c_i x_i$ *has only positive coefficients* $c_1, \ldots, c_n$, *then by making at most* $v/2 + n$ *evaluations of* h *at vertices of the cube* Q, *at most* $v + n$ *comparisons of values of* h, *and at most* $v + n$ *relative* $(1 + \varepsilon)$-*comparisons of values of* h, *one can construct a set of vertices* $N = \langle a_1, \ldots, a_t \rangle$ *containing at most* $2n(1+1/\log(1+\varepsilon))$ *vertices such that for any nonzero vertex* b *of the cube* Q *missing in* N, *there exist* $a_i, a_j \in N$ *satisfying the relations* $f(a_i) \leqslant f(b) \leqslant f(a_j)$ *and* $f(a_j)/f(a_i) \leqslant 1 + \varepsilon$.

3. Approximate solution of the knapsack problem

Let a real function f be specified on the vertices of the cube Q. Let N_1 be a nondecreasing sequence of vertices, and a the first vertex in N_1. Delete from N_1 all vertices distinct from a whose images under the mapping f occur in the right δ-neighborhood of $f(a)$. Recall that a number x belongs to the right δ-neighborhood of a number y if $0 \leqslant x - y \leqslant \delta$. Let b be the second vertex of the sequence N_2 thus obtained. Delete from N_2 all vertices distinct from b whose images under the mapping f occur in the right δ-neighborhood of b. Consider then the third vertex of the sequence N_3 thus obtained, and so on. The process ends when, after deleting the right δ-neighborhood of a vertex, that vertex is the last one remaining in the sequence. The sequence of vertices thus obtained is called the δ-*skeleton* of N_1.

The following lemma is obvious.

LEMMA 12. *The distance between the images of two vertices of the* δ-*skeleton of a nondecreasing sequence is greater than* δ.

LEMMA 13. *If a nondecreasing sequence* N_1 *contains* m *vertices, then its* δ-*skeleton can be found by at most* m δ-*comparisons of values of* f.

PROOF. We prove the lemma by induction on m. For $m = 1$, the δ-skeleton of N_1 coincides with N_1, and the lemma is evident. Let it be true for $m \leqslant k - 1$; we prove it for $m = k$. Let $N = \langle a_1, \ldots, a_k \rangle$.

Consider the differences $f(a_2)-f(a_1)$, $f(a_3)-f(a_1)$, and so on, until we get a difference greater than δ (if $f(a_k)-f(a_1) \leq \delta$, then the δ-skeleton of N_1 equals $\langle a_1 \rangle$, and the lemma is evident). Let this difference be $f(a_i)-f(a_1)$. Since N_1 is a nondecreasing sequence, the values of f at the vertices that follow a_i are not contained in the right δ-neighborhood of $f(a_1)$, and thus $N_2 = \langle a_1, a_i, \ldots, a_k \rangle$. The value of f at the vertex a_1 can be omitted in further comparisons required to find N_3, N_4, and so on, since $f(a_1) < f(a_i)$. Hence, the further construction of the δ-skeleton of N_1 is reduced to the construction of the δ-skeleton of the sequence $N_2' = \langle a_i, \ldots, a_k \rangle$, which can be accomplished, according to the induction hypothesis, by at most $(k-(i-1))$ δ-comparisons. Since we have performed $(i-1)$ δ-comparisons to pass to N_2, the lemma is proved.

The proof of this lemma immediately yields one more statement.

LEMMA 14. *Let the δ-skeleton of a nondecreasing sequence $N = \langle a_1, \ldots, a_k \rangle$ equal $\langle a_{i_1}, \ldots, a_{i_s} \rangle$. Then:*

1) $i_1 = 1$;

2) *the images of all vertices from N whose indices are not less than i_j, but less than i_{j+1}, lie in the right δ-neighborhood of $f(a_{i_j})$, $(j < s)$;*

3) *the images of all vertices from N whose indices are not less than i_s lie in the right δ-neighborhood of $f(a_{i_s})$.*

In addition to the function f, we fix another real function g on the vertices of the cube Q. Let N be a nondecreasing sequence of vertices, and let a function ψ map N to Q^0. Let L be the δ-skeleton of N. We specify a function φ on the vertices from N according to the following rule.

RULE 1. Let a be a vertex from L. Consider all vertices from N whose images under the mapping f are contained in the right δ-neighborhood of $f(a)$. We set the value of φ at a equal to the value of ψ at the vertex where the function $g \circ \psi$ attains its maximum (the circle stands for the superposition of functions, i.e. $F_1 \circ F_2(x) = F_1(F_2(x))$ for any F_1, F_2).

The function φ defined above is said to be the δ-*majorant* of ψ. The definition of the δ-majorant immediately yields

LEMMA 15. *Let φ be a δ-majorant of ψ. If $|f(b) - f \circ \psi(a)| \leq \varepsilon$ for any vertex b from a nondecreasing sequence, then $|f(b) - f \circ \varphi(a)| \leq \delta + \varepsilon$ for any vertex a from the δ-skeleton of this sequence.*

Lemma 14 shows that for any vertex b from a nondecreasing sequence, there exists a vertex from its δ-skeleton such that $f(b)$ lies in the right δ-neighborhood of its image. Together with the definition of δ-majorants, this yields

LEMMA 16. *If a number x lies in the right ε-neighborhood of the image of a vertex b from a nondecreasing sequence, then there exists a vertex a in the δ-skeleton of this sequence such that $0 \leq x - f(a) \leq \varepsilon + \delta$, $g \circ \varphi(a) \geq g \circ \psi(b)$.*

Let $N = \langle a_1, \ldots, a_k \rangle$ be a nondecreasing sequence, $L = \langle a_{i_1}, \ldots, a_{i_s} \rangle$ its δ-skeleton. By Lemma 14, it is easy to see that for $j < s$ there exist $i_{j+1} - i_j$ vertices whose images lie in the right δ-neighborhood of $f(a_{i_j})$, and $k - (i_s - 1)$ vertices whose images lie in the right δ-neighborhood of $f(a_{i_s})$. Therefore, the values of a δ-majorant of a function $\psi: N \to Q^0$ can be found by at most

$$(i_2 - i_1) + (i_3 - i_2) + \cdots + (k - i_s + 1) = k - i_1 + 1 = k$$

comparisons of values of the function g (the values are taken at the vertices from the set $\{\psi(a_1), \ldots, \psi(a_k)\}$). This is implied by the following well-known fact: it suffices to make at most r comparisons to find the maximum among any r numbers. Hence, we have

LEMMA 17. *Let L be the δ-skeleton of N. To find a δ-majorant of a function $\psi: N \to Q^0$, one needs no more comparisons of values of g than the number of elements in N.*

A nondecreasing sequence of vertices N from a set M with a function $\psi: N \to Q^0$ defined on this sequence is called a (k, δ)-*covering* for M, provided the mapping f possesses the following properties:
1) the distance between the images of two arbitrary vertices from N does not exceed δ;
2) for any vertex b from M, there exists a vertex a from N such that $f(b)$ lies in the right $k\delta$-neighborhood of its image;
3) if the image of a vertex b from M lies in the right $k\delta$-neighborhood of the image of a vertex a from N, then $g(b) \leq g \circ \psi(a)$;
4) for any vertex a from N, we have $|f(a) - f \circ \psi(a)| \leq k\delta$.

LEMMA 18. *Let N, together with a function $\psi: N \to Q^0$ defined on N, be a (k, δ)-covering sequence for the set of vertices Q_m. Let d be the vertex with the minimal index among the vertices from the sequence N, where the function $g \circ \psi$ exceeds a number x. Then for any vertex $b \in Q_m$, the relation $g(b) < g \circ \psi(d)$ is implied by the relation $f(b) < f(d)$.*

PROOF. Let $f(b) < f(d)$. According to property 2) of the definition of (k, δ)-covering sequences, $f(b)$ lies in the right $k\delta$-neighborhood of the image of a vertex q from N. We have $f(q) \leq f(b) < f(d)$. Since the sequence N is nondecreasing, the vertex q precedes the vertex d. Comparing this with property 3) of the definition of (k, δ)-covering sequences, and taking account of the choice of d, we derive the lemma.

If $N = \langle a_1, \ldots, a_t \rangle$ is a sequence of vertices of the cube Q, and ψ is a function that maps N to Q^0, then the function φ defined on the k-shift of N by

$$\varphi(\{k\}) = \{k\}, \qquad \varphi(a_i \cup \{k\}) = \psi(a_i) \cup \{k\}, \quad i = 1, \ldots, t,$$

is said to be the *k-shift of* ψ. As in §2, consider the particular case when $f = \log h$, where $h = \sum c_i x_i$ with $c_i, \ldots, c_n > 0$. The function g is assumed to be represented as $g = \sum a_i x_i$ with arbitrary $a_1, \ldots, a_n$, and δ is assumed to be given by $\delta = \log(1+\varepsilon)$ with positive ε. Since the function g is linear, Lemma 9 yields

LEMMA 19. *Let* $N \subseteq Q_k$, $\psi: N \to Q^0$, $a \in N$, $b \in Q_k$, *and let* ψ' *be the* $(k+1)$-*shift of* ψ. *Put* $a' = a \cup \{k+1\}$, $b' = b \cup \{k+1\}$. *Then* $|f(a) - f(b)| \geq |f(a') - f(b')|$, $|f(a) - f \circ \psi(a)| \geq |f(a') - f \circ \psi(a')|$, *and if* $g(b) \leq g \circ \psi(a)$, *then* $g(b') \leq g \circ \psi'(a')$.

In our case, it follows from the proof of Lemma 1 that the values of f at the vertices from Q_k belong to at most $k(1 + 1/\log(1+\varepsilon))$ nonintersecting segments of length $\log(1+\varepsilon)$. Comparing this with property 1) in the definition of (k, δ)-covering sequences, we infer

LEMMA 20. *A* $(k, \log(1+\varepsilon))$-*covering sequence of the set of vertices* Q_k *contains at most* $k(1+1/\log(1+\varepsilon))$ *vertices.*

LEMMA 21. *Let* $v_k = k(k+1)(1+1/\log(1+\varepsilon))$. *For the set* Q_k, *one can find a* $(k-1, \log(1+\varepsilon))$-*covering sequence consisting of the set of vertices* N_k *and a function* $\psi_k: N_k \to Q^0$ *defined on this sequence by performing at most* $v_k/2 + k$ *evaluations of* f *and* g *at vertices of the cube* Q, *at most* $(k+v_k)$ *comparisons of values of* f *and* g, *and at most* $(v_k + k) \log(1+\varepsilon)$-*comparisons of values of* f.

PROOF. We prove the lemma by induction on k. For $k = 1$ it is obvious, since Q_1 contains only one vertex $\{1\}$.

Let the lemma be true for $k \leq m$; we prove it for $k = m+1$. Let N_m, together with a function ψ_m defined on N_m, be an $(m, \log(1+\varepsilon))$-covering sequence for Q_m. Consider the $(m+1)$-shift of N_m, denoted by N_m^*, and the $(m+1)$-shift of ψ_m, denoted by ψ_m^*. Let us evaluate f at the vertices from N_m^* and g at the vertices from $\psi_m^*(N_m)$. By Lemma 20, this requires $1 + m(1 + 1/\log(1+\varepsilon))$ evaluations of f and the same number of evaluations of g. A number of comparisons of values of f that does not exceed the total number of points in N_m and N_m^*, i.e., does not exceed

$$1 + 2m(1 + 1/\log(1+\varepsilon)),$$

is known to allow the construction of the nondecreasing sequence N_{m+1}^*, consisting of the vertices from both N_m and N_m^* (and no others). Let us define a function ψ_{m+1}^* on N_{m+1}^* by

$$\psi_{m+1}^*(a) = \begin{cases} \psi_m(a), & \text{if } a \in N_m, \\ \psi_m^*(a), & \text{if } a \in N_m^*. \end{cases}$$

By Lemma 13 we can construct the sequence N_{m+1}, the $\log(1+\varepsilon)$-skeleton of N_{m+1}^*, by $1 + 2m(1+1/\log(1+\varepsilon))$ $\log(1+\varepsilon)$-comparisons. According to Lemma 17, we can determine a $\log(1+\varepsilon)$-majorant of the function ψ_{m+1}^* on N_{m+1} by at most $1 + 2m(1 + 1/\log(1+\varepsilon))$ comparisons of values of g. Using the induction hypothesis, Lemmas 12, 15, 16, and 19, and the relation $Q_{m+1} = Q_m \cup Q_m^*$, we infer that the sequence N_{m+1}, along with the $\log(1+\varepsilon)$-majorant of the function ψ_{m+1}^* defined on N_{m+1}, is the one required. Together with the above estimates of the number of evaluations and comparisons of f and g required to pass from N_m to N_{m+1}, this proves the lemma.

Let us introduce θ by the equation $1+\theta = (1+\varepsilon)^{n-1}$. Making the same remarks that followed the transition from Lemma 11 to Theorem 2, and accounting for Lemma 18 and the relations

$$\log(1+\varepsilon) = \frac{\log(1+\theta)}{n-1}, \qquad 1+\varepsilon = (1+\theta)^{1/(n-1)},$$

we obtain, on the basis of Lemma 21, the following theorem.

THEOREM 3. *Let $v = n(n+1)(1 + (n-1)/\log(1+\theta))$ $(\theta > 0)$, and let h_0 be the minimum of $h = \sum c_i x_i$ subject to the condition $g = \sum a_i x_i \geqslant b$ with zero-one variables $x_1, \ldots, x_n$. Then, with $c_1, \ldots, c_n > 0$, one can find a collection q such that $g(q) \geqslant b$ and $h(q)/h_0 \leqslant 1+\theta$ by performing at most $v/2 + n$ evaluations of h and g at collections of zeros and ones, at most $v + n$ relative $(1+\theta)^{1/(n-1)}$-comparisons of values of h, at most $v + n$ comparisons of values of g, at most $v + n$ comparisons of values of h, and at most $n(1 + (n-1)/\log(1+\theta))$ comparisons of values of g with the number b on collections of zeros and ones.*

REMARK 1. It is easy to see that the above algorithm applied to an unsolvable knapsack problem indicates the absence of a collection q satisfying the relation $g(q) \geqslant b$.

REMARK 2. In all the above reasoning, we disregarded the collection $x_1 = \cdots = x_n = 0$, which therefore requires a separate verification.

REMARK 3. Clearly, for $a_i \leqslant 0$, it is advantageous to set the value of x_i equal to zero. Such an initial zero-setting reduces the number of variables and thus makes the algorithm more efficient.

On the Growth of Coefficients in an Integral Linear Aggregation

L. G. BABAT

1. Introduction

Consider a system of linear equations with integer coefficients in a domain Ω:

$$\sum_{i=1}^{n} a_{i1} x_i = b_1, \ldots, \sum_{i=1}^{n} a_{ik} x_i = b_k. \qquad (1)$$

It is known that under certain conditions on Ω this system admits integer linear aggregation in this domain. This means that one can find an equation

$$\sum_{1}^{n} d_i x_i = d_0 \qquad (2)$$

with integral coefficients that has the same solutions as the initial system on the set $\Omega \cap \mathbb{Z}^n$, where $\mathbb{Z}^n$ is, as usual, the set of n-dimensional integral vectors (see, for example, [2]). However, for all the known ways of linear integral aggregation, the norm of the vector $D = (d_1, \ldots, d_n)$ grows sharply with the growth of Ω. The rate of the growth depends exponentially on the rank of the aggregated system (1); without loss of generality, this rank is assumed to be equal to k.

The question arises, whether there exists a method of integral linear aggregation that does not lead to a rapid growth of the coefficients of the aggregated equation. The following theorem confirming the negative answer to this question was formulated in [1].

THEOREM 1. *Let an n-dimensional ball of radius $R > R_0$ be inscribed in Ω so that the coordinates of its center $(x_1, \ldots, x_n)$ are integers satisfying*

1991 *Mathematics Subject Classification.* Primary 68R10.

Translation of Mathematical Methods for Solving Economic Problems, no. 8 (N. P. Fedorenko and E. G. Gol'shteĭn, editors; Suppl. to Èkonom. i Mat. Metody), "Nauka", Moscow, 1979, pp. 34–43; MR **81f**:90067.

(1). *Then the fact that equation* (2) *aggregates system* (1) *in* Ω *implies that*

$$\|D\| = \sqrt{\sum_0^n d_i^2} \geq C^* R^{k-1},$$

where C^* and R_0 are positive constants determined by the rows of the matrix of system (1) *and independent of* Ω.

In the present article we provide a proof of this theorem. Based on this proof, it is easy to construct examples of systems for which $R_0 = C^* = 1$. However, it is also not difficult to construct systems with small C^* and large R_0. Therefore, it makes sense to provide a new formulation of the aggregation problem: describe efficiently those systems, significant for applications, that are well aggregated in sufficiently wide domains, i.e. systems with small C^* and large R_0. At present, there is no such description.

Let $X^0 = (x_1^0, \ldots, x_n^0)$ be an integral vector from Ω satisfying (1). Let us shift the space by the vector $-X^0$. Let Ω be transformed into Ω_0. Clearly, equation (2) aggregates system (1) in Ω if and only if the equation

$$\sum_1^n d_i x_i = 0 \tag{3}$$

aggregates the system

$$\sum_{i=1}^n a_{i1} x_i = 0, \ldots, \sum_{i=1}^n a_{ik} x_i = 0 \tag{4}$$

in Ω_0. Therefore, Theorem 1 is a corollary of the following Theorem 2, which will be proved below.

THEOREM 2. *Let equation* (3) *aggregate system* (4) *in* Ω_0. *Then there exist positive constants R_0 and C^*, determined by the rows of the matrix of system* (4) *and independent of Ω_0, such that $\|D\| \geq C^* R^{k-1}$ if a ball of radius $R \geq R_0$ centered at the origin can be inscribed in Ω_0.*

2. Preliminary analysis

Treating the ith row of the matrix of system (4) as a vector A_i, $i = 1, \ldots, k$, rewrite the system as

$$(A_1, X) = 0, \ldots, (A_k, X) = 0, \tag{5}$$

where $X = (x_1, \ldots, x_n)$, and parentheses denote the inner product. Similarly, using the vector $D = (d_1, \ldots, d_n)$, rewrite equation (3) in the form

$$(D, X) = 0. \tag{6}$$

Consider the space H of solutions of system (5), whose rank equals k by the assumption. Applying the standard procedure of constructing the fundamental set of solutions, we find vectors $V_1, \ldots, V_{n-k}$ that constitute a basis in H. Since $A_1, \ldots, A_k$ belong to $\mathbb{Z}^n$, i.e. are integer vectors, the quantities $V_1, \ldots, V_{n-k}$ are rational. Let us apply the orthogonalization procedure to obtain rational vectors that also constitute a basis in H. Multiplying these vectors by suitable constants, we obtain a basis in H consisting of integral orthogonal vectors.

Consider the set of all integral orthogonal bases in H. The above reasoning implies that this set is nonempty. Therefore, with R greater than some constant $\varkappa$, a ball S of radius R centered at the origin contains an integer orthogonal basis of H. Let this basis consist of vectors $Y_1, \ldots, Y_{n-k}$. If $S \subseteq \Omega_0$, then $Y_1, \ldots, Y_{n-k} \in \Omega_0$. However, equation (6) has the same integral solutions in Ω_0 as system (5); therefore,

$$(D, Y_i) = 0, \qquad i = 1, \ldots, n - k. \tag{7}$$

Any vector from H can be represented as a linear combination of $Y_1, \ldots, Y_{n-k}$. Comparing this with (7), we obtain the following lemma.

LEMMA 1. *Let equation* (6) *aggregate system* (5) *in* Ω_0. *Then there exists a constant* $\varkappa > 0$ *such that* $(D, X) = 0$ *for any* $X \in H$ *if a ball of radius* $R \geqslant \varkappa$ *centered at the origin can be inscribed in* Ω_0.

Along with H, consider the subspace L spanned by the rows $A_1, \ldots, A_k$ of the matrix of system (5). It is known that H and L are the orthogonal complements to each other, and therefore any n-dimensional vector can be uniquely represented as the sum of a vector from H and a vector from L. Let us represent the vector D as $D = X + Y$, where $X \in H$ and $Y \in L$. Performing the scalar multiplication by X and using the fact that for $(Y, X) = 0$, we derive the following statement from Lemma 1.

LEMMA 2. *Let equation* (6) *aggregate system* (5) *in* Ω_0. *Then there exists a constant* $\varkappa > 0$ *such that* $D \in L$ *if a ball of radius* $R \geqslant \varkappa$ *centered at the origin can be inscribed in* Ω_0.

Let us introduce some notation. For a set Φ of vectors, denote by $\mathrm{pr}(\Phi)$ the orthogonal projection of Φ to L. In particular, if Φ is a vector, then $\mathrm{pr}(\Phi)$ is its orthogonal projection to L.

It is easy to see that if $Y \in L$, then $(Y, X) = (Y, \mathrm{pr}(X))$ for any X. Therefore,

$$(A_i, X) = (A_i, \mathrm{pr}(X)), \qquad i = 1, \ldots, k. \tag{8}$$

If the vector D belongs to L as well, then, besides (8), the following equality holds:

$$(D, X) = (D, \mathrm{pr}(X)). \tag{9}$$

Relations (8) and (9) indicate that with $D \in L$ equation (6) aggregates system (5) in the domain Ω_0 if and only if it has the same solutions in the set $\mathrm{pr}(\Omega_0 \cap \mathbb{Z}^n)$ as this system. However, the set $\mathrm{pr}(\Omega_0 \cap \mathbb{Z}^n)$ lies in the subspace L orthogonal to the subset H of solutions of (5), and hence system (5) has no nontrivial solutions in the set $\mathrm{pr}(\Omega_0 \cap \mathbb{Z}^n)$. Comparing this with Lemma 2, we obtain the following lemma.

LEMMA 3. *Let equation* (6) *aggregate system* (5) *in* Ω_0. *Then there exists a constant* $\varkappa > 0$ *such that equation* (6) *has no nontrivial solutions in the set* $\mathrm{pr}(\Omega_0 \cap \mathbb{Z}^n)$ *if a ball of radius* $R \geqslant \varkappa$ *centered at the origin can be inscribed in* Ω_0.

Consider the set of integral vectors having nonzero projections to L. This set is nonempty, since $A_1, \ldots, A_k$ are integral nonzero vectors lying in L, by the definition of this subspace. Hence the following lemma holds.

LEMMA 4. *For R greater than a constant γ, the ball of radius R centered at the origin contains an integral vector having a nonzero orthogonal projection to L.*

Put $R_0 = \max(\varkappa, \gamma)$, where $\varkappa$ and γ are the constants from Lemmas 3 and 4. Let a ball of radius $R \geqslant R_0$ be inscribed in Ω_0. In this case, according to Lemma 4, the set $\mathrm{pr}(\Omega_0 \cap \mathbb{Z}^n)$ contains a nonzero vector X. Applying Lemma 3, we see that if equation (6) aggregates system (5) in Ω_0, then $(D, X) \neq 0$, which means that D is a nonzero vector. Combining this fact with Lemmas 2 and 3, we arrive at the following statement.

LEMMA 5. *Let equation* (6) *aggregate system* (5) *in* Ω_0. *Then there exists a constant R_0 such that the vector D belongs to L, is nonzero and not orthogonal to any nonzero vector from the set* $\mathrm{pr}(\Omega_0 \cap \mathbb{Z}^n)$ *if a ball of radius* $R \geqslant R_0$ *centered at the origin can be inscribed in* Ω_0.

For further considerations, we need the following technical lemma.

LEMMA 6. *For any basis $E_1, \ldots, E_k$ of the subspace L, one can find a constant g such that the relations $|x_i| \leqslant g\|X\|$, $i = 1, \ldots, k$, hold true for $X = \sum x_i E_i$.*

PROOF. Let us decompose X with respect to some orthonormal basis $W_1, \ldots, W_k$ of the space L:

$$X = \sum v_i W_i, \qquad \|X\|^2 = \sum v_i^2. \tag{10}$$

Consider the transition matrix $[M]$ from the basis $W_1, \ldots, W_k$ to the basis $E_1, \ldots, E_k$. We have $(x_1, \ldots, x_k) = (v_1, \ldots, v_k)[M]$ and $\sum x_i^2 \leqslant (\sum v_i^2)\|[M]\|^2$. Comparing the second relation with (10), we see that

$$\sum x_i^2 \leqslant \|X\|^2 \|[M]\|^2,$$

whence $|x_i| \leqslant \|X\| \cdot \|[M]\|$. Then, setting $g = \|[M]\|$, we derive the assertion of the lemma.

With the help of Lemma 6, we establish the following important property of the set $\mathrm{pr}(\mathbb{Z}^n)$.

LEMMA 7. *There exists a constant q such that for any vector $X \in \mathrm{pr}(\mathbb{Z}^n)$ one can find a vector $Y \in \mathbb{Z}^n$ satisfying the conditions $\mathrm{pr}(Y) = X$ and $\|Y\| \leq q\|X\|$.*

PROOF. The projections of arbitrary vectors V and W are known to satisfy $\mathrm{pr}(\alpha V + \beta W) = \alpha \, \mathrm{pr}(V) + \beta \, \mathrm{pr}(W)$ for any numbers α and β. Using this relation, it is easy to see that $\mathrm{pr}(\mathbb{Z}^n)$ is a discrete lattice. The vectors $A_1, \ldots, A_k$ belong to $L \cap \mathbb{Z}^n$, and therefore they lie in the lattice $\mathrm{pr}(\mathbb{Z}^n)$. This means that $\dim \mathrm{pr}(\mathbb{Z}^n) = \dim L = k$. Let us fix a basis $E_1, \ldots, E_k$ of the lattice $\mathrm{pr}(\mathbb{Z}^n)$. By definition, the lattice $\mathrm{pr}(\mathbb{Z}^n)$ consists of the orthogonal projections of integral vectors to L, and therefore there exist integral vectors $U_1, \ldots, U_k$ such that $\mathrm{pr}(U_i) = E_i$, $i = 1, \ldots, k$. Let $X \in \mathrm{pr}(\mathbb{Z}^n)$. Since $E_1, \ldots, E_k$ is a lattice basis, one has $X = \sum x_i E_i$, where $x_1, \ldots, x_k$ are integers. Consider the vector $Y = \sum x_i U_i$. Clearly, it is integer, and $\mathrm{pr}(Y) = X$. Let us estimate the norm of Y. We find that $\|Y\| = \|\sum x_i U_i\| \leq \sum |x_i| \|U_i\| \leq \max_i \|U_i\| (\sum |x_i|)$. Applying Lemma 6 to this relation, we get $\|Y\| \leq kg\|X\| \max_i \|U_i\|$. Setting now $q = kg \max_i \|U_i\|$, we derive the assertion of the lemma.

Lemma 7 evidently yields the following fact. If S_1 is a ball of radius R centered at the origin, and S is the ball of radius R/q centered at the origin, then $\mathrm{pr}(S_1 \cap \mathbb{Z}^n) \supseteq S \cap \mathrm{pr}(\mathbb{Z}^n)$. If, in addition, $S_1 \subseteq \Omega_0$, then $\mathrm{pr}(S_1 \cap \mathbb{Z}^n) \subseteq \mathrm{pr}(\Omega_0 \cap \mathbb{Z}^n)$, and thus the following lemma holds.

LEMMA 8. *There exists a constant q that satisfies the following conditions: if a ball of radius R centered at the origin can be inscribed in Ω_0, then $\mathrm{pr}(\Omega_0 \cap \mathbb{Z}^n) \supseteq \mathrm{pr}(\mathbb{Z}^n) \cap S$, where S is the ball of radius R/q centered at the origin.*

Denote by L_D the intersection of L with the subspace orthogonal to the vector D. Since D is an integral vector, one has $D \in \mathrm{pr}(\mathbb{Z}^n)$ if $D \in L$. Comparing this with Lemmas 5 and 8, we find that Theorem 1 is a corollary of the following statement.

THEOREM 3. *Let D be a nonzero vector belonging to $\mathrm{pr}(\mathbb{Z}^n)$. If no nonzero vectors from $\mathrm{pr}(\mathbb{Z}^n)$ are contained in the intersection of L_D with a ball S of radius R centered at the origin, then $\|D\| \geq cR^{k-1}$, where c is a positive constant independent of D and S.*

A proof of this theorem is presented below.

3. Proof of Theorem 3

By the assumptions of Theorem 3, D is a nonzero vector from $\mathrm{pr}(\mathbb{Z}^n)$; however, $\mathrm{pr}(\mathbb{Z}^n) \subseteq L$, and thus D is a nonzero vector from L. This means that the dimension of the subspace L_D consisting of vectors orthogonal to D equals
$$\dim L_D = \dim L - 1 = k - 1.$$

Consider the set $\Delta = L_D \cap \mathrm{pr}(\mathbb{Z}^n)$. As we indicated in the proof of Lemma 7, $\mathrm{pr}(\mathbb{Z}^n)$ is a discrete lattice, and therefore, the set Δ is a discrete lattice as well, as the intersection of a discrete lattice and a subspace. In order to estimate the dimension of this lattice, we establish the following statement.

LEMMA 9. *The subspace L_D contains $k-1$ linearly independent integral vectors.*

PROOF. Consider the following equation with respect to the unknowns $\xi_1, \ldots, \xi_k$:

$$\sum_{1}^{k}(D, A_i)\xi_i = 0. \tag{11}$$

Since $D, A_1, \ldots, A_k$ are integral vectors, the coefficients of this equation are integers. Therefore, it has a fundamental system of solutions consisting of $k-1$ rational k-dimensional vectors. Multiplying these vectors by the corresponding constants, we obtain $k-1$ integral linearly independent vectors $\Xi_i = (\xi_{i1}, \ldots, \xi_{ik})$, $i = 1, \ldots, k-1$, satisfying (11). Put

$$X_i = \sum_{j=1}^{k} \xi_{ij} A_j, \qquad i = 1, \ldots, k-1.$$

It follows from (11) that $X_1, \ldots, X_{k-1}$ are orthogonal to D, while from the condition $A_1, \ldots, A_k \in L$ we infer that $X_1, \ldots, X_{k-1} \in L$. Consequently, $X_1, \ldots, X_{k-1} \in L_D$. Moreover, the vectors $X_1, \ldots, X_{k-1}$ are integral, because $A_1, \ldots, A_k$ are integral by assumption, and $\Xi_1, \ldots, \Xi_{k-1}$ are integral by construction. The only thing left is to show that $X_1, \ldots, X_{k-1}$ are linearly independent. Observe that

$$\sum_i \lambda_i X_i = \sum_i \lambda_i \left(\sum_j \xi_{ij} A_j\right) = \sum_j A_j \left(\sum_i \lambda_i \xi_{ij}\right).$$

Since $A_1, \ldots, A_k$ are linearly independent, this means that if $\sum \lambda_i X_i$ is the zero vector, then $\sum_i \lambda_i \xi_{ij} = 0$ for all $j = 1, \ldots, k$. This means, however, that the vector $\sum_i \lambda_i \Xi_i$ is zero, whence, by the linear independence of $\Xi_1, \ldots, \Xi_{k-1}$, we conclude that $\lambda_1 = \cdots = \lambda_{k-1} = 0$, exactly what we wanted to prove.

Using Lemma 9 and the fact that any integral vector from L belongs to $\mathrm{pr}(\mathbb{Z}^n)$, we infer

$$\dim \Delta = \dim L = k - 1.$$

By the assumptions of Theorem 3, no point from Δ distinct from the origin lies in the ball S of radius R centered at the origin. However, the intersection of the ball S and the subspace L_D is a $(k-1)$-dimensional ball, i.e. a convex body symmetric with respect to the origin. Therefore, by the Minkowski lemma, the following fact is true.

LEMMA 10. *Under the assumptions of Theorem 3, the volume V of a cell of the lattice Δ is not less than the volume of a $(k-1)$-dimensional ball of radius R divided by 2^{k-1}.*

Consider a basis $F_1, \ldots, F_{k-1}$ of the lattice Δ. A cell of the lattice Δ is a parallelepiped built on $F_1, \ldots, F_{k-1}$. The volume of such a parallelepiped is known to be equal to

$$V = \|[F_1 \times \cdots \times F_{k-1}]\|,$$

where $[F_1 \times \cdots \times F_{k-1}]$ denotes the vector product of $F_1, \ldots, F_{k-1}$. Comparing this relation with Lemma 10, we obtain the assertion of the lemma.

LEMMA 11. *Under the assumptions of Theorem 3, a basis $F_1, \ldots, F_{k-1}$ of the lattice Δ satisfies the relation $\|[F_1 \times \cdots \times F_{k-1}]\| \geq QR^{k-1}$, where Q is a constant independent of the lattice Δ.*

The vectors $F_1, \ldots, F_{k-1}$ are linearly independent and lie in the $(k-1)$-dimensional subspace L_D of the space L. Therefore, their vector product is a nonzero vector lying in L and orthogonal to L_D. This means that $[F_1 \times \cdots \times F_{k-1}]$ is a vector proportional to D, i.e. there exists a number α such that

$$\alpha[F_1 \times \cdots \times F_{k-1}] = D. \tag{12}$$

Comparing this formula with Lemma 11, we find Theorem 3 to be a corollary of the following theorem.

THEOREM 4. *Let D be a nonzero vector of the lattice $\mathrm{pr}(\mathbb{Z}^n)$, and let $F_1, \ldots, F_{k-1}$ be a basis of the lattice $\Delta = L_D \cap \mathrm{pr}(\mathbb{Z}^n)$. Then there exists a constant $\beta > 0$ independent of D and Δ such that $|\alpha| > \beta$ for α satisfying (12).*

Theorem 4 is proved in the next section.

4. Proof of Theorem 4

Let us apply the standard orthogonalization procedure to the vectors $A_1, \ldots, A_k$. Since $A_1, \ldots, A_k$ are integral, it results in k rational vectors. Multiplying these vectors by the corresponding constants, we get a basis of the subset L consisting of integral orthogonal vectors $B_1, \ldots, B_k$.

Since $B_1, \ldots, B_k \in L$, it is evident that for any vector we have

$$(B_i, W) = (B_i, \mathrm{pr}(W_i)), \quad i = 1, \ldots, k. \tag{13}$$

By the definition of the lattice $\mathrm{pr}(\mathbb{Z}^n)$, for any vector X from the lattice, there exists an integral vector Y such that $\mathrm{pr}(Y) = X$. Comparing this with (13) and taking into account that $B_1, \ldots, B_k$ are integral, we obtain the following lemma.

LEMMA 12. *If* $X \in \mathrm{pr}(\mathbb{Z}^n)$, *then the inner product* (B_i, X) *is an integer*, $i = 1, \ldots, k$.

Together with the basis $B_1, \ldots, B_k$ of the subspace L made up of integral orthogonal vectors, we need also the orthonormal basis consisting of the vectors $B_i/\|B_i\|$, $i = 1, \ldots, k$, and the basis consisting of the orthogonal rational vectors $B_i/\|B_i\|^2$, $i = 1, \ldots, k$. The following lemma is obvious.

LEMMA 13. *The ith coordinate of a vector* $X \in L$ *with respect to the orthonormal basis* $B_1/\|B_1\|, \ldots, B_k/\|B_k\|$ *equals* $(B_i, X)/\|B_i\|$. *With respect to the rational basis* $B_1/\|B_1\|^2, \ldots, B_k/\|B_k\|^2$, *this coordinate equals* (B_i, X), $i = 1, \ldots, k$.

Consider a basis $F_1, \ldots, F_{k-1}$ of the lattice Δ. Let $[F]_N$ be the matrix whose rows are the coordinates of the vectors $F_1, \ldots, F_{k-1}$ with respect to the basis $B_1/\|B_1\|, \ldots, B_k/\|B_k\|$. This basis is orthonormal, and thus $[F_1 \times \cdots \times F_{k-1}]$ is given, with respect to this basis, by

$$[F_1 \times \cdots \times F_{k-1}] = \sum_{i=1}^{k} (-1)^{i-1} \det[F_i]_N \frac{B_i}{\|B_i\|}, \qquad (14)$$

where $[F_i]_N$ is the minor of the matrix $[F]_N$ obtained by deleting the ith column. Along with the matrix $[F]_N$, consider the matrix $[F]$ whose rows are the coordinates of the vectors $F_1, \ldots, F_{k-1}$ with respect to the basis $B_1/\|B_1\|^2, \ldots, B_k/\|B_k\|^2$. It follows from Lemma 13 that the matrix $[F]$ is obtained from the matrix $[F]_N$ by multiplying the ith column of $[F]_N$ by $\|B_i\|$, $i = 1, \ldots, k$. Hence,

$$\frac{(\det[F_i]_N)}{\|B_i\|} = \det[F_i] \left(\prod_{j=1}^{k} \|B_j\| \right)^{-1},$$

where $[F_i]$ is the minor of the matrix $[F]$ obtained by deleting the ith column, $i = 1, \ldots, k$. Substituting this relation into (14), we obtain

$$[F_1 \times \cdots \times F_{k-1}] = \left(\prod_{j=1}^{k} \|B_j\| \right)^{-1} \sum_{i=1}^{k} \det[F_i] B_i. \qquad (15)$$

Let us decompose the vector D from Theorem 4 with respect to the basis $B_i/\|B_i\|^2$, $i = 1, \ldots, k$. According to Lemma 13, we have

$$D = \sum_{i=1}^{k} \frac{(B_i, D)}{\|B_i\|^2} B_i. \qquad (16)$$

By the assumptions of Theorem 4, D satisfies (12). Together with (15) and (16), this yields:

$$\alpha \left(\prod_{j=1}^{k} \|B_j\| \right)^{-1} \det[F_i] = \frac{(B_i, D)}{\|B_i\|^2}, \qquad i = 1, \ldots, k.$$

Multiplying these relations by $\prod_1^k \|B_j\|$, we get

$$\alpha\left(\prod_{j=1}^k \|B_j\|\right) \det[F_i] = (B_i, D) \|B_1\|^2 \cdots \|B_{i-1}\|^2 \|B_{i+1}\|^2 \cdots \|B_k\|^2, \qquad (17)$$
$$i = 1, \ldots, k.$$

By the assumptions of Theorem 4, $D \in L$. Comparing this with Lemma 12 and taking into account that $B_1, \ldots, B_k$ are integral, and using (17), we infer the following lemma.

LEMMA 14. *Under the assumptions of Theorem 4, the number*

$$\alpha\left(\prod_{j=1}^k \|B_j\|\right) \det[F_i]$$

is an integer for any $i = 1, \ldots, k$.

In what follows we assume a fixed basis $E_1, \ldots, E_k$ of the lattice $\mathrm{pr}(\mathbb{Z}^n)$. Clearly, $E_1, \ldots, E_k$ is at the same time a basis in L.

It follows from Lemmas 13 and 12 that the transition matrix from the basis $E_1, \ldots, E_k$ to the basis $B_i/\|B_i\|^2$, $i = 1, \ldots, k$, is integral. This means that the inverse matrix consists of rational elements. Therefore, the following lemma holds.

LEMMA 15. *If $[P]$ is the transition matrix from the basis $B_1/\|B_1\|^2, \ldots, B_k/\|B_k\|^2$ to the basis $E_1, \ldots, E_k$, then there exists a constant $C \neq 0$ such that all the elements of the matrix $C[P]$ are integers.*

We need the following auxiliary theorem.

THEOREM 5. *If C is a constant satisfying Lemma 15, and M is an arbitrary square minor of order m of the matrix $[F]$, then the number*

$$\alpha\left(\prod_1^k \|B_j\|\right) C^{k-(m+1)} \det(M)$$

is an integer.

We shall prove this theorem by induction, decreasing successively the order of minors under consideration.

By construction, the matrix $[F]$ has k columns and $k-1$ rows. Hence, the square minors of maximal order are obtained by deleting a column; these minors were denoted above by $[F_1], \ldots, [F_k]$. Lemma 14 shows that the assertion of the theorem holds for these minors. Let it be true for all minors of order $m+1$, where $m+1 \leqslant k-1$; let us prove it for minors of order m. To avoid complicated notation, we provide the proof for the minor at the upper left corner of $[F]$. For the other minors the proof is similar. Denote by $[M]$ the minor of order m standing at the upper left corner of $[F]$. Let

v be the number of the row missing in $[M]$. Such a row does exist, since $m < m + 1 \leqslant k - 1$. In our case, we can assume, without loss of generality, that $v = m + 1$. Let us introduce matrices $[M_1], \ldots, [M_k]$ by the rule

$$[M_i] = \begin{bmatrix} & & (B_i, F_1) \\ & [M] & \vdots \\ & & (B_i, F_m) \\ (B_1, F_{m+1}) \cdots (B_m, F_{m+1}) & (B_i, F_{m+1}) \end{bmatrix}, \quad (18)$$

$i = 1, \ldots, k$. It is easy to see that, for $i > m$, $[M_i]$ is a minor of the matrix $[F]$ of order $m + 1$, and $[M_i]$ is a matrix having two identical columns for $i \leqslant m$. Together with the inductive hypothesis, this yields

LEMMA 16. *The number* $\alpha(\prod_{j=1}^{k} \|B_j\|) C^{k-(m+2)} \det([M_i])$ *is an integer for* $i > m$ *and vanishes for* $i \leqslant m$.

Let us introduce the numbers $S_1, \ldots, S_k$ by setting

$$S_t = \alpha \left(\prod_{j=1}^{k} \|B_j\| \right) C^{k-(m+2)} \sum_{i=1}^{k} \det[M_i] p_{it} C, \quad (19)$$

where $p_{1t}, \ldots, p_{kt}$ are the elements of the ith column of the matrix $[P]$ introduced in Lemma 15. The following lemma is derived from Lemmas 15 and 16.

LEMMA 17. *The numbers* $S_1, \ldots, S_k$ *are integers.*

Consider the vector $X = \sum_{1}^{k} S_t E_t$. The vectors $E_1, \ldots, E_k$ constitute a basis of the lattice $\mathrm{pr}(\mathbb{Z}^n)$. Hence, according to Lemma 17,

$$X = \sum_{1}^{k} S_t E_t \in \mathrm{pr}(\mathbb{Z}^n). \quad (20)$$

Basing on (18), it is easy to see that the following relation holds:

$$\sum_{i=1}^{k} \det[M_i] p_{it} = \det[N], \quad (21)$$

where $[N]$ is the matrix

$$[N] = \begin{bmatrix} & & \sum_{i=1}^{k}(B_i, F_1) p_{it} \\ & [M] & \vdots \\ & & \sum_{i=1}^{k}(B_i, F_m) p_{it} \\ (B_1, F_{m+1}) \ldots (B_m, F_{m+1}) & \sum_{i=1}^{k}(B_i, F_{m+1}) p_{it} \end{bmatrix}.$$

Decomposing the vectors $F_1, \ldots, F_{k-1}$ with respect to the basis $E_1, \ldots, E_k$, we get

$$F_j = \sum_r f_{jr} E_j, \qquad j = 1, \ldots, k-1.$$

From Lemma 13 and the definition of the matrix $[P]$ given in Lemma 15, we deduce that $\sum_{i=1}^{k} (B_i, F_j) p_{it} = f_{lt}$, $l = 1, \ldots, m+1$, and therefore $[N]$ can be rewritten as

$$[N] = \begin{bmatrix} & & & f_{1t} \\ & [M] & & \cdots \\ & & & f_{mt} \\ (B_1, F_{m+1}) & \cdots & (B_m, F_{m+1}) & f_{m+1,t} \end{bmatrix}.$$

Denote by $[N_i]$ the minor obtained from $[N]$ by deleting the last column and the ith row. Expanding $[N]$ with respect to the last column, we get

$$\det[N] = (-1)^m \sum_{i=1}^{m+1} (-1)^{i-1} \det[N_i] f_{it}. \tag{22}$$

Here it is easily seen that

$$[N_{m+1}] = [M]. \tag{23}$$

Comparing (22) with (21) and (19), we obtain

$$S_t = (-1)^m \alpha \left(\prod_{j=1}^{k} \|B_j\| \right) C^{k-m+1} \sum_{i=1}^{m+1} \det[N_i] f_{it} (-1)^{i-1}, \quad t = 1, \ldots, k.$$

Substituting these quantities $S_1, \ldots, S_k$ into $X = \sum_{1}^{k} S_t E_t$, we get

$$X = (-1)^m \alpha \left(\prod_{j=1}^{k} \|B_j\| \right) C^{k-(m+2)} \sum_{i=1}^{m+1} (-1)^{i-1} \det[N_i] F_i. \tag{24}$$

Hence we infer that X belongs to the subspace L_D, which contains the vectors $F_1, \ldots, F_{m+1}$. Comparing this with (20), we see that X belongs to the lattice $\Delta = L_D \cap \operatorname{pr}(\mathbb{Z}^n)$. But $F_1, \ldots, F_{k-1}$ is a basis of the lattice Δ, and therefore the coordinates of X with respect to this basis must be integers. Along with (24) and (23), this indicates that $\alpha(\prod_{j=1}^{k} \|B_j\|) C^{k-m+1} \det[M]$ is an integer, which completes the induction step of the proof of Theorem 5.

Having proved Theorem 5, we return to Theorem 4. Since D is a nonzero vector, $\alpha \neq 0$ by (12). The constant C introduced in Lemma 15 does not vanish either. Therefore, to prove Theorem 4, it suffices to establish the following lemma.

LEMMA 18. *For the constant C introduced in Lemma 15, the number*

$$\alpha \left(\prod_{1}^{k} \|B_j\| \right) C^{k-1}$$

is an integer.

PROOF. Consider the numbers $S_1, \ldots, S_k$ determined by

$$S_t = \alpha \left(\prod_{j=1}^{k} \|B_j\| \right) C^{k-2} \sum_{i=1}^{k} (B_i, F_1) p_{it} C, \qquad (25)$$

where $p_{1t}, \ldots, p_{kt}$ are the elements of the ith column of the matrix $[P]$ defined in Lemma 15. The numbers $(B_1, F_1), \ldots, (B_k, F_1)$ represent the elements of the first row of the matrix $[P]$, and therefore, according to Theorem 5 and Lemma 15, the numbers $S_1, \ldots, S_k$ are integers. Consider the vector $X = \sum_1^k S_t E_t$. Since $E_1, \ldots, E_k$ is a basis of the lattice $\mathrm{pr}(\mathbb{Z}^n)$, and $S_1, \ldots, S_k$ are integers, we have

$$X = \sum_1^k S_t E_t \in \mathrm{pr}(\mathbb{Z}^n). \qquad (26)$$

Let us decompose the vector F_1 with respect to the basis $E_1, \ldots, E_k$:

$$F_1 = \sum_1^k f_t E_t. \qquad (27)$$

By the definition of the matrix $[P]$ and according to Lemma 13, it follows that $\sum_1^k (B_i, F_1) p_{it} = f_t$, $t = 1, \ldots, k$. Substituting this in (25), we get

$$S_t = \alpha \left(\prod_{j=1}^{k} \|B_j\| \right) C^{k-1} f_t, \qquad t = 1, \ldots, k.$$

Using this relation and (26), we obtain

$$X = \alpha \left(\prod_{j=1}^{k} \|B_j\| \right) C^{k-1} F_1 \qquad (28)$$

for the vector $X = \sum_1^k S_t E_t$. Hence $X \in L_D$, because $F_1 \in L_D$. Comparing this with (26), we find that X belongs to the lattice $\Delta = L_D \cap \mathrm{pr}(\mathbb{Z}^n)$. But $F_1, \ldots, F_{k-1}$ is a basis of Δ, so the coordinates of X with respect to this basis must be integers. By comparing with (28), we derive the assertion of the lemma.

The author is thankful to A. A. Fridman for the formulation of the problem and help in this work.

REFERENCES

1. L. G. Babat, *On integral linear aggregation*, Èkonom. i Mat. Metody **13** (1977), 599–601. (Russian)
2. A. A. Fridman, *Boolean methods and their applications*, Proc. Second Winter School Math. Programming and Related Problems, Part I, Tsentral. Èkonom.-Mat. Inst. Akad. Nauk SSSR, Moscow, 1969, pp. 202–309. (Russian)

A Fast Algorithm for Constructing a Maximum Flow Through a Network

B. V. CHERKASSKIĬ

This paper deals with the classical problem of constructing a maximum one-commodity flow through a network. Several algorithms have been suggested to solve this problem, namely, the Ford–Fulkerson algorithm [5], the Dinits algorithm [4], the Karzanov algorithm [7], and the author's algorithm [3]. The last three algorithms require $O(n^2 p)$, $O(n^3)$, and $O(n^2 \sqrt{p})$ operations, respectively, where n is the number of vertices in the network, and p the number of arcs. However, the work on this problem should not be considered complete either in theory or in practice. In theory, it would be of great interest to find an algorithm requiring $O(np)$ operations and obtain lower bounds of the number of operations distinct from the trivial one, $O(p)$. In practice, we need simple and fast codes to handle problems of large dimensions.

This paper suggests one more algorithm for constructing a maximum flow through a network. We call it AWB, the Algorithm Without Balance. It requires $O(n^3)$ operations, quite like the Karzanov algorithm (KA), but more than the Algorithm With Decomposition (AWD) from [3]. However, at the cost of dropping some operations inherent in the Karzanov algorithm, AWB becomes simpler than KA, let alone AWD. We also provide the results of an experimental comparison of the speed of procedures implementing the above algorithms.

1. Basic notions

Let $G = \{N, U, c\}$ be a given network, where N is the set of its vertices, U is the set of its arcs, and c is a positive capacity function specified on

---

1991 *Mathematics Subject Classification.* Primary 68R10, 90B10.
Translation of Combinatorial Methods in Flow Problems, no. 3 (A. V. Karzanov, editor), Vsesoyuz. Nauchno-Issled. Inst. Sistem. Issled., Moscow, 1979, pp. 90–96.

©1994 American Mathematical Society
0065-9290/94 $1.00 + $.25 per page

the arcs. Let $A(i)$ and $B(i)$ be the sets of the arcs leaving and entering i, respectively. Denote by s the source, and by t the sink.

The mathematical formulation of the maximum flow problem is well known (see, for example, [5] or [1]), and therefore we omit it here. We suppose also that the reader is familiar with the maximum flow—minimum cut theorem.

Let a nonnegative function $f(u)$ be specified on the arcs of the network. The *deficit* of a vertex i with respect to this function is the difference between the incoming and outcoming flows at this vertex:

$$df_f(i) = \sum_{u \in B(i)} f(u) - \sum_{u \in A(i)} f(u).$$

The vertices having a nonzero deficit are called *deficient*.

Let the residual capacity network G_f (as defined in [1]) be constructed on the basis of the function f (in the definition of the network G_f, the function f is not necessarily a flow). The *one-way directory of the shortest paths* is the network $M = \{N_M, U_M, c_M\}$, where N_M is the set of the directory vertices (exactly those vertices from which a directed path in the network G_f leads to the sink), U_M is the set of the directory arcs (exactly those arcs of the network G_f that enter at least one of the shortest (in number of arcs) paths leading from a vertex $i \in N_M$ to the sink), and c_M is the capacity of the directory arcs in the network G_f. The one-way directory of the shortest paths will be called just the *directory*.

A nonnegative integer $R(i)$ called the *rank* of the vertex can be assigned to any directory vertex i. The rank of a vertex i is the length (the number of arcs) of the path that leads in the directory from i to t. The rank of t is assumed zero. Clearly, if $u = (i, j)$ is a directory arc, then $R(i) = R(j) + 1$. The definition of the directory indicates also that at least one arc leaves each directory vertex (except t). Let $A_M(i)$ be the set of directory arcs leaving i.

2. Description of the algorithm

Before the algorithm starts, one must specify a function $f(u)$ on the arcs of the network. It is required that $df_f(i) \geq 0$ for $i \neq s$, i.e. that the deficits of all vertices, except the source, with respect to this function be nonnegative. The source deficit is assumed infinite. For example, one can take $f(u) \equiv 0$ as the initial function satisfying all the requirements.

ALGORITHM. 1. *Constructing the directory.* We construct the one-way directory of the shortest paths. (The algorithm of constructing the directory is described in [6] and [2]. Actually, there is no need to reconstruct the directory each time. Some corrections will suffice, which affects the estimate of the complexity of the algorithm.) We then proceed to step 2.

Denote by D_l the set of the deficient vertices of rank l.

2. *Completion.* Let k be the maximal rank of the deficient vertices. If we find no deficient vertices in the directory, then we proceed to step 3.

Consider the set D_k and a vertex $i \in D_k$. We take the first arc $u = (i, j)$ from the list $A_M(i)$ and start to process it.

We put $\varepsilon = \min(c_M(u), df_f(i))$, increase $f(u)$ by ε, and decrease $c_M(u)$ by ε. At this, $df_f(i)$ decreases by ε, while $df_f(j)$ increases by ε.

If $df_f(i)$ remains positive, then we take the next arc from the list $A_M(i)$ and start to process it. If $df_f(i)$ becomes zero or the list $A_M(i)$ is exhausted, then we finish with the vertex i and go over to the next vertex from D_k. When the entire set D_k is examined, then we start examining D_{k-1}.

Having examined the set D_1, we finish the completion and proceed to step 1.

3. *Correction.* We apply the chain expansion to the function $f(u)$ just constructed. (The chain expansion algorithm is described in [1]). All the chains that result from the expansion originate at the source. Some chains end at the sink, while the rest of them end at inner vertices with zero deficits. We set the flows on the chains ending at inner vertices equal to zero, and reduce $f(u)$ on the corresponding arcs. At this, the algorithm stops.

3. Justification

We now prove that the algorithm constructs a maximum flow. First, we introduce some notation and describe some properties of the network G_f needed in the proofs below.

Denote by $f_{k+1}(u)$ the function obtained as a result of the kth completion. Denote by $f_{\text{fin}}(u)$ the function resulting from the final completion. Denote by M_k the directory obtained as a result of the kth construction. We drop the subscripts M and f for all objects associated with this directory, i.e. we write N_k instead of N_{M_k}, df_k instead of df_{f_k}, and so on.

The network G_f possesses the following property. Provided $c_f(i, j) = 0$, if an arc u leads from i to j in the initial network, then $f(u) = c(u)$, while if u leads from j to i, then $f(u) = 0$.

Let $\bar{u}$ denote the arc opposite to u. By the properties of the network G_f, if $u \in G_k$ and $u \notin G_{k-1}$, then $\bar{u} \in U_{k-1}$ and in the kth completion the function f increases on the arc $\bar{u}$.

Recall that the rank of a vertex, $R_k(i)$, is the distance from i to the sink in the kth directory. If $i \notin N_k$, then we put $R_k(i) = \infty$.

LEMMA 1. *The distance from each vertex to the sink does not decrease under completions, i.e.* $R_{k+1}(i) \geq R_k(i)$.

PROOF. Clearly, the assertion of the lemma holds for the vertices of rank 1. Let the assertion of the lemma hold for the vertices with $R_{k+1}(i) \leq l$. We prove that if $R_{k+1}(i) = l+1$, then $R_k(i) \leq l+1$. Since $R_{k+1}(i) = l+1$, there exists an arc $u \in U_{k+1}$ leading from i to some vertex j with $R_k(j) = l$. If $u \in G_k$, then $R_k(i) \leq R_k(j) + 1$ and the inequality $R_k(j) \leq R_{k+1}(j)$ implied by the inductive hypothesis yields $R_k(i) \leq R_{k+1}(j) + 1 = l+1$. If $u \notin G_k$, then $\bar{u} \in U_k$ by the properties of the network G_f. Hence $R_k(i) = R_k(j) - 1$, and $R_{k+1}(i) = R_{k+1}(j) + 1 \geq R_k(j) + 1 = R_k(i) + 2$.

LEMMA 2. *If for some vertex i there exists an arc $u = (i, j)$ leaving i, $u \in U_{k+1}$, $u \notin U_k$, then $R_{k+1}(i) > R_k(i)$.*

PROOF. If $u \in G_k$ and $u \notin U_k$, then $R_k(i) \leq R_k(j)$. However, $R_{k+1}(i) = R_{k+1}(j) + 1 \geq R_k(j) + 1 > R_k(i)$. If $u \notin G_k$, then $\bar{u} \in U_k$. Hence $R_k(i) = R_k(j) - 1$ and $R_{k+1}(i) \geq R_k(i) + 2 > R_k(i)$.

LEMMA 3. *Let $i \in N_{k+1}$ and let the deficit of i be strictly positive. Then $R_{k+1}(i) > R_k(i)$.*

PROOF. If $df_{k+1}(i) > 0$, then at the kth completion the list of arcs was exhausted in an attempt to increase f_k on the arcs leaving i. This means that the network G_{k+1} contains no arc from U_k beginning at i. Hence the list $A_{k+1}(i)$ consists of arcs that do not belong to U_k. Then, by Lemma 2, $R_{k+1}(i) > R_k(i)$. The lemma is proved.

Let n be the number of vertices in the network.

THEOREM 1. *The number of completions performed by the above algorithm does not exceed n^2.*

PROOF. Lemma 1 implies that the distance from any vertex to the sink does not decrease. Since no vertex can be at a finite distance exceeding n from the sink, its distance can be increased at most n times. Hence, the total number of distance increments cannot exceed n^2. If no deficient vertices are left after a completion, then no more completions follow. By Lemma 3, other completions increase at least one vertex-to-sink distance. Hence, the total number of completions does not exceed n^2.

COROLLARY. *The algorithm is finite.*

Denote by L_k the set $N \setminus N_k$.

LEMMA 4. *If $i \in L_k$ and $j \in N_k$, then $c_k(i, j) = 0$.*

PROOF. If $j \in N_k$, then a directed path Γ_j from j to t exists in the network G_k. Should the assertion of the lemma be violated, then the path $\Gamma_i = ((i, j), \Gamma_j)$ would lead from i to t, which contradicts the fact that $i \notin N_k$.

COROLLARY. *In the initial network the cut (L_k, N_k) is saturated.*

LEMMA 5. *If $df_{\mathrm{fin}}(i) > 0$ before the correction, then $i \notin N_{\mathrm{fin}}$.*

PROOF. If after the construction of a directory we proceed to the correction, then no directory vertex is deficient.

COROLLARY. *The source s does not belong to N_{fin}.*

THEOREM 2. *The algorithm constructs a maximum flow.*

PROOF. By Theorem 1, the number of completions is finite and at some step we proceed to the correction. The corollary of Lemma 4 claims that the cut $(L_{\text{fin}}, N_{\text{fin}})$ is saturated before the correction. Let us prove that the chains at which the flow vanishes under the correction cannot follow the arcs of this cut. Indeed, by Lemma 5, the beginning and the end of each eliminated chain lie in the set L_{fin}. If such a chain follows the arcs of the cut, then it must pass from N_{fin} to L_{fin} at least once. However, this is impossible, since, due to the fact that the cut is saturated, the flows on its arcs are directed from L_{fin} to N_{fin}. Hence, the cut $(L_{\text{fin}}, N_{\text{fin}})$ remains saturated after the correction. The correction algorithm implies that all the deficits of inner vertices become zero after the correction. Thus, the correction results in a flow function that saturates a cut. By the corollary of Lemma 5, the source s belongs to L_{fin}. By the definition of the directory, the sink t belongs to N_{fin}, i.e. the saturated cut $(L_{\text{fin}}, N_{\text{fin}})$ separates the source from the sink. Now the maximum flow—minimum cut theorem implies that a maximum flow has been constructed.

4. Complexity bounds

Let us assume that the information on the network and the directory is arranged in such a way that a finite number $O(1)$ of operations is required to proceed from one arc leaving a given vertex to the next such arc. Such a possibility is offered, for example, by the list representation of the network and directory [1]. Recall that n denotes the number of vertices in the network, and p the number of arcs.

We now estimate the number of operations required by completions. Let us estimate first the number of operations involved in processing arcs. The description of the completion algorithm implies that an arc at a given distance from the sink cannot be saturated more than once. Since arc-to-sink distances do not exceed n, each arc can be saturated at most n times. Hence the total number of saturation events does not exceed np and the number of operations required by the saturation does not exceed $O(np)$.

While processing the arcs leaving a vertex, all the arcs except perhaps the last one become saturated. Thus, at each completion, at most one arc with a nonzero and nonsaturating flow appears at any vertex. The total number of such arcs cannot exceed n at a completion. By Theorem 1, the total number of completions does not exceed n^2. Hence the total number of nonsaturation events does not exceed n^3 and the total number of operations spent on nonsaturations does not exceed $O(n^3)$.

Observe that the number of operations required for processing vertices during one completion does not exceed $O(n)$, because each vertex is processed at most once. Hence, the total number of operations required for processing vertices does not exceed $O(n^3)$.

Let us estimate now the number of operations spent on constructing the directories. If we had to construct the directory each time anew, then we would need $O(n^2 p)$ operations, because the construction of each directory requires $O(p)$ operations [1]. However, we can improve this estimate if we use the algorithm of modifying the current directory suggested by Dinits [1]. Let us describe this algorithm briefly as it applies to our case.

0. Prior to the first completion, we construct the directory in the conventional way.

1. In the process of completion, we form vertex sets K_l by including in them the vertices of rank l whose list of leaving arcs is exhausted. The vertices from K_l are declared as "candidates" for the $(l+1)$st rank.

2. We start processing the sets K_l with that of minimal index. Let $i \in K_l$. We delete from the lists of leaving arcs all directory arcs that lead from the vertices of rank $(l+1)$ to i. If the list of the arcs leaving some vertex is exhausted, then we include this vertex in K_{l+1}.

We start processing the arcs of the network G_f that leave i. If at least one of these arcs leads to a vertex of rank l, then we put the rank of i equal to $l+1$ and form the list of the arcs leaving i consisting of the arcs leading to vertices of rank l. If we find no arc leading to vertices of rank l, then we include i in K_{l+1} and start processing the next vertex from K_l. When the set K_l is processed, we proceed to processing K_{l+1}, and so on, until we process K_{n-1}. The vertices included in K_n need no processing. They acquire infinite ranks and do not take part in subsequent completions.

The above algorithm allows one to modify some parts of the directory instead of reconstructing it each time anew. Due to this fact, the complexity of processing directories does not exceed $O(np)$, as indicated in [1].

The complexity of the correction is determined by the complexity of the chain expansion procedure and, as shown in [1], does not exceed $O(np)$.

Thus, we have proved that the overall complexity of our algorithm is determined by the complexity of completions and can be estimated as $O(n^3)$.

5. An experimental comparison of the speed of different algorithms for constructing a maximum flow

To accomplish the experiment, the author used the programming language PL/1 to develop procedures that implement the five algorithms of constructing a maximum flow, namely, FFFL for the Ford–Fulkerson algorithm (in the Edmonds–Karp modification), DFLW for the Dinits algorithm, KFL for the Karzanov algorithm, CHFL for the algorithm AWD, and CHFL1 for the algorithm AWB suggested in the present paper. The experiment showed that the algorithm of modifying the directory employed by AWB can be improved to gain speed in the following way. If no vertices of rank l are found in the directory while processing the set K_l, then any further modification of the directory is terminated. All vertices in the sets K_q, $q \geq l$, that are not processed yet, as well as the directory vertices distant from the sink by more than l, acquire infinite ranks and are not processed further.

In the course of the experiment, we solved problems with dimensions ranging from 100 vertices and 500 arcs to 1000 vertices and 5000 arcs. The networks were constructed with the help of a random number generator. The tail and the head of an arc were random values distributed uniformly from 1 to n. Moreover, in problems with p arcs, we replaced the random tail of the first $\lfloor p/40 \rfloor + 1$ arcs by the source, while the heads of the same number of the last arcs were replaced by the sink. Arc capacities were random values distributed uniformly from 10 to 500. The capacities of the source and sink arcs added were made fivefold.

A special code controlled the correctness of the procedure performance, flow restrictions, and the saturation of the minimum cut. For chosen network dimensions (Table 1), each procedure (except FFFL, which appeared too slow for large problems) was applied to solve three identical problems on a DOS operated ES-1022 computer.

Table 1

Solution time (in seconds) averaged for three problems with given network dimensions

Number of vertices	Number of arcs	FFFL	DFLW	KFL	CHFL	CHFL1
100	500	17.0	4.2	5.3	5.6	2.7
200	1000	68.3	9.6	11.8	13.0	7.4
300	1500	159.7	14.3	17.7	19.4	9.4
400	2000	285.1	20.0	24.8	27.1	19.7
500	2500	—	24.9	30.0	32.7	22.7
600	3000	—	41.0	48.0	55.0	19.1
700	3500	—	41.7	50.9	56.0	24.5
800	4000	—	46.7	56.6	61.8	34.5
900	4500	—	57.3	69.7	76.8	38.4
1000	5000	—	54.5	66.9	72.4	43.5

Table 1 indicates that the running time of FFFL grows with the network dimensions considerably faster than that of the other procedures. This allows us to infer that the directory technique applied to the problem of constructing a maximum flow appears to be very efficient not only in theory, but in practice as well. We can regard the rate of growth of the running time required by the other procedures approximately linear (for the problems tested). It is by far less than that given by upper running time estimates, though these estimates are tight (sharp) in the sense that a series of extreme examples exists for each of the algorithms under investigation that confirms the tightness.

The analysis of the process of solving the problems generated has shown that these problems appear "too simple" for the theoretically effective, but sophisticated algorithms that have been developed. The Karzanov algorithm and especially the algorithm AWD are created as if they were expected to construct a flow in long and complicated directories. In the real problems solved here, the maximal length of directories did not exceed 30, which devaluated all the efforts of the authors to make their algorithms work more effectively. In our experiment, simpler algorithms rose to the forefront, namely, the Dinits algorithm, whose work in one directory is not complicated, and AWB, which passes from a step directory to a current one and skips the balancing implied by KA.

We have not managed to reveal peculiarities of different algorithms involving directories in this experiment. It would be interesting to continue experiments with more sophisticated network generators and understand how difficult practical problems are. However, on the basis of our limited practical experience, we can recommend the procedure implementing AWB for solving problems of large dimensions. The RAM space required to handle information arrays grows linearly with network dimensions. This space (in bytes) can be calculated by the formula $S = k_1 p + k_2 n$. The coefficients k_1 and k_2 for the above procedures are listed in Table 2.

Table 2

Coefficients	FFFL	DFLW	KFL	CHFL	CHFL1
k_1	16	18	24	26	20
k_2	6	8	14	14	16

References

1. G. M. Adel'son-Vel'skiĭ, E. A. Dints, and A. V. Karzanov, *Flow algorithms*, "Nauka", Moscow, 1975. (Russian)
2. B. V. Charkasskiĭ, *Multiterminal two-commodity problems*, Studies in Discrete Optimization (A. A. Fridman, editor), "Nauka", Moscow, 1976, pp. 261–290. (Russian)
3. _____, *An algorithm for constructing a maximal flow through a network requiring $O(n^2\sqrt{p})$ operations*, Mathematical Methods for Solving Economic Problems, no. 7 (N. P. Fedorenko and E. G. Gol'shteĭn, editors; Suppl. to Ėkonom. i Mat. Metody), "Nauka", Moscow, 1977, pp. 117–126. (Russian)
4. E. A. Dinits, *An algorithm for solving the maximum flow problem in a network with a power estimate*, Dokl. Akad. Nauk SSSR **194** (1970), 754–757; English transl. in Soviet Math. Dokl. **11** (1970).
5. L. R. Ford, Jr., and D. R. Fulkerson, *Flows in networks*, Princeton Univ. Press, Princeton, NJ, 1962.
6. V. P. Grishukhin, *Polyhedra associated with structures and minimax combinatorial problems*, Graphs, Hypergraphs, and Discrete Optimization Problems, "Znanie", Kiev, 1977, pp. 14–16. (Russian)
7. A. V. Karzanov, *Finding the maximum flow through a network by the method of preflows*, Dokl. Akad. Nauk SSSR **215** (1974), 49–52; English transl. in Soviet Math. Dokl. **15** (1974).

On the Extremality of the Rank Function of a Connected Semimodular Lattice

V. P. GRISHUKHIN

1.

The following situation, illustrated by the examples below, is typical for finding faces and vertices (extremal points) of polyhedra.

The faces of the travelling salesman polyhedron are described by direction vectors in the space indexed by the edges of the complete graph on the vertex set V. To each vector $h = \{h_{ij} : i, j \in V\}$ of this kind, i.e., to each face, there corresponds a weighted subgraph G that is the support of the vector h. An edge (ij) is contained in the above graph if and only if $h_{ij} \neq 0$; in this case its weight equals h_{ij}.

Extreme rays of the metric cone on a finite set V are also defined by direction vectors $\{h_{ij} : i, j \in V\}$ whose supports define certain subgraphs of the complete graph. Naturally, not every weighted subgraph determines a face of the travelling salesman polyhedron, or an extreme ray of the metric cone. Therefore, it is of interest to distinguish certain classes of graphs that generate faces or extreme rays.

The situation is similar for extreme rays of the cone of monotone submodular set functions defined on the Boolean algebra $\mathscr{B}(V) = 2^V$ of all subsets of the set V. For any subset $T \subseteq V$, there exists a unique maximal subset $T_f \supseteq T$ such that $f(T_f) = f(T)$. The mapping $T \to T_f$ takes the Boolean algebra $\mathscr{B}(V)$ to the lattice L_f consisting of f-closed sets. The restriction of f to L_f, denoted by $\bar{f}$, is a strictly monotone submodular function; that is, it belongs to the cone $\mathscr{C}_0(L_f)$ of monotone submodular functions on L_f.

It was shown in [2] that an $f \in \mathscr{C}_0(V)$ is extremal, that is, lies on an extreme ray of the cone $\mathscr{C}_0(V)$, if and only if $\bar{f} \in \mathscr{C}_0(L_f)$ is extremal. However, there exist finite lattices L such that the cones $\mathscr{C}_0(L)$ have no strictly

---

1991 *Mathematics Subject Classification.* Primary 90B10.
Translation of Problems of Discrete Optimization and Methods for Their Solution, Tsentral. Èkonom.-Mat. Inst. Akad. Nauk SSSR, Moscow, 1987, pp. 218–223.

monotone extremal functions. For example, there are no such functions in $\mathscr{C}_0(V)$. Therefore, the following problem arises: describe the class $\mathscr{K}$ of finite lattices L such that the corresponding $\mathscr{C}_0(L)$ possesses strictly monotone extremal functions. The above examples show that the class $\mathscr{K}$ defines extreme rays of the cone $\mathscr{C}_0(V)$.

In this paper we show that connected semimodular lattices belong to the class $\mathscr{K}$.

2.

Let L be a finite lattice. Two intervals b_1/a_1 and b_2/a_2 are said to be *perspective* if either $a_1 \wedge b_2 = a_2$, $a_1 \vee b_2 = b_1$, or $a_2 \wedge b_1 = a_1$, $a_2 \vee b_1 = b_2$. The transitive closure of the perspectivity relation is an equivalence (called the *projectivity*) on the set of simple intervals of the lattice L. A lattice L is said to be *connected* if all the simple intervals belong to the same projectivity class.

Let L be a semimodular (from above) lattice. In a semimodular lattice, all the chains connecting an arbitrary element $a \in L$ with $0(L)$ have the same length $h(a)$; it is equal to the number of simple intervals in each of these chains, and is called the *rank*, or the *height*, of a. The height h, regarded as a function on a lattice, is submodular on L. Let $h_P(a)$ be the number of simple intervals of some chain connecting a with $0(L)$ that belong to the class P. The proof that $h_P(a)$ is independent of the choice of the chain for modular lattices presented in [1], Chapter X, §4, requires only the semimodularity condition, and hence can be easily modified to become valid for semimodular (from above) lattices.

PROPOSITION 1. *The function h_P is submodular on L for any projectivity class P.*

PROOF. Let $a, b \in L$ be incomparable. Choose a maximal chain $l = \{a_0 = a \wedge b \prec a_1 \prec \cdots \prec a_k = a\}$ connecting $a \wedge b$ with a. Let $b_1 = a_1 \vee b$. Since L is semimodular, b_1 covers b, i.e., $b_1 \succ b$. Let a_i be the minimal element of the chain l not less than b_1; then $a_i \wedge b_1 = a_{i-1}$ and $b_2 = a_i \vee b_1 \succ b_1$. Evidently, the simple intervals b_1/b and b_2/b_1 are perspective to the simple intervals $a_1/a \wedge b$ and a_i/a_{i-1}, respectively. Moving ahead in the same manner, we see that there exists a chain l_1 connecting b with $a \vee b$ such that any simple interval of l_1 is perspective to a simple interval of the chain l. Since $h_P(a \vee b) - h_P(b)$ is the number of simple intervals of l_1 belonging to the class P, we obtain $h_P(a \vee b) - h_P(b) \leq h_P(a) - h_P(a \wedge b)$, or, in other words, $h_P \in \mathscr{C}_0(L)$.

Since $h = \sum_P h_P$, and an extremal function cannot be represented as the sum of other monotone submodular functions, we obtain the following.

COROLLARY 2. *If a semimodular lattice is not connected, then its rank function is not extremal.*

It turns out that the converse statement is also true.

PROPOSITION 3. *The rank function of a connected semimodular lattice L is extremal.*

PROOF. Consider the following equations:

$$f(a) - f(b) = f(p) - f(p_*)$$

for all $p \in \mathscr{F}(a/b)$ and for all simple intervals a/b that are not irreducible, that is, $a/b \neq p/p_*$. Here $\mathscr{F}(a/b)$ is the set of $\vee$-nonfactorable elements $p \in L$ such that $p \vee b = a$, $p \wedge b = p_*$. Since the lattice L is connected, we obtain the uniqueness, up to a factor, of the solution of the above equations for which the differences $f(a) - f(b) = t$ are equal for all simple intervals; in other words, f is proportional to the rank function: $f = t \cdot h$.

Let us show that h is the unique strictly monotone extremal function on a semimodular lattice. We say that two simple perspective intervals a_1/b_1 and a_2/b_2 form a *square* (in the Hasse diagram) if either $a_1 \succ a_2$ and $b_1 \succ b_2$, or $a_2 \succ a_1$ and $b_2 \succ b_1$. The four elements $(a, b, a \wedge b, a \vee b)$ form a square if all the four intervals $a/a \wedge b$, $b/a \wedge b$, $a \vee b/a$, $a \vee b/b$ are simple. For a semimodular lattice, the simplicity of the intervals $a/a \wedge b$ and $b/a \wedge b$ implies that $(a, b, a \wedge b, a \vee b)$ is a square.

Hien Quang Nguyen showed in [3] that the system of submodularity inequalities

$$f(a) + f(b) - f(a \wedge b) - f(a \vee b) \geq 0 \tag{1}$$

for a Boolean algebra is equivalent to the subsystem of the same system consisting of the inequalities corresponding to quadruples $(A \cup i, A \cup j, A, A \cup \{i, j\})$, $i, j \notin A \subset V$, that form squares.

This is not the case for semimodular lattices. For example, for the simplest nontrivial semimodular lattice (Figure 1) having three squares, the function

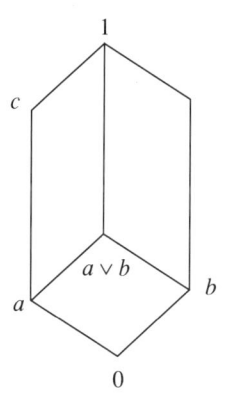

FIGURE 1. The simplest semimodular lattice.

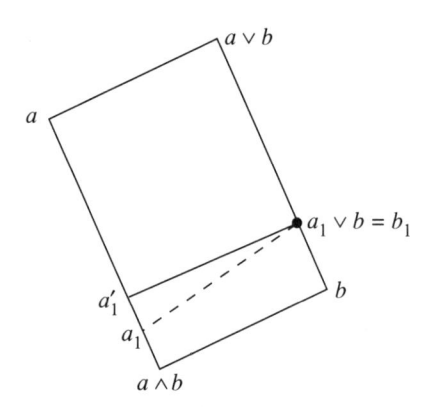

FIGURE 2

f taking values $f(0) = 0$, $f(a) = f(b) = 2$, $f(a \vee b) = 3$, $f(c) = f(d) = -2$, $f(1) = -3$ is not submodular, though the submodularity inequalities hold for all the squares.

However, the following statement is true.

LEMMA 4. *For a monotone f to be submodular on a semimodular lattice L, it suffices that the inequalities (1) hold only for squares (the remaining inequalities following by monotonicity).*

PROOF. Let f be monotone. Let us show that inequality (1) for a quadruple $(a, b, a \wedge b, a \vee b)$ that is not a square is implied by the same inequalities for certain squares.

First of all, we assume that b covers $a \wedge b$, that is, $b \succ a \wedge b$, and prove the assertion of the lemma by induction on $k = h(a) - h(a \wedge b)$. Evidently, $k > 1$, since for $k = 1$, i.e., for $a \succ a \wedge b$, the semimodularity implies that $(a, b, a \wedge b, a \vee b)$ is a square. Let an element $a_1 < a$ cover $a \wedge b$. Then $b_1 = a_1 \vee b$ covers b. Put $a_1' = a \wedge b_1$ (see Figure 2) and consider inequalities (1) for the quadruples $(a, b, a \wedge b, a \vee b)$, $(a, b_1, a \wedge b_1 = a_1', a \vee b_1)$, $(a_1', b, a \wedge b, a_1' \vee b)$:

$$f(a) + f(b) - f(a \wedge b) - f(a \vee b) \geq 0,$$
$$f(a) + f(b_1) - f(a_1') - f(a \vee b) \geq 0,$$
$$f(a_1') + f(b) - f(b_1) - f(a \wedge b) \geq 0.$$

It is easy to see that the first of the above inequalities is the sum of the other two. Since $h(a) - h(a_1') < k$, the second inequality is implied by square inequalities according to the inductive hypothesis. Since f is monotone and $a_1' \geq a_1$, we obtain $f(a_1') \geq f(a_1)$. Hence, the third inequality is yielded by the inequality $f(a_1) + f(b) - f(b_1) - f(a \wedge b) \geq 0$ for the square $(a_1, b, a \wedge b, b_1)$.

Assume now that b does not cover $a \wedge b$. In this case, we prove the lemma by induction on $m = h(b) - h(a \wedge b)$. Let b_1 be the element covered by b such that $b_1 \succ a \wedge b$, and hence, $h(b_1) - h(a \wedge b) = m - 1$. Evidently, $a \wedge b_1 = a \wedge b$. If $a \vee b_1 = a \vee b$, then, since $f(b) \geqslant f(b_1)$, inequality (1) is implied by the inequality $f(a) + f(b_1) - f(a \wedge b_1) - f(a \vee b_1) \geqslant 0$, which follows from square inequalities by the inductive hypothesis.

If $a \vee b_1 \neq a \vee b$, then $(a \vee b_1) \vee b = a \vee b$ and $(a \vee b_1) \wedge b = b_1$. In this case, the inequalities $f(a) + f(b_1) - f(a \wedge b) - f(a \vee b_1) \geqslant 0$ and $f(a \vee b_1) + f(b) - f(b_1) - f(a \vee b) \geqslant 0$, which are implied by square inequalities according to the inductive hypothesis, are to be added to obtain inequality (1).

The above lemma implies that to any basis subsystem of (1) defining an extremal function there corresponds a basis subsystem of square inequalities. However, the rank function satisfies such a system. Since any basis system possesses a unique solution, up to a factor, we see that the rank function is the unique solution for any basis system.

Thus, the above reasoning yields

THEOREM 5. *A semimodular lattice possesses a strictly monotone extremal function if and only if it is connected; in this case, such a function is unique and coincides with the rank function of the lattice.*

In conclusion, we want to mention that any connected semimodular lattice is simple. Yet, not every simple semimodular lattice is connected. In Figure 3 we present a simple disconnected semimodular lattice having two projectivity classes: $-$ and $=$.

For modular and geometric lattices, the notions of simplicity and connectivity coincide [1]. Since any geometric lattice is the lattice of closed sets of a connected matroid, Theorem 5 implies the following statement due to Hien Quang Nguyen [3]:

COROLLARY 6. *The rank function of a matroid is extremal if and only if the matroid is connected.*

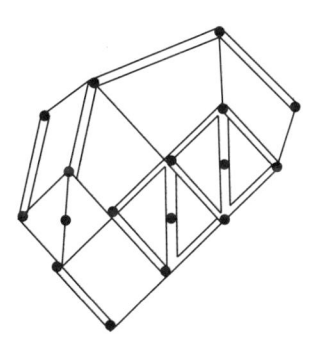

FIGURE 3. A simple disconnected semimodular lattice with two projectivity classes.

References

1. Garrett Birkhoff, *Lattice theory*, new ed., Amer. Math. Soc., Providence, RI, 1967.
2. V. P. Grishukhin, *Generators of the cone of submodular matrices*, Tsentral. Èkonom.-Mat. Inst. Akad. Nauk SSSR, Moscow, 1982. (Russian)
3. Hien Quang Nguyen, *Semimodular functions and combinatorial geometries*, Trans. Amer. Math. Soc. **238** (1978), 355–383.

On Polynomial Solvability Conditions for the Simplest Plant Location Problem

V. P. GRISHUKHIN

It is well known that plant location problems are very important and often occur in applications; for references see the highly informative book [3]. In the present paper, we consider two main cases when the simplest plant location problem (SPLP) can be solved in polynomial time; one of them occurs when the incidence matrix of the problem is totally balanced, and the other when it is connected with respect to a tree.

The classical SPLP has the following form:

$$\sum_{i \in I} c_i x_i + \sum_{i \in I, j \in J} c_{ij} x_{ij} \to \min, \tag{1}$$

$$\sum_{i \in I} x_{ij} = 1, \quad 0 \leqslant x_{ij} \leqslant x_i, \quad x_i \in \{0, 1\}. \tag{2}$$

Here $c_i \geqslant 0$ is the cost of establishing the ith facility, while $c_{ij} \geqslant 0$ is the transportation cost from the ith facility to the jth consumer.

For a fixed $(0, 1)$-vector $\{x_i : i \in I\}$ with support $X = \{i \in I : x_i = 1\}$, the SPLP decomposes into $n = |J|$ simple problems

$$\sum_{i \in X} c_{ij} x_{ij} \to \min, \quad \sum_{i \in X} x_{ij} = 1, \quad x_{ij} \geqslant 0, \quad i \in X \quad (j \in J).$$

These problems have the following $(0, 1)$-solutions:

$$x_{ij} = \begin{cases} 1 & \text{for } i = i_j \in \operatorname{Arg\,min}_{i \in X} c_{ij}, \\ 0 & \text{for } i \neq i_j. \end{cases}$$

Hence the optimal value of the SPLP for the above fixed $\{x_i\}$ equals

$$f(X) = \sum_{i \in X} c_i + \sum_{j \in J} \min_{i \in X} c_{ij},$$

1991 *Mathematics Subject Classification.* Primary 68R10.
Translation of Economico-Mathematical Modelling of Discrete Processes, Tsentral. Èkonom.-Mat. Inst. Akad. Nauk SSSR, Moscow, 1988, pp. 52–65.

and the SPLP itself is equivalent to the minimization problem for $f(X)$.

The function $f(X)$ is a supermodular set function; using $(0, 1)$-variables x_i (see [3]), one can rewrite $f(X)$ in the following pseudo-Boolean form:

$$f(x) = \sum_{i \in I} c_i x_i + \sum_{j \in J} \sum_{k=1}^{m} \Delta_{jk} \prod_{i \in A_{jk}} (1 - x_i). \qquad (3)$$

Here A_{jk} is the subset of the set I obtained in the following way. The natural nondecreasing ordering of the elements in the jth column of the matrix $\|c_{ij}\| = C$,

$$c_{i_1 j} \leqslant c_{i_2 j} \leqslant \cdots \leqslant c_{i_m j},$$

induces an ordering $\{i_1, \ldots, i_m\}^j$ of the set I. The set $A_{jk} = \{i_1, \ldots, i_k\}^j$ consists of the first k elements of the above ordering, while

$$\Delta_{jk} = c_{i_{k+1} j} - c_{i_k j}, \quad 1 \leqslant k \leqslant m-1, \qquad \Delta_{jm} = 0.$$

A. A. Ageev [1] was apparently the first to observe that the minimization problem for the function $f(x)$ in Boolean variables x_i is equivalent to the following covering problem:

$$\sum_{i \in I} c_i x_i + \sum_{j \in J} \sum_{k=1}^{m} \Delta_{jk} z_{jk} \to \min,$$

$$\sum_{i \in A_{jk}} x_i + z_{jk} \geqslant 0, \qquad x_i, z_{jk} \in \{0, 1\}. \qquad (4)$$

This fact is easy to understand upon observing that the optimal values of the variables x_i and z_{jk} for problem (4) satisfy the relation

$$z_{jk} = \prod_{i \in A_{jk}} (1 - x_i).$$

Substituting this expression for z_{jk} in the objective function in (4), one obtains exactly the function of x from (3).

Let us rewrite the constraints of problem (4) in the matrix form:

$$A^T x + z \geqslant 1,$$

where $A = A(C) = (a_{i, jk})$ is the incidence matrix for the SPLP; its (jk)th column a_{jk} is the incidence vector of the set A_{jk}.

Whether the SPLP is solvable in polynomial time or not depends mostly on the properties of its incidence matrix.

If A is a balanced matrix, then the polyhedron of the linear relaxation for (4) possesses only integer vertices. This is equivalent to the fact that the optimum for the SPLP is attained at an integer vertex of the linear relaxation for (1),(2). In this case one can use any polynomial linear programming algorithm to obtain an integer optimal solution, provided the algorithm is endowed with a procedure for finding a basis solution. Unfortunately, there are no polynomial algorithms to determine whether a $(0, 1)$-matrix is balanced or not.

DEFINITION. A $(0, 1)$-matrix is said to be *balanced* if it has no odd-length cycle submatrices. A square $(0, 1)$-matrix is called a *cycle matrix* if it has no equal rows or columns, and each of its rows and columns contains exactly two units. The number of rows is said to be the *length* of the cycle.

If a balanced matrix does not contain even-length cycle submatrices, it is said to be *totally balanced*. This property can be verified in polynomial time [5]. The algorithm is based on the following constructive property Q, which is equivalent to the property of being totally balanced. A $(0, 1)$-matrix possesses *property Q* (is a *Q-matrix*) if any of its submatrices contains either a row having exactly one nonzero entry, or two comparable columns. Since the initial definition is symmetric with respect to rows and columns, one can interchange rows and columns in the above formulation as well.

A Q-matrix can be reduced to the standard form by transposing its rows and columns. An $m \times n$ matrix A in standard form possesses the following property: given any i, $1 \leqslant i \leqslant m-1$, delete the first $i-1$ rows; then all the columns of the truncated matrix that contain unity in the first position (in other words, those that initially contained unity in the ith position) are comparable.

The dual problem for the linear relaxation of the covering problem (4) with a totally balanced constraint matrix in standard form is solved by the greedy algorithm. Once a dual solution is obtained, it is easy to find an integer solution of the initial problem. V. A. Trubin [6] was first to discover this fact.

He arrived at totally balanced matrices (in fact, at matrices with property Q) while studying the system of subtrees of a tree T with the vertex set $V(T) = I$. Yet it is easy to show that any totally balanced matrix is the incidence matrix for the vertex sets of the subtrees of some tree. To do this, it suffices, for instance, to use the simple polynomial algorithm for recognition of such tree-matrices described in [2].

In this algorithm, it is assumed that the row set of the matrix tested spans a tree T (i.e., is the vertex set for this tree), while each of its columns is the incidence vector for the vertex set of some subtree in T. First, all the columns having a single nonzero entry are deleted; they correspond to one-vertex subtrees. Next, the algorithm tests whether a given row corresponds to a terminal vertex of the tree T, and if it does, the row is deleted. This test relies on the following simple observation: all non-one-vertex subtrees that contain a terminal vertex of T must contain also the only vertex adjacent to it. In other words, the row that corresponds to a terminal vertex must be contained in the row corresponding to the adjacent vertex. It is easy to see that, according to property Q, a totally balanced matrix having no one-element columns always possesses comparable rows.

It is interesting to note that the pseudo-Boolean polynomial $f(x)$ in (3) that corresponds to the SPLP with a totally balanced incidence matrix $A(C)$ can be minimized with the greedy algorithm as well. Totally balanced polynomials generalize the regular polynomials defined in [3]. For a regular polynomial, the family of subsets A_{jk} is the family of the vertex sets of subchains of the chain formed by the set I. An algorithm for minimizing a regular polynomial is based on the following fact: the set I always possesses an element (e.g., an extreme one) such that all the sets A_{jk} containing this element are comparable by inclusion. This property for totally balanced systems can be deduced easily from property Q, while for matrices in the standard form it is evident; this enables us to apply the algorithm of V. I. Beresnev [3] to totally balanced polynomials.

This algorithm is a realization of the ordinary pseudo-Boolean programming algorithm applied to a totally balanced polynomial $f(x)$. In any pseudo-Boolean programming algorithm, a variable x_i is chosen as distinguished, and the polynomial $f(x)$ is represented as $f(x) = x_i f_1^i(x) + f_2^i(x)$, where the polynomials f_1^i and f_2^i do not depend on x_i.

Let x^* be the value of x that minimizes $f(x)$. Evidently, if $f_1^i(x^*) > 0$, then one necessarily has $x_i^* = 0$, while if $f_1^i(x^*) < 0$, then $x_i^* = 1$. These properties define x_i^* implicitly as a Boolean function of the other variables x_k^*. For a totally balanced polynomial, this function can be written down explicitly.

Let us rewrite the polynomial (3) in the variables $y_i = 1 - x_i$:

$$f(y) = c_0 - \sum_{i \in I} c_i y_i + \sum_{B \in \mathscr{A}} \Delta(B) \prod_{i \in B} y_i,$$

where $c_0 = \sum_{i \in I} c_i$, $\mathscr{A}$ is the family of all distinct A_{jk}'s, and $\Delta(B) = \sum \{\Delta_{jk} : A_{jk} = B\}$. Let the set I be ordered in accordance with the ordering of the rows of A written in the standard form. As it was mentioned above, all the sets $B \in \mathscr{A}$ containing the first element of I form the family $\mathscr{A}_1$ whose members are included one into another:

$$\mathscr{A}_1 = \{B \in \mathscr{A} : B \ni 1\} = \{B_1 \subset B_2 \subset \cdots \subset B_k\},$$

where $k = |\mathscr{A}_1|$.

Let us separate the terms of $f(y)$ that contain the variable y_1:

$$f(y) = c_0 + y_1 \left(-c_1 + \sum_{B \in \mathscr{A}_1} \Delta(B) \prod_{i \in B - \{1\}} y_i \right)$$
$$- \sum_{i \in I - \{1\}} c_i y_i + \sum_{B \in \mathscr{A} - \mathscr{A}_1} \Delta(B) \prod_{i \in B} y_i$$
$$= y_1 f_1 + f_2.$$

Let us find an explicit formula for the first component y_1^* of the vector y^* at which the polynomial $f(y)$ attains its minimum.

First of all, let us observe that

$$-c_1 + \sum_{l=1}^{k} \Delta(B_l) < 0, \quad k = |\mathscr{A}_1|,$$

yields $y_1^* = 1$. Therefore, we can assume that $\sum_{l=1}^{k} \Delta(B_l) \geq c_1$. For $B_l \in \mathscr{A}_1$, put $B_l' = B_l - \{1\}$, $1 \leq l \leq k$. Let q, $1 \leq q \leq k$, be an index such that $-c_1 + \sum_{l=1}^{q-1} \Delta(B_l) < 0$, yet $-c_1 + \sum_{l=1}^{q} \Delta(B_l) \geq 0$. Let us consider the function $1 - \prod_{i \in B_q'} y_i$ and prove that there exists a minimizing vector $y^* = \{y_i^* : i \in I\}$ such that $y_1^* = 1 - \prod_{i \in B_q'} y_i^*$.

Let q_p be the minimal index l of the sets B_l that contain elements $p \in I$ satisfying $y_p^* = 0$. If there are no such sets, we assume $q_p = k+1$. Therefore,

$$\prod_{i \in B_l'} y_i^* = 1 \text{ for } l \leq q_p - 1, \qquad \prod_{i \in B_l'} y_i^* = 0 \text{ for } l \geq q_p,$$

and

$$f_1(y^*) = -c_1 + \sum_{l=1}^{q_p-1} \Delta(B_l).$$

Thus, if $q_p \leq q$, then $f_1(y^*) < 0$ and $1 - \prod_{i \in B_q'} y_i^* = 1$, whereas if $q_p > q$, then $f_1(y^*) > 0$ and $1 - \prod_{i \in B_q'} y_i^* = 0$, in accordance with the relation $y_1^* = 1 - \prod_{i \in B_q'} y_i^*$.

Substituting $1 - \prod_{i \in B_q'} y_i$ for y_1 in f, we obtain a polynomial in fewer variables. It is easy to verify that this polynomial is totally balanced, and the procedure described above can be applied once more.

We remark here that the following two facts are equivalent: (a) the polynomial $f(x)$ corresponds to the supermodular set function $f(x)$, and (b) the polynomials $f_1^i(x)$, $1 \leq i \leq m$, are monotone nonincreasing in x. Hence, the inequality $f_1^i(0) \leq 0$ yields the inequalities $f_1^i(x) \leq 0$ for all x; therefore, we can set $x_i^* = 1$. Similarly, if $f_1^i(1) \geq 0$, then $f_1^i(x) \geq 0$ for all x, and hence $x_i^* = 0$. Thus, we have arrived exactly at the famous elimination rules of the successive computations method due to V. Cherenin.

Another important case of polynomially solvable SPLP arises when the matrices $C = \|c_{ij}\|$ are connected with respect to a tree; this notion was introduced by E. Gimadi [4]. Below this property is described in somewhat different terms than in Gimadi's original paper. Let J be the column set of the matrix C. For any ordered pair of indices $i, k \in I$, introduce the sets

$$J_{ik}^C = \{j \in J : c_{ij} < c_{kj}\}.$$

DEFINITION. A matrix C is said to be *connected (with respect to a tree)* if there exist a tree T with the vertex set $V(T)$ and an embedding $\varphi : J \to V(T)$ such that each pair of subsets of vertices $\varphi(J_{ik}^C)$ and $\varphi(J_{ki}^C)$ is separated by an edge in T.

It is easy to show that if a matrix C is connected with respect to a tree, then the incidence matrix $A(C)$ is connected with respect to a certain (in general, another) tree. Therefore, we can restrict ourselves with studying of $(0,1)$-matrices. For brevity, the matrices connected with respect to a tree are occasionally just called *connected*.

We emphasize that the main difficulty in testing whether a matrix is connected with respect to a tree lies in finding not only the tree T, but also the corresponding embedding $\varphi: J \to V(T)$. For a given tree and a given embedding of J into $V(T)$, it is rather easy to detect connectivity; this takes $O(mn^2)$ operations with $n = |J|$, $m = |I|$.

In particular, it is easy to detect connectivity with respect to a tree T for a $(0,1)$-matrix whose rows are incidence vectors for vertex sets of subtrees of T. In this case, the connectivity property with respect to a tree coincides with the property of having no forks, introduced in [2]. A family $\mathscr{F}$ of subtrees of a tree T *contains no forks* if for any two subtrees T_1, $T_2 \in \mathscr{F}$, the differences $V(T_1) \setminus V(T_2)$ and $V(T_2) \setminus V(T_1)$ generate connected subtrees of T. The paper [2] contains a polynomial algorithm for solving the covering problem whose matrix A is the incidence matrix for a family of subtrees containing no forks.

One can easily verify, using the algorithm of recognition of $(0,1)$-tree-matrices [2], that any $(0,1)$-matrix A containing a column consisting only of units is the incidence matrix of subtrees for some tree $T(A)$. Since a column consisting only of units can be added to a $(0,1)$-matrix without violation of its connectivity, it is natural to assign to each $(0,1)$-matrix A the canonical tree $T(A1)$, where $A1$ is obtained from A by adding a column consisting only of units. Unfortunately, a connected matrix A need not be connected with respect to $T(A1)$.

The following necessary (yet not sufficient) condition for an arbitrary matrix C to be connected is known: for any pair of partitions $V_1 \cup V_2 = V_1' \cup V_2' = V(T)$, $V_1 \cap V_2 = V_1' \cap V_2' = \varnothing$, of the vertex set $V(T)$ of the tree T defined by edges of T, there exists a nonintersecting pair of subsets V_i and V_j' belonging to different partitions, i.e., $V_i \cap V_j' = \varnothing$. According to this condition, if a matrix C is connected, then for any two pairs of indices (ik) and (ml) at least one of the intersections

$$J_{ik}^C \cap J_{ml}^C, \quad J_{ki}^C \cap J_{ml}^C, \quad J_{ik}^C \cap J_{lm}^C, \quad J_{ki}^C \cap J_{lm}^C \qquad (5)$$

is empty. This condition turns out to be sufficient for strict families, defined below.

DEFINITION. A family of pairs of nonintersecting sets is said to be *strict* if each pair of sets in this family is a partition of the set J.

For a strict family of pairs $\{J_{ik}^C, J_{ki}^C\}$, the following two facts are equivalent: (a) at least one of the intersections in (5) is empty, and (b) the family is *parallel*, that is, for $J_1 \in \{J_{ik}^C, J_{ki}^C\}$, $J_2 \in \{J_{lm}^C, J_{ml}^C\}$, exactly one of the following conditions holds:

$$J_1 \subseteq J_2; \quad J_2 \subseteq J_1; \quad J_1 \cup J_2 = J; \quad J_1 \cap J_2 = \emptyset.$$

LEMMA 1. *The above necessary connectivity condition is also sufficient for matrices C with strict families $\{J_{ik}^C, J_{ki}^C\}$.*

PROOF. It suffices to find a tree T and an embedding $\varphi: J \to V(T)$ such that each pair of sets J_{ik}^C and J_{ki}^C is separated by an edge in T. An algorithm for reconstructing a tree from a parallel set of partitions (cuts) is well known, and so we describe it only briefly.

Choose $j_0 \in J$, and take the set not containing j_0 out of any pair (J_{ik}^C, J_{ki}^C). Then, for the representatives (just chosen) J_1 and J_2 of any two pairs, exactly one of the following options holds:

$$J_1 \subseteq J_2; \quad J_2 \subseteq J_1; \quad J_1 \cap J_2 = \emptyset.$$

An arc e_i is assigned to each representative. In the tree T to be constructed, the tails of two arcs e_1 and e_2 are adjacent if $J_1 \cap J_2 = \emptyset$ and the family does not contain a set J_3 such that either

$$J_1 \subseteq J_3, \quad J_3 \cap J_2 = \emptyset,$$

or

$$J_2 \subseteq J_3, \quad J_3 \cap J_1 = \emptyset.$$

The head of e_1 and the tail of e_2 are adjacent if $J_1 \supseteq J_2$, and there is no set J_3 such that $J_1 \supset J_3 \supset J_2$.

For such a construction, a subset of J (possibly empty) corresponds to each vertex of the tree. Exactly one arc enters every vertex of T, except for the root. If an arc e_i enters a vertex v, and arcs $e_{i_1}, e_{i_2}, \ldots, e_{i_k}$ leave this vertex, then the set corresponding to v is given by $J(v) = J_i - \bigcup_{j=1}^k J_{i_j}$. The set $J(r) = J - \bigcup_{j=1}^k J_{i_j}$ corresponds to the root r, where e_{i_j} are the arcs incident to r; recall that no arcs enter the root. If a set $J(v)$ of cardinality $|J(v)| > 1$ corresponds to a vertex v, then the vertex v in T must be replaced by an arbitrary tree with $|J(v)|$ vertices assigned arbitrarily to the elements $j \in J(v)$. In such a way we obtain the embedding of J into the vertex set of the newly obtained tree.

Lemma 1 is the base of an algorithm that detects connectivity of arbitrary matrices. In this algorithm, a given family of pairs is extended, in every possible way, to a strict family. Each strict family thus obtained is then tested for parallelism. If a parallel strict family is detected, then, by Lemma 1, a tree is assigned to this family.

The above algorithm for detecting connectivity is of branch-and-bound type. The main elimination test is as follows: among all the extensions of a given pair (J_{ik}^C, J_{ki}^C), only those are chosen that are parallel to all the nonextended pairs. This relies on the following statement.

PROPOSITION 1. *Given two pairs* (J_{ik}^C, J_{ki}^C) *and* (J_{lm}^C, J_{ml}^C), *let exactly one of the intersections in* (5) *be empty, for instance, let* $J_{ik}^C \cap J_{lm}^C = \varnothing$. *Then any two partitions that extend the above pairs contain the pairs*

$$(J_{ik}^C, J_{ki}^C \cup J_{lm}^C) \quad \text{and} \quad (J_{lm}^C, J_{ml}^C \cup J_{ki}^C). \tag{6}$$

PROOF. Indeed, let (J'_{ik}, J'_{ki}) and (J'_{lm}, J'_{ml}) be the extending partitions. Since there are three nonempty sets among the four possible intersections of the above sets, we see that $J'_{ik} \cap J'_{lm} = \varnothing$. Since $J_{ki} = J \setminus J_{ik}$, $J_{ml} = J \setminus J_{lm}$, we obtain $J'_{ki} \supseteq J'_{lm}$ and $J'_{ml} \supseteq J'_{ik}$; in other words, J'_{ki} and J'_{ml} contain the second members of the pairs from (6).

Remarkably, connected (0,1)-matrices possess almost the same property that is characteristic for totally balanced matrices.

PROPOSITION 2. *Any* $(0, 1)$-*matrix connected with respect to a tree contains no cycle submatrices of length* > 3.

The proof of this proposition follows immediately from Lemma 2 below and the following fact: if a matrix is connected with respect to a tree, each of its submatrices is connected with respect to the same tree.

LEMMA 2. *A cycle matrix of length* $n \geqslant 4$ *is not connected with respect to any tree.*

PROOF. Let $J = \{1, 2, \ldots, n\}$. Below all sums of the form $i_1 + i_2$ are considered modulo n. Without loss of generality, we can assume that the elements of a cycle matrix A have the following form:

$$a_{ii} = a_{i, i+1} = 1, \quad 1 \leqslant i \leqslant n,$$
$$a_{ij} = 0, \quad \text{otherwise};$$

here n is the length of the cycle. Since $n \geqslant 4$, there exist two rows i and $i+2$ such that

$$J_{i, i+2}^A = \{i+2, i+3\}, \quad J_{i+2, i}^A = \{i, i+1\}.$$

Suppose that there exist a tree T and an embedding $J \to V(T)$ such that A is connected with respect to T. Let us denote vertices in the image of J by the same indices $i, j \in J$. Since A is connected with respect to T, there exists a partition $V_1 \cup V_2 = V(T)$, $V_1 \cap V_2 = \varnothing$, defined by an edge of T, and satisfying the relations $\{i, i+1\} \subseteq V_1$, $\{i+2, i+3\} \subseteq V_2$. Let $j_1 \in V_2$ be the maximal element in the set $J \cap V_2$. This means that $j_1 + 1 \in V_1$. Evidently, the pair (j_1, j_1+1) differs from the pairs $(i, i+1)$, $(i+1, i+2)$, $(i+2, i+3)$, and besides, $\{i+1, i+2\} \cap \{j_1, j_1+1\} = \varnothing$. Hence,

$$J_{i+1, j_1}^A = \{j_1, j_1+1\}, \quad J_{j_1, i+1}^A = \{i+1, i+2\},$$

and, by the assumption, there exists a partition $V'_1 \cup V'_2 = V(T)$, $V'_1 \cap V'_2 = \varnothing$ generated by an edge in T such that $\{j_1, j_1+1\} \subseteq V'_1$, $\{i+1, i+2\} \subseteq V'_2$.

Since
$$V_1 \cap V_1' \ni j_1+1, \quad V_1 \cap V_2' \ni i+1, \quad V_2 \cap V_1' \ni j_1, \quad V_2 \cap V_2' \ni i+2,$$
the pair of the partitions (V_1, V_2) and (V_1', V_2') is not parallel, and so at least one of the above partitions is not generated by an edge, a contradiction.

Since the family of pairs (J_{ik}^A, J_{ki}^A) generated by the cycle matrix with $n \geq 5$ is parallel, Lemma 2 implies that the necessary connectivity condition (5) is not sufficient.

Since totally balanced matrices do not contain cycle submatrices at all, Lemma 2 leads naturally to the following conjecture: the class of connected matrices contains the class of totally balanced matrices. However, this is not the case.

The properties of a (0,1)-matrix to be totally balanced or connected are independent in the sense that neither of them implies the other.

Connected matrices are not necessary balanced, since they can contain a cycle submatrix of length 3, which is not balanced or connected.

Similarly, there exist disconnected totally balanced matrices. Observe, that the rows of any totally balanced matrix A are the incidence vectors of the vertex sets of subtrees of some tree $T(A)$. However, the tree $T(A)$ itself is not defined uniquely by the matrix A. Among the trees $T(A)$, a tree can exist such that A is connected with respect to it. Even if A is disconnected with respect to any tree of type $T(A)$, there can exist another tree, with respect to which A turns out to be connected. For such a tree, the rows of A do not correspond to connected subtrees.

Consider the following totally balanced matrix A_σ introduced in [5]:

$$A_\sigma = \begin{array}{c|cccccc} & 1 & 2 & 3 & 4 & 5 & 6 \\ \hline a & 1 & 0 & 0 & 0 & 1 & 0 \\ b & 0 & 0 & 1 & 0 & 1 & 0 \\ c & 0 & 1 & 0 & 0 & 0 & 1 \\ d & 0 & 0 & 0 & 1 & 0 & 1 \\ e & 1 & 1 & 0 & 0 & 1 & 1 \\ f & 0 & 0 & 1 & 1 & 1 & 1 \end{array}$$

Applying the above algorithm of detecting (0,1)-tree-matrices, it is easy to see that the rows (as well as the columns) of A_σ are the incidence vectors of subchains of the tree represented in the figure.

PROPOSITION 3. *The matrix A_σ is not connected with respect to any tree.*

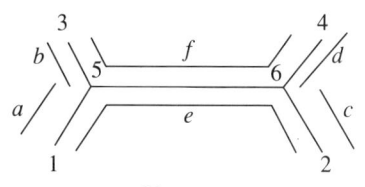

FIGURE

PROOF. Let us write down the pairs of the form $\{J_{ik}^C, J_{ki}^C\}$ for which both sets have exactly two elements:

$$J_{ac} = \{2, 6\}, \quad J_{ca} = \{1, 5\}, \quad J_{bc} = \{2, 6\}, \quad J_{cb} = \{3, 5\},$$
$$J_{ef} = \{3, 4\}, \quad J_{fe} = \{1, 2\}, \quad J_{ad} = \{4, 6\}, \quad J_{da} = \{1, 5\},$$
$$J_{bd} = \{4, 6\}, \quad J_{db} = \{3, 5\}.$$

Let there exist a tree T with the vertex set $V(T)$ such that A_σ is connected with respect to T. Then there exists a partition (V_1, V_2) such that $J_{fe} = \{1, 2\} \subseteq V_1$ and $J_{ef} = \{3, 4\} \subseteq V_2$. There are four possibilities for arranging the elements $5, 6 \in J$ within the sets V_1 and V_2:

1) $\{1, 2, 5, 6\} \subseteq V_1$, $\{3, 4\} \subseteq V_2$;
2) $\{1, 2, 6\} \subseteq V_1$, $\{3, 4, 5\} \subseteq V_2$;
3) $\{1, 2, 5\} \subseteq V_1$, $\{3, 4, 6\} \subseteq V_2$;
4) $\{1, 2\} \subseteq V_1$, $\{3, 4, 5, 6\} \subseteq V_2$.

It is easy to see that the first partition is not parallel to the pair $J_{bd} = \{4, 6\}$, $J_{db} = \{3, 5\}$, the second partition is not parallel to the pair $\{4, 6\}$, $\{1, 5\}$, the third partition is not parallel to the pair $\{2, 6\}$, $\{3, 5\}$, and the fourth partition is not parallel to the pair $\{2, 6\}$, $\{1, 5\}$. Hence, there is no tree T admitting all the pairs.

Unfortunately, no polynomial algorithm for detecting whether an arbitrary matrix is connected with respect to a tree is known at present.

References

1. A. A. Ageev, *On the complexity of minimization problems for polynomials in Boolean variables*, Upravlyaemye Sistemy No. 23 (1983), 3–11. (Russian)
2. I. Bárány, J. Edmonds, and L. Wolsey, *Packing and covering a tree by subtrees,*, Combinatorica **6** (1986), 221–233.
3. V. L. Beresnev, È. Kh. Gimadi, and V. T. Dement'ev, *Extremal standardization problems*, "Nauka", Novosibirsk, 1978. (Russian)
4. È. Kh. Gimadi, *Efficient algorithm for solving the plant location problem for serving regions connected with respect to an acyclic network*, Upravlyaemye Sistemy No. 23 (1983), 12–23. (Russian)
5. A. J. Hoffman, A. W. J. Kolen, and M. Sakarovitch, *Totally-balanced and greedy matrices*, SIAM J. Algebraic Discrete Methods **6** (1985), 721–730.
6. V. A. Trubin, *Efficient algorithm for solving the plant location problem on tree-like networks*, Dokl. Akad. Nauk SSSR **231** (1976), 547–550; English transl. in Soviet Math. Dokl. **17** (1976).

Minimal Mean Weight Cuts and Cycles in Directed Graphs

A. V. KARZANOV

1. Introduction

In discrete optimizations, a sort of problem arises in which the goal is to find a set with the minimum (or maximum) mean weight of an element among the sets of a given type. More formally, the problem is posed as follows. Let E be a finite set whose elements $e \in E$ have real weights $c(e) \in \mathbb{R}$, and let $\mathscr{F} \subset 2^E$ be a collection of nonempty subsets of E; in applications, the sets in $\mathscr{F}$ usually have a certain combinatorial sense and can be given implicitly. By the *weight* of a subset $E' \subseteq E$ we mean the sum $c(E') := \sum_{e \in E'} c(e)$, and by its *mean weight* the ratio $m_c(E') := c(E')/|E'|$. It is required to find a set $F \in \mathscr{F}$ having the minimum (or the maximum) mean weight. Since we do not impose any restrictions on the sign of c, it is sufficient to consider only the minimization problem. We call it the *minimum mean* problem and denote it by $\mathscr{M}(\mathscr{F}, c)$. The value of the minimum is denoted by $m^* = m^*(\mathscr{F}, c)$.

In this paper we are interested in the minimum mean problem for some natural families of cuts and circuits of directed graphs. For some of them, the problem turns out to be intractable (NP-hard), while for some other it is solvable in polynomial time. Besides, we consider an analog for the minimum mean problem for linear cones and develop a general method to solve it; implementations of this method give rise to polynomial algorithms that solve $\mathscr{M}(\mathscr{F}, c)$ for a certain class of sets $\mathscr{F}$, including the set of all cuts and the set of all circuits of a digraph.

Let us start by specifying our terminology and notation. Consider a digraph $G = (V, E)$ with possible multiple edges. Given an edge $e \in E$, denote by $h(e)$ ($t(e)$) the vertex $v \in V$ such that e enters (respectively,

---

1991 *Mathematics Subject Classification.* Primary 05C38, 90B10.
Translation of Qualitative and Approximate Methods in the Study of Operator Equations (V. S. Klimov, editor), Yaroslav. Gos. Univ., Yaroslavl, 1985, pp. 72–83.

leaves) v. By a *path* (a *circuit*) in G we mean an alternating sequence $P = (v_0, e_0, v_1, e_1, \ldots, e_{k-1}, v_k)$ ($k \geq 1$) of vertices and edges such that $\{t(e_i), h(e_i)\} = \{v_i, v_{i+1}\}$ and all the vertices $v_0, \ldots, v_k$ are distinct (respectively, all except $v_0 = v_k$); in the rest of the paper, it will be convenient to identify P with its edge-set $\{e_0, \ldots, e_{k-1}\}$. P is said to be a *directed path* (*circuit*) if $t(e_i) = v_i$ for all i. Denote by $\mathscr{C}^+ = \mathscr{C}^+(G)$ the set of all directed circuits in G, and by $\mathscr{C}_e^+ = \mathscr{C}_e^+(G)$ the set of directed circuits containing a fixed edge $e \in E$. Given $X \subseteq V$, denote by $\partial^+ X$ the set of edges e in G leaving X and entering $V \setminus X$, that is, $t(e) \in X$ and $h(e) \in V \setminus X$; the set $\partial^+(V \setminus X)$ is denoted by $\partial^- X$. A nonempty set $\partial^+ X$ is called a *cut*; if, moreover, $\partial^- X = \varnothing$, it is called a *directed cut*. For $s, t \in V$ we call $\partial^+ X$ an (s, t)-*cut* if $s \in X \not\ni t$.

Denote by $\mathscr{R} = \mathscr{R}(G)$ and $\mathscr{R}^+ = \mathscr{R}^+(G)$ the sets of all cuts and all directed cuts of G, respectively. For $(s, t) \in V$, let $\mathscr{R}_{st}$ denote the set of (s, t)-cuts, and $\mathscr{R}_{st}^+$ denote the set of directed (s, t)-cuts. For $x \in \mathbb{R}^E$, put $|x| := \sum_{e \in E} |x(e)|$.

The minimum mean directed circuit problem $\mathscr{M}(\mathscr{C}^+, c)$ is the most popular one among all minimum mean problems. The original approach to this problem (see, for example, [1], §8.4.1, for a survey) was based on the routine "bisection method"; this approach yields an algorithm whose running time is $\log |c|$ times a polynomial in $|E|$ (assuming that c is integer-valued). Later on, strongly polynomial algorithms exploiting specific combinatorial techniques were found; we particularly mention an elegant algorithm, of complexity $O(|V| \cdot |E|)$, due to Karp [3].

The "bisection" approach relies on two simple observations, which can be outlined, for an arbitrary $\mathscr{F}$ and integer-valued c, as follows. The first one is that the minimum mean problem remains, in essence, the same after adding a scalar to c, that is, if c is replaced by a new function c^a, where $a \in \mathbb{R}$ and $c^a(e) := c(e) + a$ for $e \in E$; obviously, $m_{c^a}(E') = m_c(E') + a$ for any $E' \subseteq E$, whence $m^*(\mathscr{F}, c^a) = m^*(\mathscr{F}, c) + a$. We say that c^a is the *shift* of c by a. Since $m^* = m^*(\mathscr{F}, c)$ is a fraction with denominator at most $|E|$ (as c is integer-valued), finding m^* is reduced to $O(|E| \log |c|)$ verifications of the inequality $m^*(\mathscr{F}, c) < -a$ for properly chosen numbers a. The second observation is that the above inequality is valid if and only if there exists $F \in \mathscr{F}$ such that $c^a(F) < 0$. Thus, $\mathscr{M}(\mathscr{F}, \cdot)$ is polynomially equivalent to the problem $\mathscr{O}(\mathscr{F}, \cdot)$. Here, for $c' \in \mathbb{R}^E$, the problem $\mathscr{O}(\mathscr{F}, c')$ is:

Decide whether $\mathscr{F}$ contains a set F of negative weight $c'(F) < 0$.

The polynomial equivalence between $\mathscr{M}(\mathscr{F}, \cdot)$ and $\mathscr{O}(\mathscr{F}, \cdot)$ immediately implies the above mentioned result for $\mathscr{F} = \mathscr{C}^+$, since the negative circuit problem $\mathscr{O}(\mathscr{C}^+, \cdot)$ is well known to be polynomially solvable. On the other hand, such an equivalence enables us to establish the NP-hardness of the problems $\mathscr{M}(\mathscr{C}_e^+, \cdot)$, $\mathscr{M}(\mathscr{R}_{st}, \cdot)$, and $\mathscr{M}(\mathscr{R}, \cdot)$. To see this, consider the following problem MAX$(\mathscr{F}, k)$:

Given $\mathscr{F} \subset 2^E$ and a natural k, decide whether $\mathscr{F}$ contains a set F of cardinality $|F|$ at least k.

Assume that all the sets in $\mathscr{F}$ contain a common element $e \in E$. Then MAX$(\mathscr{F}, k)$ turns into the problem $\mathscr{O}(\mathscr{F}, c)$ with $c(e) = k - 1$ and $c(e') = -1$ for $e' \in E \setminus \{e\}$. Therefore, if MAX$(\mathscr{F})$ is intractable, the same must be true for $\mathscr{O}(\mathscr{F})$ and $\mathscr{M}(\mathscr{F})$ as well. The problems MAX$(\mathscr{C}_e^+)$ and MAX$(\mathscr{R}_{st})$ are NP-complete (the former is a version of the directed Hamiltonian cycle problem, while the latter is a version of the maximum (s, t)-cut problem; without loss of generality, one can assume that G has an edge from s to t).

The fact that $\mathscr{M}(\mathscr{R}_{st})$ is intractable yields the same for $\mathscr{M}(\mathscr{R})$. Indeed, consider the problem $\mathscr{O}(\mathscr{R}_{st}, c)$. For each vertex $v \in V \setminus \{s, t\}$, add to G new edges e_v and e'_v such that $t(e_v) = s$, $h(e_v) = t(e'_v) = v$, $h(e'_v) = t$, and an edge e^0 from s to t; denote the resulting graph by G'. Put $c(e_v) := c(e'_v) := b$ and $c(e^0) := -(|V| - 2)b$, where $b \geq |c|$. It is easy to see that for any $X \subset V$ one has

$$c(\partial_{G'}^+ X) = c(\partial_G^+ X) \quad \text{if } s \in X \not\ni t,$$
$$c(\partial_{G'}^+ X) \geq 0 \qquad \text{otherwise.}$$

Therefore, the problem $\mathscr{O}(\mathscr{R}_{st}, c)$ is reduced to the problem $\mathscr{O}(\mathscr{R}(G'), c)$, and the proof is completed. One can show that $\mathscr{M}(\mathscr{R})$ and $\mathscr{M}(\mathscr{R}_{st})$ are NP-hard even for the class of symmetric graphs G and symmetric $c \in \mathbb{Z}^E$, that is, for the case when there exists a bijection $\gamma \colon E \to E$ such that γ^2 is the identity mapping, and for any $e \in E$ one has $t(e) = h(\gamma(e))$, $c(e) = c(\gamma(e))$.

In §2 we present a strongly polynomial algorithm for solving the minimum mean directed cut problem $\mathscr{M}(\mathscr{R}^+, c)$; it uses a relationship between the mean and the total weight minimization problems, and solves the latter for the case $\mathscr{F} = \mathscr{R}^+$ with the help of flow techniques. That section will also contain strongly polynomial algorithms for $\mathscr{M}(\mathscr{R}_{st}^+, c)$ with arbitrary G and c and for $\mathscr{M}(\mathscr{R}, c)$ when G is symmetric and c is skew-symmetric. In §3, we develop a general method for solving the mean weight problem for convex cones; this method forms the core of the paper. When applied to $\mathscr{M}(\mathscr{R}^+, c)$, the method yields a strongly polynomial algorithm different from the one described in §2. As another illustration of the method, we derive a new strongly polynomial algorithm for the minimum mean directed circuit problem (however, its running time is worse than that for Karp's algorithm [3] mentioned above).

2. Relations to the weight minimization problem

We explain the idea of the algorithm by considering arbitrary $\mathscr{F} \subset 2^E$ and $c \in \mathbb{R}^E$. Denote by MIN$(\mathscr{F}, c')$ the problem of finding a set $F \in \mathscr{F}$ having the minimum possible weight $c'(F)$. Let us show that the problem

$\mathcal{M}(\mathcal{F}, c)$ can be reduced to at most $|E|$ problems $\text{MIN}(\mathcal{F}, c')$.

First of all, we shift c by a sufficiently large negative number a (e.g., by $a := -\max_{e \in E} c(e)$) in order to guarantee the inequality $m^*(\mathcal{F}, c) \leq 0$ for the function $c := c^a$ just obtained. Assume that c is the current function for a given iteration, and $F \in \mathcal{F}$ is a minimum-weight set. If $c(F) < 0$, then we make the shift $c \mapsto c^a$ with $a = -c(F)/|F|$ and proceed to the next iteration. But if $c(F) = 0$, then, obviously, F is the required solution for $\mathcal{M}(\mathcal{F}, c)$ with the final function c, whence F is optimal for the initial function c as well.

Let us show that the above process terminates in at most $|E|+1$ iterations. Denote by c', c'' the current function c at two consecutive iterations, and by F', F'' the solutions for the problems $\text{MIN}(\mathcal{F}, c')$, $\text{MIN}(\mathcal{F}, c'')$, respectively. Evidently, one has

$$c'(F') < 0, \qquad c'(F') \leq c'(F''), \qquad c''(F') \leq c''(F'') = 0.$$

If $c''(F'') < 0$, then the relations

$$c''(F') = c'(F') + a|F'|, \qquad c''(F'') = c'(F'') + a|F''|, \qquad a = -\frac{c'(F')}{|F'|} > 0,$$

yield $|F''| < |F'|$, and the required bound on the number of iterations follows.

For $\mathcal{F} = \mathcal{R}^+$ and $m^*(\mathcal{F}, c) \leq 0$, the problem $\text{MIN}(\mathcal{F}, c)$ is equivalent, in a sense, to the problem of minimizing a modular function on a distributive lattice, which is known to be polynomially solvable. Let us describe the algorithm for the problem $\text{MIN}(\mathcal{R}^+, c)$ in combinatorial terms. For a vertex $v \in V$, put $d(v) := \text{div}_c(v)$, where

$$\text{div}_x(v) := \sum_{e \in E,\, t(e)=v} x(e) - \sum_{e \in E,\, h(e)=v} x(e) \quad \text{for } x \in \mathbb{R}^E.$$

It is easy to see that for any directed cut $\partial^+ X$ one has $c(\partial^+ X) = d(X)$ ($:= \sum_{v \in X} d(v)$). Therefore, the problem $\text{MIN}(\mathcal{R}^+, c)$ is equivalent to finding the closed vertex subset X having the least possible weight $d(X)$ (a set $X \subseteq V$ is said to be *closed* if $\partial^- X = \varnothing$; since both $\varnothing$ and V are closed, and $d(\varnothing) = d(V) = 0$, we are forced to impose the restriction $m^*(\mathcal{F}, c) \leq 0$). The latter problem can be solved by a strongly polynomial flow-type algorithm [5] (see also [4]). This results in a strongly polynomial algorithm for the initial problem $\mathcal{M}(\mathcal{R}^+, c)$ as well.

To design a strongly polynomial algorithm for the problem $\mathcal{M}(\mathcal{R}^+_{st}, c)$, we reduce the problem $\text{MIN}(\mathcal{R}^+_{st}, c')$ to the problem $\text{MIN}(\mathcal{R}^+(G'), c')$. The graph G' is obtained from G by adding the edge e^0 and edges e_v, e'_v for each $v \in V \setminus \{s, t\}$ in the same way as it was done in the proof of intractability for $\mathcal{M}(\mathcal{R})$ (see the Introduction); for the added edges, we put $c'(e) := 0$. Then the set of directed cuts in G' corresponds to the set of directed (s, t)-cuts in G, and the weights of the corresponding cuts are the same, as required.

To conclude this section, consider the problem $\mathscr{M}(\mathscr{R}, c)$ for the case when G is a symmetric graph and c is a skew-symmetric function, that is, $c(e) = -c(\gamma(e))$, $e \in E$, where γ is the edge symmetry for G. Observe that the problem $\mathscr{O}(\mathscr{R}, c)$ for G and c as above is trivial, yet this fact does not help us to solve the problem $\mathscr{M}(\mathscr{R}, c)$, since shifts destroy the skew symmetry for c. Construct a new graph G' by replacing each pair of edges $\{e, e'\}$, $e' = \gamma(e)$, of G by a pair of edges $\{e^1, e^2\}$ and a vertex v such that $t(e^1) = t(e)$, $t(e^2) = t(e')$, $h(e^1) = h(e^2) = v$; put $c'(e^1) := -c'(e^2) := c(e)$. One can see that each cut $\partial^+ X$ in G corresponds to a set of directed cuts $\partial^+ X'$ in G' such that $c(\partial^+ X) = c'(\partial^+ X')$, $|\partial^+ X| \leqslant |\partial^+ X'|$, and there is an X' that turns the last inequality into an equality. Since $m^*(\mathscr{R}, c)$ is evidently nonpositive, the above reasons imply that a solution for $\mathscr{M}(\mathscr{R}^+(G'), c')$ provides a solution for $\mathscr{M}(\mathscr{R}, c)$. The problem $\mathscr{M}(\mathscr{R}_{st}, c)$ with G and c as above is solved similarly for the case $m^*(\mathscr{R}_{st}, c) \leqslant 0$.

3. Minimum mean problem for cones

In what follows we make no distinction between the set $\mathbb{R}^E$ of all real-valued functions on E and the linear space $\mathbb{R}^{|E|}$ whose coordinates correspond to the elements of E.

Given $q \in \mathbb{R}^E$, we introduce the following notation: q^+ (q^-) is the vector with components $\max\{q(e), 0\}$ (respectively, $\min\{q(e), 0\}$), $e \in E$,

$$S(q) := \{e \mid c(e) \neq 0\}, \quad S^+(q) := \{e \mid c(e) > 0\},$$
$$S^-(q) := \{e \mid c(e) < 0\}, \quad S^0(q) := \{e \mid c(e) = 0\}.$$

Consider a vector $c \in \mathbb{R}^E$ and a set Q of nonzero vectors from the nonnegative orthant $\mathbb{R}^E_+ = \{x \in \mathbb{R}^E \mid x \geqslant 0\}$ such that

$$\bigcup_{x \in Q} S(x) = E. \tag{1}$$

The minimum mean problem for Q and c (denoted by $\mathscr{M}(Q, c)$) consists in finding a vector $x \in Q$ minimizing the function $m_c(x) = cx/|x|$, where $cx := \sum_{e \in E} c(e)x(e)$; the minimum value of $m_c(x)$ (if it exists) is denoted by $m^* = m^*(Q, c)$.

Denote by $\mathscr{L} = \mathscr{L}(Q)$ and $\mathscr{K} = \mathscr{K}(Q)$ the linear and the nonnegative linear hulls of vectors from Q, respectively. Evidently, for $x, x' \in \mathbb{R}^E_+ \setminus \{0\}$ and $\lambda > 0$ one has

$$m_c(\lambda x) = m_c(x), \quad m_c(x + x') \geqslant \min\{m_c(x), m_c(x')\}.$$

The equality in the latter relation occurs if and only if $m_c(x) = m_c(x')$; hence $m^*(Q, c) = m^*(\mathscr{K}, c)$, and moreover, whenever an optimal vector for $\mathcal{M}(\mathcal{K}, c)$ is represented as a positive linear combination of vectors from Q, each of these vectors is an optimal solution for $\mathcal{M}(Q, c)$.

DEFINITION. The set Q is said to be *regular* if $\mathcal{K}$ coincides with the set $\mathcal{L}^+ := \mathcal{L} \cup \mathbb{R}^E_+$.

Similar to $\mathcal{M}(\mathcal{F}, c)$, the problem $\mathcal{M}(Q, c)$ remains equivalent under shifts of the function c by any number a. Another transformation that preserves cx and $m_c(x)$ for any $x \in \mathcal{L}^+$ is $c \mapsto c + y$, where y is an arbitrary vector from the orthogonal complement $\mathcal{L}^\perp$ to $\mathcal{L}$ in $\mathbb{R}^E$. Evidently, $m^*(Q, c) \geq \min_{e \in E} c(e)$. The following theorem is due to B. A. Papernov (an unpublished manuscript, 1983).

THEOREM. *If $Q \subset \mathbb{R}^E_+$ is a regular set and $c \in \mathbb{R}^E$, then there exists $y^* \in \mathcal{L}^\perp(Q)$ such that $c(e) + y^*(e) \geq m^*(Q, c)$ for any $e \in E$ (and hence, if $x \in Q$ is an optimal solution for $\mathcal{M}(Q, c)$, then $S(x) \subseteq \{e \in E \mid c(e) + y^*(e) = m^*(Q, c)\}$). In other words,*

$$m^*(Q, c) = \max_{y \in \mathcal{L}^\perp(Q)} \min_{e \in E} \{c(e) + y(e)\}.$$

We propose a method to solve the problem $\mathcal{M}(\mathcal{L}^+, c)$, which, in particular, implies the proof of the above theorem. Evidently, without loss of generality, one may assume that

$$\bigcup_{x \in \mathcal{L}^+} S(x) = E. \qquad (2)$$

First, we shift c by a sufficiently large negative number in order to guarantee the inequality $m^*(\mathcal{L}^+, c) < 0$. The algorithm consists of iterations. At each iteration, one transforms the current c by adding to c a vector $y \in \mathcal{L}^\perp$ (this does not affect m^*); next, c can be shifted by some number $a > 0$ (this increases m^* by a). If a shift of c occurs at an iteration, a vector x^* such that $cx^* = 0$ for the new function c is provided along with c. In the iteration process, the inequality $m^*(\mathcal{L}^+, c) \leq 0$ holds and the following monotonicity property is maintained: at each iteration, the set $S^-(c)$ does not increase, and $c(e)$ is monotone nondecreasing while $e \in E$ belongs to $S^-(c)$. Eventually, the current vector c becomes everywhere nonnegative. Then the relation $cx^* = 0$ for the final c and x^* obviously implies $m_c(x^*) = 0 \leq m^*(\mathcal{L}^+, c)$; therefore x^* is an optimal solution for the problem $\mathcal{M}(\mathcal{L}^+, c)$ (for the final, initial, and all the intermediate c's). The sum of all the y's obtained provides the required vector y^*, while the sum of all the a's provides the value m^*.

Let us describe an iteration. At each iteration, an element $r \in S^-(c)$ is fixed, and the following pair of dual linear problems is solved:

$$y(r) \to \max, \quad y \in \mathcal{L}^\perp, \quad c^+(e) + y(e) \geq 0, \quad e \in E \setminus \{r\}; \qquad (3)$$

$$c^+x \to \min, \quad x \in \mathcal{L}^+, \quad x(r) = 1. \qquad (4)$$

To verify that (3) and (4) are dual to each other, one must rewrite (3) in the standard form by replacing the condition $y \in \mathscr{L}^\perp$ by equalities $z^i y = 0$, $i = 1, \ldots, N$, where $\{z^1, \ldots, z^N\}$ is a basis in $\mathscr{L}$, and perform simple transformations of the standard dual problem. Since $y = 0$ is a feasible solution for (3), and (4) has a feasible solution (because of (2)) as well, one obtains that (3) and (4) have optimal solutions y and x with $y(r) = c^+ x$. Put $c' := c + y$. The complementary slackness conditions for (3) and (4) have the following form:

$$e \in E \setminus \{r\}, \ c^+(e) + y(e) > 0 \Rightarrow x(e) = 0;$$

whence for $e \in S(x) \setminus \{r\}$ we deduce

$$c'(e) = 0 \quad \text{if } c(e) \geqslant 0, \qquad (5)$$
$$c'(e) = c(e) \quad \text{for } c(e) < 0. \qquad (6)$$

If $c'(r) \geqslant 0$, take the new c to be c', while if $c'(r) < 0$, put $a := -c'x/|x|$, $c := (c')^a$, $x^* := x$ (note that (5) and (6) yield $c'(e) \leqslant 0$ for each $e \in S(x)$; hence $a > 0$). Therefore, the equality $cx^* = 0$ is valid for x^* and the new c. Observe that the constraints for y in (3) and the positivity of a imply the monotonicity property. If $S^-(c)$ is still nonempty, proceed to the next iteration obeying the following rule: if $c(r) < 0$, the same element r must be fixed at the next iteration.

Finiteness of the above algorithm. We say that an iteration is *essential* if the set $S^-(c)$ at this iteration decreases; by the monotonicity property, there are at most $|E|$ essential iterations. Hence it suffices to verify that the number of consecutive unessential iterations is finite. Below we prove that this number is bounded from above both by the number η of generators (extreme rays) for the cone $\mathscr{L}^+$ and by $q^2|E|$, where q is the maximum entry among the minimal integer vectors defining the extreme rays of $\mathscr{L}^+$ (q is sometimes called the *fractionality* of the cone). Consider two consecutive unessential iterations. According to the above rule, the same element r is fixed at both iterations. Put $D = S^+(c) \cup S^0(c)$. Let c_i, c_i', x_i, y_i, a_i, $i = 1, 2$, be the corresponding objects for the first ($i = 1$) and the second ($i = 2$) of the iterations in question; then $c_2 = (c_1')^{a_1}$. Let us prove that $x_1(D) > x_2(D)$. Evidently, $c_2(e) \geqslant a_1$ for each $e \in D$, and (5) yields $c_2(e) = a_1$ for $e \in S(x_1) \cap D$. The relations $c_2 x_1 = c_2^+ x_1 + c_2^- x_1 = 0$ imply $|c_2^- x_1| = c_2^+ x_1 = a_1 x_1(D)$; this relation together with $x_1(r) = 1$ yields $|c_2(r)| \leqslant a_1 x_1(D)$. On the other hand, one has $|c_2(r)| > y_2(r) = c_2^+ x_2 \geqslant a_1 x_2(D)$. Hence $a_1 x_1(D) > a_1 x_2(D)$, and thus $x_1(D) > x_2(D)$. Next, let $x_i = \lambda_i^1 z_i^1 + \cdots + \lambda_i^{m(i)} z_i^{m(i)}$, $i = 1, 2$, where all the λ_i^j are positive and the z_i^j are extreme vectors of $\mathscr{L}^+$ such that $z_i^j(r) = 1$. Clearly, each z_i^j is an optimal solution of (4) (for $c = c_i$). So we may consider z_1^1 and z_2^1 instead of x_1 and x_2, respectively, whence $z_1^1(D) > z_2^1(D)$. Now, to complete the proof, observe that the number of distinct values of $z(D)$ for extreme vectors z satisfying $z(r) = 1$ exceeds neither η nor $q^2|E|$, as we asserted above.

Let us return to the problem $\mathcal{M}(\mathcal{F}, c)$. It is equivalent to $\mathcal{M}(Q(\mathcal{F}), c)$, where $Q(\mathcal{F})$ is the set of incidence vectors of the sets in $\mathcal{F}$. For a regular $Q(\mathcal{F})$, the above method yields an optimal solution of this problem; moreover, the number of iterations does not exceed $|E|^2$, since in this case $q = 1$ (in fact, one can obtain easily from the above proof that the total number of iterations is bounded by $\rho|E|$ with $\rho := \max_{F \in \mathcal{F}} |F|$). Therefore, to design an efficient algorithm for $\mathcal{M}(\mathcal{F}, c)$, we need efficient algorithms for (3) and (4).

Well-known results in linear algebra imply the following: $Q(\mathcal{F})$ is regular if and only if $\mathcal{L} = \mathcal{L}(Q(\mathcal{F}))$ is the set of solutions for $Ax = 0$ with a unimodular $m \times |E|$-matrix A, where m is the corank of $\mathcal{L}$. Next, it is known that the spaces $\mathcal{L}$ for $\mathcal{F} = \mathcal{R}^+$ and $\mathcal{F} = \mathcal{C}^+$ can be defined by the following unimodular matrices A: the incidence matrix of fundamental cycles and edges of G in the first case, and that of vertices and edges of G in the second case. Therefore, the sets $Q(\mathcal{R}^+)$ and $Q(\mathcal{C}^+)$ are regular (in contrast to the sets $Q(\mathcal{R}^+_{st})$ and $Q(\mathcal{C}^+_e)$, which, in general, fail to be regular).

Let us illustrate the above method for the case $\mathcal{F} = \mathcal{R}^+$; this will give us a strongly polynomial algorithm different from that outlined in §2. Assume that G does not contain directed circuits (this corresponds to condition (1)). It is easy to see that $\mathcal{L}^\perp$ is the space of *circulations* for G, that is, of functions $y \in \mathbb{R}^E$ such that $\operatorname{div}_y(v) = 0$ for all $v \in V$. We choose $r \in S^-(c)$ and put $t := t(r)$, $s := h(r)$. Observe that (3) is the problem of finding a circulation y having the maximal value at the edge r subject to lower capacity restrictions $y(e) \geqslant -c^+(e)$. This is reduced to the standard problem of finding a maximum flow from s to t in the graph G' obtained from G by adding to each $e \in E \setminus \{r\}$ a new edge e' from $h(e)$ to $t(e)$; the edge e' has upper capacity $c^+(e)$, while e is of capacity ∞ (all the lower capacities are zero). Let y' be a maximum flow from s to t in G', and let $\partial^+_{G'} X'$ be a minimum (s, t)-cut in G'. Put $X := V \setminus X'$. Then each edge in G' from X' to X is saturated by y', and $y'(e) = 0$ for each edge from X to X'. This easily implies that $\partial^+_G X$ is a directed cut in G, and its incidence vector provides an optimal solution x for (4). The above proof of the finiteness of the method is specified as follows: while processing a fixed edge r, each newly obtained cut $\partial^+ X$ contains strictly fewer edges e satisfying $c(e) \geqslant 0$ than the previous one. Thus the algorithm for the problem $\mathcal{M}(\mathcal{R}^+, c)$ consists in finding $|E|^2$ maximum flows.

Finally, consider the problem $\mathcal{M}(\mathcal{C}^+, c)$. Condition (1) means the strong connectivity for G. It is easy to see that $\mathcal{L}^\perp$ is the space of functions $y_\pi \in \mathbb{R}^E$ generated by "potential" functions $\pi \in \mathbb{R}^V$, that is, $y(e) = \pi(h(e)) - \pi(t(e))$, $e \in E$. Let $r \in S^-(c)$ be an edge from t to s. Then (3) turns into the problem of finding a function y_π such that $\pi(s) - \pi(t)$ is maximal, provided $\pi(t(e)) - \pi(h(e)) \leqslant c^+e$ for all $e \in E$; in other words, it is the problem of finding a shortest directed path from t to s in the graph whose edges are oriented reverse to the edges of G and have lengths c^+. The incidence vector for the corresponding directed path from s to t, extended by the edge r, is an optimal solution for (4). During consecutive unessential iterations (for a fixed edge r), each newly obtained path contains strictly fewer edges e with $c(e) \geqslant 0$ than the previous one. Therefore the algorithm terminates in at most $|V| \cdot |E|$ iterations, and hence its running time equals $O(|V|^3|E|)$ (assuming that an $O(|V|^2)$-algorithm is applied for determining shortest paths, like Dijkstra's [2]).

References

1. Nicos Christofides, *Graph theory: an algorithmic approach*, Academic Press, New York, 1975.
2. E. W. Dijkstra, *A note on two problems in connection with graphs*, Numer. Math. **1** (1959), 269–271.
3. Richard M. Karp, *A characterization of the minimum cycle mean in a digraph*, Discrete Math. **23** (1978), 309–311.
4. A. V. Karzanov, *On closed subsets of a directed graph*, Zh. Vychisl. Mat. i Mat. Fiz. **24** (1984), 1903–1906; English transl. in USSR Comput. Math. and Math. Phys. **24** (1984).
5. Jean-Claude Picard, *Maximal closure of a graph and applications to combinatorial problems*, Management Sci. **22** (1975/76), 1268–1272.

An Algorithm for Determining a Maximum Packing of Odd-Terminus Cuts, and Its Applications

A. V. KARZANOV

1. Introduction

Throughout the paper by a *graph* we mean a finite undirected graph without loops and multiple edges. The vertex-set and the edge-set of a graph H are denoted by VH and EH, respectively; an edge with end vertices u and v is denoted by uv. A *chain*, or an *st-chain*, of a graph is a subgraph L in it such that $VL = \{s = v_0, v_1, \ldots, v_k = t\}$ and $EL = \{v_{i-1}v_i \mid i = 1, \ldots, k\}$. A connected subgraph all of whose vertices are of valency 2 is called a *circuit*.

We deal with a connected graph G whose edges $e \in EG$ have nonnegative rational-valued weights $l(e) \in \mathbb{Q}_+$ (*lengths* of edges), in which a subset $T \subseteq VG$ of *even* cardinality $|T|$, called the set of *terminals*, is specified.

A subgraph J in G is called a *T-join* if the set of odd valency vertices of J coincides with T (such a definition slightly differs from that introduced in [11], since we allow circuits in J). Clearly, a T-join can be represented as the union of pairwise edge-disjoint chains and circuits so that the ends of these chains are distinct and form the set T. Originally, T-joins appeared in connection with the so-called "Chinese postman problem" (see [9] and [2]) that consists in determining a closed route of minimum length in G passing through each edge at least once. It is easy to see that the length of such a route is equal to $l(EG) + l(EJ)$, where J is a minimum length T'-join for the set T' of odd valency vertices of G. (For a subset $S' \subseteq S$ and a mapping $g: S \to \mathbb{Q}$, $g(S')$ denotes $\sum(g(e) \mid e \in S')$.)

There is a minimax relation between T-joins and packings of special cuts of G. More precisely, for $X \subseteq VG$, let $\partial X = \partial^G X$ denote the set of edges of G with one end in X and the other in $VG \setminus X$. We say that $X \subset V$ is an *odd-terminus set* if $|X \cap T|$ is odd; the cut ∂X for such an X is

---

1991 *Mathematics Subject Classification.* Primary 05B40, 68R05.
Translation of Studies of Applied Graph Theory (A. S. Alekseev, editor), "Nauka", Novosibirsk, 1986, pp. 126–140.

usually called a *T-cut* [12]. Let $D(G, T)$ denote the set of odd-terminus sets for G and T. When $V = T$, we say that $X \in D(G, T)$ is an *odd* set. For a collection $D' \subseteq 2^{VG}$ of subsets of vertices in G, we call a mapping $f: D' \to \mathbb{Q}_+$ an *l-packing* of D' if the corresponding "cuts" weighted by f satisfy the packing condition:

$$\lambda^f(e) := \sum (f(X) \mid X \in D', e \in \partial^G X) \leqslant l(e) \quad \text{for all} \quad e \in EG. \quad (1)$$

Given a D', an *l*-packing f is *maximum* if the value

$$1 \cdot f := \sum (f(X) \mid X \in D')$$

is the greatest possible.

It is a simple fact that a subgraph J of G is a *T*-join if and only if $|EJ \cap \partial X|$ is odd for all $X \in D(G, T)$. This implies, for an *l*-packing $f: D(G, T) \to \mathbb{Q}_+$ and a *T*-join J, that

$$1 \cdot f \leqslant \sum_{X \in D(G,T)} f(X)(|\partial X \cap EJ|) = \sum_{e \in EJ} \sum (f(X) \mid X \in D(G, T), e \in \partial X)$$
$$= \sum_{e \in EJ} \lambda^f(e) \leqslant l(EJ).$$

Edmonds and Johnson proved that there exist f and J for which the inequalities in this expression hold with equality.

THEOREM 1 [3]. $\max\{1 \cdot f\} = \min\{l(EJ)\}$, *where f runs over the l-packings of $D(G, T)$ and J runs over the T-joins in G.*

The proof of this theorem given in [3] follows from an algorithm developed there to find optimal f and J. An analysis of this algorithm shows that it can be implemented with running time $O(n^4)$, where $n := |VG|$. Whenever l is integer-valued, the algorithm determines an optimal f which turns out to be *half-integral* (an independent proof of the existence of a half-integral optimal packing of T-cuts appeared in [8]).

The latter result was strengthened by Seymour as follows. We say that $l \in \mathbb{Z}_+^{EG}$ is *cyclically even* if the length $l(EC)$ of every circuit C in G is even ($\mathbb{Z}_+$ is the set of nonnegative integers).

THEOREM 2 [12]. *If l is cyclically even, then the equality in Theorem 1 is attained for an integral l-packing f.*

Note that the proof of Theorem 2 given in [12] is "nonconstructive". On the other hand, in the case of cyclically even l, the algorithm in [3] guarantees only a half-integral, rather than integral, optimal packing of T-cuts.

The problem of determining optimal f and J is denoted by $\mathscr{P}(G, T, l)$. In the present paper, we describe an algorithm to solve $\mathscr{P}(G, T, l)$ for $l \in \mathbb{Q}_+^{EG}$ in running time $O(pnm + p^4)$, where $m := |EG|$ and $p := |T|$

(§§2 and 3). In §4, a modification of this algorithm is developed, which is based on a dynamic data structure and has running time $O(pm \log n + p^3 \log p)$. The algorithm, as well as its modification, determines an integral optimal f whenever l is cyclically even. This gives an alternative proof of Theorem 2. The algorithm uses the reduction method developed in [6], §5, to solve a certain larger class of cut packing problems. Namely, the problem $\mathscr{P}(G, T, l)$ in question is reduced to the "smaller" problem $\mathscr{P}(K_T, T, h)$. Here K_T is the complete graph with the vertex-set T, and $h(st)$ is the distance $\text{dist}_l(s, t)$ between terminals $s, t \in T$ in the graph G with length l of edges, that is, $\text{dist}_l(s, t) := \min\{l(EL) \mid L \text{ is an } st\text{-chain in } G\}$. The fact that h is a metric implies that $\mathscr{P}(K_T, T, l)$ is, in essence, a variant of the minimum weight perfect matching problem, as we explain in §3; therefore, it can be solved by use of alternating chains techniques.

Two applications of the problem in question are well known.

I. Suppose that U is a subset of edges of G, and $\mathscr{X}$ is the collection of all sets $X \subset VG$ such that $|\partial X \cap U| = 1$. Seymour studied the problem of the existence of an l-packing $f' : \mathscr{X} \to \mathbb{Q}_+$ satisfying the equality $\lambda^{f'}(e) = l(e)$ for all $e \in U$ (the problem $\mathscr{A}(G, U, l)$).

THEOREM 3 [10]. $\mathscr{A}(G, U, l)$ *has a solution if and only if the inequality*

$$l(EC \cap U) \leq l(EC \setminus U) \qquad (2)$$

holds for any circuit C in G; in other words, when the graph G whose edges are weighted as $w(e) := l(e)$ for $e \in EG \setminus U$ and $w(e) := -l(e)$ for $e \in U$ has no negative circuit.

Problem $\mathscr{A}(G, U, l)$ is immediately reduced to the problem $\mathscr{P}(G, T, l)$ with T the set of vertices in G covered by an odd number of edges from U. More precisely, let f and J be optimal solutions for the latter problem. Since U generates a T-join for given T, one has $l(EJ) \leq l(U)$. If $l(EJ) < l(U)$, then $\mathscr{A}(G, U, l)$ has no solution (since the subgraph induced by the edge set $(EJ \setminus U) \cup (U \setminus EJ)$ obviously contains a circuit C, which violates (2)). But if $l(EJ) = l(U)$, then f determines a solution of $\mathscr{A}(G, U, l)$ (since $1 \cdot f = l(U)$ easily implies that: (i) $|\partial X \cap U| = 1$ whenever $X \in D(G, T)$ and $f(X) > 0$, and (ii) $\lambda^f(e) = l(e)$ for all $e \in U$). Theorem 2 implies also that if l is cyclically even and $\mathscr{A}(G, U, l)$ is solvable, then it has an integral solution [10] (note that, as it was shown in [7], §8, this stronger version of Theorem 3 can be derived directly from Theorem 3 itself).

Thus, the algorithm can be applied to solve the problem $\mathscr{A}(G, U, l)$ and, as a consequence, to recognize a negative circuit in an undirected edge-weighted graph. In the latter case, the running time of the algorithm is $O(\min\{pm \log n, pn^2\} + p^3 \log p)$, where p is the number of vertices covered by an odd number of edges of negative weight; the time bound becomes smaller because we do not need to construct any packing of cuts.

II. Let there be given a planar graph G (explicitly embedded in the plane), a subset U of its edges, a vector $c \in \mathbb{Q}_+^{EG \setminus U}$ of edge *capacities*, and a vector $d \in \mathbb{Q}_+^U$ of *demands*. It is required to find a multicommodity flow $\{F_u \mid u \in U\}$ such that: (i) F_u is a flow in the graph $(VG, EG \setminus U)$ that connects the ends of the edge $u \in U$ and has value $d(u)$, and (ii) the total flow through an edge $e \in EG \setminus U$ does not exceed $c(e)$. Let G^* denote the planar graph dual to G, and U^* the set of edges of G^* corresponding to U. Define $l(e^*)$ to be $c(e)$ for $e \in EG \setminus U$ and $d(e)$ for $e \in U$, where e^* denotes the edge in G^* corresponding to $e \in EG$. One can see that the above multicommodity flow problem is equivalent to $\mathscr{A}(G^*, U^*, l)$, and hence, Theorems 2 and 3 imply the following result.

THEOREM 4 [12]. *For G, U, c, d as above, the required multicommodity flow exists if and only if the cut condition*

$$c(\partial X \setminus U) - d(\partial X \cap U) \geq 0 \qquad (3)$$

holds for any $X \subset VG$. Moreover, if c and d are integer-valued and the value of the left hand side in (3) is nonnegative and even, then the problem has an integral solution.

2. Reduction

The algorithm for solving $\mathscr{P}(G, T, l)$ consists of three stages.

The *first stage* is to determine the distances $\text{dist}_l(s, t)$ for all $s, t \in T$. It takes $O(pn^2)$ time, provided a shortest path procedure of complexity $O(n^2)$ is used, e.g., Dijkstra's method.

Let h denote the restriction of the distance function dist_l to EK_T. Observe that if l is cyclically even, so is h (h concerns K_T).

The *second stage*, the core of the algorithm, is described in §3. The aim of this stage is to solve the reduced problem $\mathscr{P}(K_T, T, h)$. More precisely, we construct an h-packing $g: D(K_T, T) \to \mathbb{Q}_+$ of *odd* sets in K_T and a T-join of a special form, namely, a perfect matching M in K_T, so that

$$1 \cdot g = h(M). \qquad (4)$$

(Recall that a *matching* in a graph is a subset of its edges all ends of which are distinct; a matching is *perfect* if it covers all vertices of the graph; clearly the subgraph induced by a perfect matching in K_T is a T-join.)

Define $Q := Q(g) := \{A \in D(K_T, T) \mid g(A) > 0\}$. The function g found at the second stage satisfies the following additional properties:

$$g \text{ is integral whenever } h \text{ is cyclically even}; \qquad (5)$$

$$\text{for any distinct } A, B \in Q, \text{ either } A \subset B, \text{ or } B \subset A, \text{ or } A \cap B = \varnothing. \qquad (6)$$

The aim of the *third stage* is to transform g and M to an l-packing $f: D(G, T) \to \mathbb{Q}_+$ and T-join J in G such that

$$1 \cdot f = 1 \cdot g \quad \text{and} \quad l(EJ) = h(M).$$

This and (4) imply $1 \cdot f = l(EJ)$, and hence, f and J give an optimal solution of $\mathscr{P}(G, T, l)$.

We now describe the third stage. A T-join J as required is formed in a natural way. More precisely, for each $st \in M$ we choose an st-chain L_{st} in G with $l(L_{st}) = h(st)$; one may take for L_{st} the shortest chain found at the first stage of the algorithm. Let J be the subgraph induced by the edges in G occurring in odd number of these chains. It is easy to see that J is a T-join, and $l(EJ) \leq h(M)$ (actually, this inequality holds with equality).

Finding a packing f as required is a bit more complicated. This is solved by the following algorithm described in [6], §5. Observe that it is important for the algorithm that Q satisfies property (6); that Q consists of odd subsets is inessential.

ALGORITHM (for constructing f). Choose a *minimal* set A in Q. Define

$$X := \{x \in VG \mid \text{dist}_l(s, x) = 0 \text{ for some } s \in A\};$$

and

$$a := \min\{g(A), \min\{l(e) \mid e \in \partial X\}\}.$$

Put $f(X) := a$, and change g and l by putting $g(A) := g(A) - a$ and $l(e) := l(e) - a$ for $e \in \partial X$ (preserving the old values of g and l on the other elements). By the definition of a, the new l and g are nonnegative. If $g(A)$ vanishes, remove A from Q. Repeat these steps until the current Q becomes empty.

Let f be the resulting function (extended by zero to the sets $X \in D(G, T)$ that do not appear in the steps of the algorithm). It follows from the definition of a that f is an l-packing (for the initial l), that is, f satisfies (1). It was shown in [6] that

$$g(A) = \sum(f(X) \mid X \in D(G, T), X \cap T = A) \qquad (7)$$

for each $A \in Q$. This implies that $1 \cdot f = 1 \cdot g$, as required.

To make our presentation self-contained, let us prove (7). The proof is divided into several claims.

CLAIM 1. $X \cap T = A$.

PROOF. If $x \in A$ then $\text{dist}_l(x, x) = 0$ implies $x \in X$, while if $x \in T \setminus A$, then for any $s \in A$ we have

$$\text{dist}_l(s, x) = h(s, x) \geq \lambda^g(sx) = \sum(g(B) \mid B \in Q, sx \in \partial B) \geq g(A) > 0,$$

and therefore, $x \notin X$.

In particular, it follows from Claim 1 that every set X occurring in the algorithm belongs to $D(G, T)$.

CLAIM 2. *Let l' and g' be the functions obtained from current l and g as a result of one step. Let $\lambda^g(pq) \leq \text{dist}_l(p, q)$ for all $p, q \in T$. Then $\lambda^{g'}(pq) \leq \text{dist}_{l'}(p, q)$ for all $p, q \in T$ (in other words, g' is an h'-packing* if g is an h-packing, where h' is the distance function in K_T with respect to l').

PROOF. Put $\lambda := \lambda^g$ and $\lambda' := \lambda^{g'}$. One has to prove that for a pq-chain L in G with $p, q \in T$,

$$l'(EL) \geq \lambda'(pq). \tag{8}$$

Apply induction on $k(L) := |EL \cap \partial X|$.

(i) If $k(L) = 0$, then $l'(EL) = l(EL) \geq \mathrm{dist}_l(p, q) \geq \lambda(pq) = \lambda'(pq)$.

(ii) Let $k(L) = 1$. Then exactly one of p and q is in A. We have $l'(EL) = l(EL) - a$ and $\lambda'(pq) = \lambda(pq) - a$, and (8) follows.

(iii) If $p, q \in A$, then, by (6) and the minimality of A, $\lambda(pq) = \lambda'(pq) = 0$.

(iv) Suppose we are in a case different from (i)–(iii). Then L contains a vertex x such that $x \in X$, $k(L') \geq 1$ and $k(L'') \geq 1$, where L' and L'' are the parts of L from p to x and from x to q, respectively. By the definition of X, there is a terminal $s \in A$ such that $\mathrm{dist}_l(sx) = 0$. Choose an sx-chain P with $l(EP) = 0$; clearly, $VP \subseteq X$. Let L_1 be a ps-chain in $L' \cup P$ and L_2 an sq-chain in $L'' \cup P$. Obviously, $k(L_1) \leq k(L') < k(L)$ and $k(L_2) \leq k(L'') < k(L)$, whence by induction $l'(EL_1) \geq \lambda'(ps)$ and $l'(EL_2) \geq \lambda'(sq)$. We have

$$l'(EL) \geq l'(EL_1) + l'(EL_2) \geq \lambda'(ps) + \lambda'(sq) \geq \lambda'(pq)$$

(the last inequality follows from the fact that $B \subset T$ and $pq \in \partial B$ imply $\{ps, sq\} \cap \partial B \neq \emptyset$), as required.

CLAIM 3. *Let the same set $A \in Q$ be chosen at the ith and jth steps of the algorithm, $i < j$, and let X_i and X_j be the sets determined on these steps, respectively. Then $X_i \subset X_j$.*

PROOF. Since l can only decrease during the algorithm, we have $X_i \subseteq X_j$. Moreover, this inclusion is strict because, after the ith iteration, $g(A)$ remains positive and therefore $l(xy)$ vanishes for some $x \in X_i$ and $y \in VG \setminus X_i$ (by the definition of a), whence $y \in X_j$.

Claims 1–3 prove the correctness of the algorithm for the third stage. Claim 3 shows that the algorithm terminates after at most $n|Q|$ steps, and that all the sets X found on the steps are distinct. Now Claim 2 implies (7).

To estimate the number of steps of the algorithm, observe that the cardinality of Q is bounded by a linear function in the number of terminals. Namely, the following claim is easily proved by induction on $|T|$.

CLAIM 4. *If $Q \subset 2^T$ satisfies (6), then $|Q| \leq 2|T| - 3$.*

Thus, the algorithm for the third stage consists of $O(pn)$ steps. Obviously, a step can be designed to take $O(m)$ operations. This gives the running time of the third stage to be $O(pnm)$. (Note also that, by Claim 3, the sets X found for the same $A \in Q$ can be stored as sets of consecutive elements in a certain ordering of vertices (depending on A). This enables us to shorten the space needed to write the output of the algorithm.)

3. Algorithm for the reduced problem

In what follows the graph K_T and its edge set EK_T are denoted briefly by K and E, respectively. The current objects of the algorithm solving the problem $\mathscr{P}(K, T, h)$ (the second stage of the algorithm for $\mathscr{P}(G, T, l)$) that we develop in this section are:

(i) a (not necessarily perfect) matching $M \subseteq E$,
(ii) a collection $D \subseteq D(K, T)$ of odd sets in K, and
(iii) an h-packing $g: D \to \mathbb{Q}_+$ satisfying (5).

An edge $e \in E$ is called *saturated* (with respect to g) if $\lambda^g(e) = h(e)$. Let M^A be the set of edges in M with both ends in a subset $A \subseteq T$. The following properties are maintained during the algorithm:

(9) All edges in M are saturated.

(10) D is *regular*; this means that (i) for any distinct $A, B \in D$, either $A \subset B$, or $B \subset A$, or $A \cap B = \varnothing$, and (ii) $|M^A| = (|A| - 1)/2$ for any $A \in D$.

Let r_A denote the only vertex in $A \in D$ not covered by M^A (the *root* of A). One can see that (9) and (10) imply the following properties of M:

(11) $|M \cap \partial A|$ is at most 1 for any $A \in D$, and is exactly 1 if and only if r_A is covered by M.

(12) If M is perfect, then $1 \cdot g = h(M)$.

The algorithm consists of $|T|/2$ iterations. Initially, one puts $M := \varnothing$ and $D := \varnothing$. An iteration starts with choosing a vertex $r \in T$ not covered by the current M. The purpose of the iteration is to transform the current M, D and g (preserving validity of (9) and (10)) in such a way that r becomes covered by M. As soon as M becomes perfect, the current g (extended by zero on $D(K, T) \setminus D$) and M turn into an optimal solution of $\mathscr{P}(K, T, h)$, by (12). Observe that (10)(i) provides the validity of (6).

We need some terminology and notation.

1) Let $\mathscr{V} = \mathscr{V}^D$ be the set whose elements are the vertices of K and the sets of D. Define a partial order $\prec$ on $\mathscr{V}$ by setting $v \prec v'$ if either $v, v' \in D$ and $v \subset v'$, or $v \in T$, $v' \in D$ and $v \in v'$ (in particular, $s \prec \{s\}$ if $s \in T$ and $\{s\} \in D$). Note that any two noncomparable elements in $\mathscr{V}$ contain no common vertex, by (10)(i). When $v \prec v'$, we say that v *precedes* v'; if, in addition, there is no v'' such that $v \prec v'' \prec v'$, we say that v *immediately precedes* v'.

2) For $S \subseteq T$, let W_S denote the set of elements $v \in \mathscr{V}$ such that v is *maximal* if either $v \in S$ or v is strictly included in S. For $S \subseteq T$ and $s \in S$, $w_S(s)$ denotes the only element v of W_S for which $s \preceq v$. Let $F_S = F_S^D$ be the multigraph whose vertex set is W_S and elements $v, v' \in W_S$ are connected by k edges, where k is the number of *saturated* edges $st \in E$

with $v = w_S(s)$ and $v' = w_S(t)$. The edge in F_S corresponding to $st \in E$ is denoted by $\tau_S(st)$. A vertex v in F_S is called *simple* if $v \in T$. When $S = T$, we write W, $w(s)$, F and $\tau(st)$ for W_S, $w_S(s)$, F_S and $\tau_S(st)$, respectively.

3) The algorithm operates with the current multigraphs F and F_A, $A \in D$. Let $M(F)$ ($M(F_A)$) denote the set of edges e of F (F_A) such that $\tau^{-1}(e)$ belongs to the matching M.

The property (11) and the fact that each set in D has odd cardinality imply the following properties:

(13) $|W|$ is even, and $|W_A|$ is odd for all $A \in D$.

(14) $M(F)$ ($M(F_A)$) is a matching in F (in F_A).

A chain L in F (F_A) is called *alternating* (with respect to M) if it contains $\lfloor |EL|/2 \rfloor$ edges in $M(F)$ ($M(F_A)$). During the algorithm the following additional property holds:

(15) For each $A \in D$ with $|A| > 1$, there is a circuit C_A in F_A passing through all the vertices of F_A and containing $(|W_A| - 1)/2$ edges in $M(F_A)$.

For $v \in W_A$, the chain in C_A that joins v and $w(r_A)$ and has an even number of edges is denoted by $L_A(v)$ (clearly, such a chain is alternating); if $|A| = 1$, we put $L_A(v) := (\{r_A\}, \varnothing)$.

ITERATION. Like the majority of matching algorithms, the main work on the iteration consists in "growing" an alternating tree. We say that a subgraph H in F is an *alternating tree rooted at* $w(r)$ if it satisfies the following conditions:

(i) $w(r) \in VH$, and H is connected and has no circuits.

(ii) For each $v \in VH$, the chain in H joining v and $w(r)$ is alternating; this chain is denoted by $L(v)$.

(iii) For each one-valency vertex v in H, $L(v)$ has even number of edges.

Let VH^+ (VH^-) denote the set of vertices of H for which $|EL(v)|$ is even (respectively, odd), and let $W^0 := W \setminus VH$.

An iteration consists of a sequence of steps. A *step* is an application of one of the Procedures P0–P5 below. At the beginning of an iteration we put $H := (\{w(r)\}, \varnothing)$ and start with P0.

PROCEDURE P0. Choose an edge e in F with ends u and v such that either (i) $u \in VH^+$ and $v \in W^0$, or (ii) $u, v \in VH^+$. Let $e = \tau(st)$, $u = w(s)$ and $v = w(t)$. In case (i), go to P1 (increasing the matching) if v is not covered by $M(F)$, and go to P2 (increasing the tree) if it is. In case (ii), go to P3 (shrinking an odd circuit). If there is no edge e as above, choose in VH^- a nonsimple vertex $A \in D$ with $g(A) = 0$ and go to P4 (destroying a nonsimple vertex). If such an A does not exist, go to P5 (changing the packing g).

PROCEDURE P1 (INCREASING M). Let $r' := t$ if $v = t$ (that is, if v is a simple vertex), and $r' := r_A$ if $v = A \in D$; then r' is not covered by M. Add the edge e and the vertex v to the alternating chain $L(u)$ in

H, forming an alternating chain L in F connecting the vertices $w(r)$ and $v = w(r')$ (these vertices are not covered by $M(F)$). If L contains a non-simple vertex $A' \in D$, we replace A' by the alternating chain with an even number of edges from the circuit $C_{A'}$, forming an alternating chain in $F^{D'}$ that connects vertices not covered by $M(F^{D'})$; here $D' := D \setminus \{A'\}$. Repeat such replacements, one by one, until an alternating rr'-chain $\tilde{L}$ in the graph K is obtained. Now change M along $\tilde{L}$ by putting $M := M \triangle E\tilde{L}$ ($X \triangle Y$ denotes the symmetric difference $(X \setminus Y) \cup (Y \setminus X)$ of sets X, Y).

Procedure P1 completes the iteration. The resulting M becomes larger and the vertices r and r' are now covered by M. One can check that (9), (10), and (15) remain true.

PROCEDURE P2 (INCREASING H). Let e' be the edge in $M(F)$ incident to v, and let v' be the other end of e'. (It follows from property (iii) of H that $v' \in W^0$.) Extend H by adding the vertices v, v' and the edges e, e'. Return to P0.

PROCEDURE P3 (SHRINKING AN ODD CIRCUIT). Let C be the circuit in the graph H' obtained by adding the edge e to H; then C is formed by e and the corresponding parts of the chains $L(u)$ and $L(v)$ in H, and it contains $(|EC| - 1)/2$ edges in $M(F)$. Form a new odd set $A := \{s' \in T \mid w(s') \in VC\}$, and put $D := D \cup \{A\}$, $g(A) := 0$ and $C_A := C$. The new tree H is obtained from H' by shrinking C into the new vertex A in the new F^D.

PROCEDURE P4 (DESTROYING A NONSIMPLE VERTEX). Let $e = \tau(st)$ and $e' = \tau(s't')$ be the edges in H incident to A, and let $e \in M(F)$. To be definite, let s and s' be in A; then $s = r_A$. Take the chain $L := L_A(w_A(s'))$ in C_A connecting $w_A(s')$ with the "root" $w_A(r_A)$ of F_A. Delete the set A from D and correct H by replacing the vertex A in it by the chain L. Return to P0.

Prior to the description of P5, let us define the following sets:

$$E^{0+} := \{st \in E \mid w(s) \in W^0, w(t) \in VH^+\};$$
$$E^{0-} := \{st \in E \mid w(s) \in W^0, w(t) \in VH^-\};$$
$$E^{++} := \{st \in E \mid w(s), w(t) \in VH^+, w(s) \neq w(t)\};$$
$$E^{--} := \{st \in E \mid w(s), w(t) \in VH^-, w(s) \neq w(t)\}.$$

PROCEDURE P5 (CHANGING g). Find the value $\varepsilon := \min\{\varepsilon^{0+}, \varepsilon^{++}, \gamma\}$, where

$$\varepsilon^{0+} := \min\{h(e) - \lambda^g(e) \mid e \in E^{0+}\};$$
$$\varepsilon^{++} := \tfrac{1}{2}\min\{h(e) - \lambda^g(e) \mid e \in E^{++}\};$$
$$\gamma := \min\{g(A) \mid A \in VH^- \cap D\}.$$

Add to D all the one-element sets $\{s\}$ such that s is a simple vertex in

VH^+. Transform g as $g(A) := g(A)+\varepsilon$ for $A \in VH^+$ and $g(A) := g(A)-\varepsilon$ for $A \in VH^- \cap D$. Return to P0.

Correctness and complexity of the algorithm.

LEMMA 3.1. *When applying* P5, *all the vertices in* VH^- *are nonsimple.*

PROOF. As it follows from the description of P0, when we go from P0 to P5, the current g satisfies

$$\lambda^g(e) < h(e) \quad \text{for any edge } e \in E^{0+} \cup E^{++} \tag{16}$$

(otherwise one must go from P0 to one of P1–P3 rather than to P5). Suppose that the lemma is not valid, and let v be a simple vertex in VH^-, that is, $v \in T$. Consider edges sv, vt in K such that $\tau(sv)$, $\tau(vt)$ are the edges in H incident to v. Since $\{v\} \notin D$ and $w(s) \neq w(t)$, there is no set $A \in D$ such that $sv, vt \in \partial A$, whence $|\{sv, vt\} \cap \partial A'| = |\{st\} \cap \partial A'|$ for any $A' \in D$. This implies $\lambda^g(st) = \lambda^g(sv) + \lambda^g(vt)$. But $\lambda^g(e) = h(e)$ for $e = sv, vt$, and now from (16) and the fact that $w(s), w(t) \in VH^+$ we conclude that $h(st) > h(sv) + h(vt)$, which is impossible because h is a metric.

Lemma 3.1 and the definition of ε imply easily that the function g resulting from Procedure P5 is an h-packing. We leave to the reader to check that M, D, g, H resulting from each of Procedures P1–P5 are correct; in particular, (9), (10), and (15) are true for them.

Suppose that the iteration is completed in a finite number of steps, and denote by N_i the number of occurrences of Procedure Pi during the iteration. Let $N := \sum_{i=0}^{5} N_i$. Then $N_1 \leq 1$ and $N_0 = \sum_{i=1}^{5} N_i$. One can see from the definition of ε in P5 that, upon applying this procedure, at least one of the two possibilities occurs: $\lambda^g(e) = h(e)$ for some $e \in E^{0+} \cup E^{++}$; or $g(A) = 0$ for some non-simple vertex $A \in VH^-$. (The case when all the sets E^{0+}, E^{++}, VH^- are empty is impossible; otherwise H would consist of a single vertex, and $VH = W$, whence $|W| = 1$ would follow, contrary to (13).) Hence P5 will be immediately followed by P0 and then by one of P1–P4. Thus, $N_5 \leq \sum_{i=1}^{4} N_i$, and therefore, N is $O(N_2 + N_3 + N_4)$. To estimate the latter quantity, introduce the sets

$$\widetilde{T} := \{s \in T \mid w(s) \in VH^+\},$$
$$\widetilde{D} := D \setminus \{A \in D \mid A \preceq v \text{ for some } v \in VH^+\}.$$

We make the following observations:

(i) After every application of P0–P5, the new set $\widetilde{T}$ contains the previous one and the new set $\widetilde{D}$ is contained in the previous one.

(ii) Every application of P2 or P3 increases $\widetilde{T}$.

(iii) Every application of P4 decreases $\widetilde{D}$.

Thus, $N_2 + N_3 \leq |T| - 1$, and N_4 does not exceed the cardinality of D at the beginning of the iteration. By Claim 4 in §2, this cardinality is $O(|T|)$. Hence, the iteration is terminated in $O(p)$ steps.

In order to estimate the running time of the iteration, we need to specify data structures used in it. The elements $A \in D$ are given as identificators (references) rather than the whole lists of their elements in T. Elements $v, v' \in \mathscr{V}$ such that v immediately precedes v' are joined by references to each other; such references define a (directed) forest on $\mathscr{V}$. Thus $\mathscr{V}$ together with the inclusion structure on its members is given by using $O(p)$ elements. The multigraphs F, F_A ($A \in D$) and the tree H are designed in the natural way.

Clearly, each of Procedures P1–P4 can be executed using $O(p)$ operations. P0 consists, in fact, in examination of vertices and edges of F (and/or F_A's) and it takes $O(p^2)$ operations. When P5 applies, we have to compute efficiently the values $\lambda^g(e)$ for $e \in E^{0+} \cup E^{++}$, required for determining ε and for correction of the edge-set of F (according to the new g). To do this, define for $v \in \mathscr{V}$ the set $\mathscr{V}(v)$ to be $\{v\} \cup \{v' \in \mathscr{V} \mid v' \prec v\}$ and the set $T(v)$ to be $T \cap \mathscr{V}(v)$; define the value

$$\rho(v) := \rho^g(v) = \sum \{g(A) \mid A \in D, v \preceq A\}.$$

One can see that $\lambda^g(st) = \rho(s) + \rho(t)$ for any $s, t \in T$ such that $w(s) \neq w(t)$. Observe also that, for fixed $v \in W$, the numbers $\rho(v')$ can be computed recursively for all $v' \in \mathscr{V}(v)$ using $O(|V(v)|)$, or $O(|T(v)|)$, operations. Hence, for $u, v \in W$, $u \neq v$, determining the values $\lambda^g(st)$ for all edges st in the set $S := \{st \in E \mid s \in T(u), t \in T(v)\}$ takes $O(|T(u)||T(v)|)$, or $O(|S|)$, operations. This gives the bound $O(p^2)$ for the running time of P5.

Thus, the iteration can be executed within $O(p^3)$ operations, whence the running time for the algorithm to solve $\mathscr{P}(K_T, T, h)$ is $O(p^4)$. This implies that the running time of the algorithm for the initial problem $\mathscr{P}(G, T, l)$ is exactly as mentioned in the Introduction.

It remains to show that whenever h is cyclically even, the algorithm finds an integral optimal h-packing. It suffices to prove that when h is cyclically even and P5 applies to an integral g, the resulting function g' will be integral as well. In other words, one has to prove that the value ε^{++} defined in the description of P5 is an integer. To see this, consider arbitrary $s, t \in T$ such that $w(s), w(t) \in VH^+$ and $w(s) \neq w(t)$. Let C be the circuit in F formed by the edge $\tau(st)$ and the corresponding parts of the chains $L(w(s))$ and $L(w(t))$. Repeatedly replacing nonsimple vertices in C by appropriate alternating chains (in a similar way as it was done in P1), we get a circuit $\widetilde{C}$ in K all of whose edges except st are saturated by g. Then

$$h(st) - \lambda^g(st) = h(E\widetilde{C}) - \lambda^g(E\widetilde{C}) = h(E\widetilde{C}) - \sum_{A \in D} g(A)|E\widetilde{C} \cap \partial A|.$$

Now the fact that $h(E\widetilde{C})$ and $|E\widetilde{C} \cap \partial A|$ are even together with the integrality of g imply that $h(st) - \lambda^g(st)$ is even. Hence, ε^{++} is an integer.

4. A faster modification

In this section we describe a faster modification of the above algorithm. It is based on certain dynamic data structures. As a consequence, the running time of the second stage becomes $O(p^3 \log p)$ (instead of $O(p^4)$) and that of the third stage becomes $O(pm \log n)$ (instead of $O(pnm)$).

More precisely, in the modification we have to search, as fast as possible, for a minimal element in a dynamic ordered set. Formally, the problem can be posed as follows. Suppose we are given a current set S whose elements $e \in S$ have rational weights $a(e)$. At moments $1, \ldots, N$ the set S is changed by removing some of its elements and inserting new ones. At some of these moments it is required to find an element e in the current set such that $a(e)$ is minimum. One approach is to design S in form of the well-known AVL-tree [1]. Then the above task can be executed with $O(\eta \log \omega)$ operations, where η is the cardinality of the initial S plus the number of elements inserted at moments $1, \ldots, N$, and ω is the maximum cardinality of a current S (note that for our purposes one can use the method developed in [1], which is simpler to implement but requires $O(\eta \log \eta)$ operations). We shall use the term "order structure" for a set S together with a design of it which enables us to execute the above task in time at most $O(\eta \log \eta)$.

First of all, we explain how to modify the third stage of the algorithm. We use the notation of §2. One may assume that the steps on which the same set $A \in Q$ is chosen go in succession. We arrange the set of edges of the cut $\partial^G X$ as an order structure $\mathscr{R} = \mathscr{R}(A)$. An element $e \in \mathscr{R}$ has a weight $q(e)$ (these weights define the ordering in $\mathscr{R}$), and there is a number d associated with $\mathscr{R}$. The numbers $q(e)$ and d are assigned so that for each $e \in \partial^G X$, the current length $l(e)$ is equal to $q(e) - d$. The structure $\mathscr{R}(A)$ is created at the beginning of the treatment of A; at this moment we put $q(e) := l(e)$, $e \in \mathscr{R}(A)$, and $d := 0$.

Suppose that i steps with a given $A \in Q$ have been executed; let $\mathscr{R}_i$ and X_i stand for $\mathscr{R}$ and X, respectively, obtained at the ith step. Then the new set X_{i+1} is constructed as follows. First, find the set Δ of elements $e \in \mathscr{R}_i$ such that $q(e) - d$ ($= l(e)$) is 0 (obviously, Δ is the set of minimal elements in $\mathscr{R}$). Second, find the set Z of vertices $x \in VG \setminus X_i$ such that there is a chain of zero length in G connecting x with a vertex incident to an edge from Δ. Then X_{i+1} is just $X_i \cup Z$.

To get the new $\mathscr{R} = \mathscr{R}_{i+1}$ corresponding to X_{i+1}, one should delete from $\mathscr{R}_i$ the edges with one end in X_i and the other in Z, and add the edges with one end in Z and the other in $V \setminus X_{i+1}$. Each latter edge e is included in $\mathscr{R}$ with weight $q(e) := l(e) + d$. According to the algorithm of §2, the number a, defined for $X = X_{i+1}$ is the minimum of values $q(e) - d$ among the elements $e \in \mathscr{R}_{i+1}$. We correct d to $d := d + a$; this corresponds to decreasing by a the lengths of the edges in ∂X_{i+1}.

Clearly, while working with the same $A \in Q$, each edge of G can be included in $\mathscr{R}$ at most once. Hence, the total number of operations to handle with $\mathscr{R}$ during this period is $O(m \log m)$, or $O(m \log n)$. This implies that the running time of the third stage is $O(pm \log n)$, as required.

Now we explain how to modify the second stage of the algorithm. In comparison with the algorithm of §3, the differences are as follows.

(i) The set E^{++} is designed as an order structure ranged by numbers $b(e)$, $e \in E^{++}$, such that $b(e) - d$ is equal to the "excess" $h(e) - \lambda^g(e)$, where d is a number attached to E^{++} as a whole. Similarly, for each $s \in T$ such that $w(s) \in VH^- \cup W^0$, there is an order structure $\mathscr{R}(s)$ consisting of the edges $st \in E$ with $w(t) \in VH^+$; these edges are ranged in $\mathscr{R}(s)$ by numbers $c(st)$ such that $c(st) - d(s)$ is equal to $h(st) - \lambda^g(st)$, where $d(s)$ is a number attached to $\mathscr{R}(s)$. These order structures are created at the beginning of each iteration, and at this moment the numbers d and $d(s)$ are assigned to be zero.

(ii) The first part of Procedure P0 is executed by the examination of minimal elements in the structures E^{++} and $\mathscr{R}(s)$ for all $s \in T$ such that $w(s) \in W^0$; we determine whether there exists an edge st among them such that the excess $b(st) - d$ (if $st \in E^{++}$) or $c(st) - d(s)$ (if $st \in \mathscr{R}(s)$) is 0. This implies that each occurrence of P0 requires running time $O(p)$. Next, when P5 applies, the number ε^{++} is determined as $(b(e) - d)/2$, and ε^{0+} is determined as $\min\{c(st) - d(s) \mid s \in T, w(s) \in W^0\}$, where e (respectively, st) is a minimal element in E^{++} (respectively, $\mathscr{R}(s)$). When changing the current function g in P5, one should correct d and $d(s)$ for $s \in T$ by $w(s) \in W^0$; namely, one has to put $d := d + 2\varepsilon$ and $d(s) := d(s) + \varepsilon$ (this corresponds to increasing $\lambda^g(e)$ by 2ε for $e \in E^{++}$ and by ε for $e \in \mathscr{R}(s)$).

(iii) Each application of Procedures P2–P4 has to be completed with the correction of the corresponding order structures. We explain how to correct them for P4, which consists in destroying a nonsimple vertex $A \in VH^-$ (for P2 and P3 the corresponding structures are easier to correct, and this is left to the reader). Let $\widehat{D} := D - \{A\}$, $\widehat{W} := (W - \{A\}) \cup W_A$ and $\widehat{H}$ be the objects obtained from D, W and H as a result of applying P4. As explained earlier, $\widehat{H}$ is formed from H by replacing the vertex A by a chain L from C_A. Define L^+ (respectively, L^-) to be the set of vertices v in L such that the part of L from $w_A(r_A)$ to v has an even (respectively, odd) number of edges. Let $W_A^0 := W_A - VL$. Then

$$V\widehat{H}^+ = VH^+ \cup L^-, \quad V\widehat{H}^- = (VH^- - \{A\}) \cup L^+, \quad \widehat{W}^0 = W^0 \cup W_A^0.$$

Using techniques described at the end of §3, determine $\rho(s)$, $\rho(t)$ and then $\lambda^g(st)$ for all $s, t \in T$ such that $s \preceq v'$ and $t \preceq v''$, where v' runs over the set $V\widehat{H}^- \cup \widehat{W}^0$ and v'' runs over L^-; by arguments of §3, this takes a number of operations proportional to the number of such edges st. Now for each $s \in T$ with $\widehat{w}(s) \in V\widehat{H}^- \cup \widehat{W}^0$, one has to insert in $\mathscr{R}(s)$ all the edges $st \in E$ for which $\widehat{w}(t) \in L^-$, and put $c(st) := h(st) - \lambda^g(st) + d(s)$ ($\widehat{w}(x)$ denotes the maximal element $v \in \widehat{W}$ such that $x \preceq v$). In addition, for each $s \in T$ with $\widehat{w}(s) \in L^-$, each element $st \in E$ such that $\widehat{w}(t) \in VH^+$ has to be transferred from $\mathscr{R}(s)$ to the structure E^{++}, with weight $b(st) := c(st) - d(s) + d$; after that the rest of $\mathscr{R}(s)$ is deleted.

It was shown in §3 that the current set $\widetilde{T} := \{s \in T \mid w(s) \in VH^+\}$ is monotonically extended during the iteration. This implies that each edge $st \in E$ can be included at most once in $\mathscr{R}(s)$ or $\mathscr{R}(t)$ and at most once in E^{++}. Hence, the total number of operations used on the iteration to support the above order structures is $O(p^2 \log p)$. This and arguments in (ii) and (iii) give the running time of the iteration to be $O(p^2 \log p)$, and the running time of the second stage to be $O(p^3 \log p)$. Thus, the modified algorithm to solve the initial problem $\mathscr{P}(G, T, l)$ has running time as mentioned in the Introduction.

References

1. G. M. Adel'son-Vel'skiĭ and E. M. Landis, *An algorithm for the organization of information*, Dokl. Akad. Nauk SSSR **146** (1962), 263–266; English transl. in Soviet Math. Dokl. **3** (1962).
2. Jack Edmonds, *The Chinese postman problem*, Oper. Res. **13** (1965), 373.
3. Jack Edmonds and Ellis L. Johnson, *Matchings, Euler tours and the Chinese postman*, Math. Programming **5** (1973), 88-124.
4. A.V. Karzanov, *A dynamic data structure to search for maximal elements in a set and its applications*, Studies in Discrete Optimization (A. A. Fridman, editor), "Nauka", Moscow, 1976, pp. 348-359. (Russian)
5. _____, *Combinatorial methods for solving cut-determined multiflow problems*, Combinatorial Methods for Flow Problems, no. 3 (A. V. Karzanov, editor), Vsesoyuz. Nauchno-Issled. Inst. Sistem. Issled., Moscow, 1979, pp. 6-69. (Russian)
6. _____, *Multicut problems and methods to solve them*, Preprint, Vsesoyuz. Nauchno-Issled. Inst. Sistem. Issled., Moscow, 1982. (Russian)
7. _____, *A generalized MFMC-property and multicommodity cut problems*, Finite and Infinite Sets (Proc. Sixth Hungarian Combinatorial Colloq., Eger, 1981; A. Hajnal, et al., editors), Vol. 2, Colloq. Math. Soc. János Bolyai, vol. 37, North-Holland, Amsterdam, 1984, pp. 443-486.
8. L. Lovász, *2-matchings and 2-covers of hypergraphs*, Acta Math. Acad. Sci. Hungar. **26** (1975), 433-444.
9. Kwan Mei-Ko, *Graphic programming using odd or even points*, Chinese Math. **1** (1962), 273-277.
10. P. D. Seymour, *Sums of circuits*, Graph Theory and Related Topics (W. Tutte Sixtieth Birthday Conf.; J. A. Bondy and U. S. R. Murty, editors), Academic Press, New York, 1978, pp. 341-355.
11. _____, *On multi-colouring of cubic graphs, and conjectures of Fulkerson and Tutte*, Proc. London Math. Soc. (3) **38** (1979), 423-460.
12. _____, *On odd cuts and planar multicommodity flows*, Proc. London Math. Soc. (3) **42** (1981), 178-192.

Translated by the author

Maximum- and Minimum-cost Multicommodity Flow Problems Having Unbounded Fractionality

A. V. KARZANOV

1. Introduction

Among the questions that are of interest in combinatorial optimization theory is the one dealing with fractionality of solutions for certain classes of linear programs. Let us give the formal setting of the problem. Assume that $\mathscr{P}$ is a set of programs of the form $Ax \leqslant b$ (*admissibility* problems), or $\max\{cx \,|\, Ax \leqslant b\}$ (*optimization* problems), with integer matrices A and integer vectors b and c; usually the set $\mathscr{P}$ (called a *class* of problems, or a *problem*) has a specific combinatorial meaning. By the *fractionality* $k(P)$ of an instance $P \in \mathscr{P}$ we mean the least natural number k such that for some admissible (respectively, optimal) solution x for P, the vector kx has only integer entries; if P has no solutions, we put $k(P) = 0$. The fractionality $k(\mathscr{P})$ for a class $\mathscr{P}$ is defined as $\max\{k(P)\,|\,P \in \mathscr{P}\}$; if $k(\mathscr{P}) = \infty$, we say that $\mathscr{P}$ has unbounded fractionality.

At present, no general theorems are known dealing with rather large classes $\mathscr{P}$ of fractionality $\geqslant 2$ (in contrast to well-known results on problems with totally unimodular constraint matrices, or on problems with submodular restrictions, which provide two representative examples of classes of fractionality 1). To find the fractionality, or even to establish that it is bounded, turns out to be a very difficult task for many combinatorial problems; one can mention, for instance, the Tutte-Seymour conjecture that the problem of finding an exact covering of the edges of an undirected graph by cycles has fractionality 2 (see [9]), which has withstood many efforts to solve it.

In the present paper, we discuss unbounded fractionality phenomena in multicommodity flow problems on undirected networks (the case of directed

---

1991 *Mathematics Subject Classification*. Primary 90B10.
Translation of Problems of Discrete Optimization and Methods for Their Solution, Tsentral. Èkonom.-Mat. Inst. Akad. Nauk SSSR, Moscow, 1987, pp. 123–135.

networks appears to be significantly simpler, as we explain below).

For our purposes, it is convenient to consider multicommodity flows as certain chain packings, rather than collections of flows (these two definitions are known to be equivalent [1]). By a *graph* we mean a finite undirected graph without loops and multiple edges; an edge with endpoints x and y is denoted by xy. By a *chain*, or an *xy-chain*, of a graph we mean a nonvoid subgraph $L = (VL, EL)$ of it with vertices $VL = \{x = x_0, x_1, \ldots, x_k = y\}$, $x_i \neq x_j$, and edges $EL = \{x_i x_{i+1} \mid i = 0, \ldots, k-1\}$; the chain L is denoted by $x_0 x_1 \cdots x_k$.

The main objects of our studies are a graph $G = (V, E)$, a distinguished subset $T \subseteq V$ of vertices (called *terminals*), a nonnegative integer-valued function $c \colon E \to \mathbb{Z}_+$ (of edge *capacities*), and a graph $H = (T, U)$ without isolated vertices (H is called a *commodity graph*, or a *scheme*). Given $x, y \in V$, we denote by $\mathscr{L}(G, xy)$ the set of all xy-chains in G. Put

$$\mathscr{L} := \mathscr{L}(G, H) := \bigcup_{st \in U} \mathscr{L}(G, st).$$

A function $f \colon \mathscr{L} \to \mathbb{R}_+$ is said to be a *multicommodity flow*, or a *multiflow*. A multiflow f is *c-admissible* if the following capacity restrictions are valid:

$$\zeta^f(e) := \sum_{L \in \mathscr{L},\, e \in EL} f(L) \leqslant c(e), \qquad e \in E.$$

Given a multiflow f, one defines its *value* between terminals s and t by $v(f, st) := \sum_{L \in \mathscr{L}(G, st)} f(L)$, and its *total value* by $v(f) := 1 \cdot f = \sum_{L \in \mathscr{L}} f(L)$.

Below we formulate three main types of multiflow problems.

DEMAND PROBLEM (to be denoted hereafter by $D(G, H, c, d)$): given a function $d \colon U \to \mathbb{Z}_+$ (flow *demands*), find a c-admissible multiflow f satisfying the condition

$$v(f, st) = d(st) \quad \text{for all } st \in U$$

(or prove that such a multiflow does not exist).

MAXIMUM MULTIFLOW PROBLEM (to be denoted hereafter by $M(G, H, c)$): find a maximum c-admissible multiflow, that is, a multiflow f with the greatest total value $v(f)$.

MINIMUM COST MAXIMUM MULTIFLOW PROBLEM (to be denoted hereafter by $C(G, H, c, a)$): given a function $a \colon E \to \mathbb{Z}_+$ (of edge *costs*), find among the maximum multiflows the one (denote it by f) having the least cost

$$\sum_{L \in \mathscr{L}(G, H)} a(EL) f(L) = \sum_{e \in E} a(e) \zeta^f(e);$$

hereafter $g(S')$, $S' \subseteq S$, stands for $\sum_{e \in S'} g(e)$ for a function g on S.

Let us classify the problems according to their schemes, thus combining in one class all the problems of a given type having the same scheme H.

Namely, given a scheme $H = (T, U)$, denote by $D(H)$ the set of all the demand problems $D(G, H, c, d)$ for arbitrary graphs $G = (V, E)$, $V \supseteq T$, and arbitrary functions $c: E \to \mathbb{Z}_+$, $d: U \to \mathbb{Z}_+$; in a similar way one defines the sets $M(H)$ and $C(H)$. The fractionalities $k(\mathscr{P})$ for the classes $\mathscr{P} = D(H)$, $\mathscr{P} = M(H)$, $\mathscr{P} = C(H)$ are denoted by $k_1(H)$, $k_2(H)$, $k_3(H)$, respectively.

The following simple proposition is true.

CLAIM 1. 1. *If a scheme H' is a subgraph of a scheme H, then $k_1(H') \le k_1(H)$.*

2. If a scheme H' is an induced subgraph of a scheme H (that is, H' is the subgraph induced by a vertex subset of H), then $k_i(H') \le k_i(H)$, $i = 2, 3$.

PROOF. Indeed, let $H' = (T', U') \subset H = (T, U)$; then the problem $D(G' = (V', E'), H', c', d')$ is equivalent to the problem $D(G, H, c, d)$ with $G = (V' \cup (T \setminus T'), E')$, $c(e) := c'(e)$ for $e \in E'$, $d(u) := d'(u)$ for $u \in U'$, $d(u) := 0$ for $u \in U \setminus U'$; this yields the first assertion. The second one is proved in a similar way.

The aim of this paper is to prove the following theorems.

THEOREM 1. *If a scheme H contains three distinct pairwise intersecting anticliques A, B, C such that $A \cap B \ne A \cap C$, then $k_2(H) = \infty$.*

THEOREM 2. *If a scheme H contains two distinct intersecting anticliques (i.e., H is not a complete multipartite graph), then $k_3(H) = \infty$.*

Recall that an *anticlique* of a graph is defined as a maximal (with respect to inclusion) independent (that is, inducing an empty graph) subset of its vertices.

We make a few comments. Denote by K_n the complete graph on n vertices, by Z_r the star with r edges (that is, the graph all of whose r edges possess a common vertex), and by $\Gamma_1 + \Gamma_2 + \cdots + \Gamma_m$ the union of pairwise disjoint graphs $\Gamma_1, \Gamma_2, \ldots, \Gamma_m$; instead of $\Gamma + \cdots + \Gamma$ (m times) we write $m\Gamma$. Let $\mathscr{A}(H)$ stand for the set of all anticliques of H.

1) It is known that (a) $k_1(H) = 1$, provided $H = Z_r$ [1]; (b) $k_1(H) = 2$, provided H is the union of two stars and $H \ne Z_r$ (according to Dinits, this fact is an easy consequence of the half-integrality theorem for two-commodity flows [2]); and (c) $k_1(H) = 2$, provided H is K_4, or H is the cycle on 5 vertices [7] (see also [10] and [8]).

Recently, the author proved that $k_1(H) = 2$, provided $H = K_5$ or H is the union of K_3 and a star [5]. On the other hand, it is shown in [8] that $k_1(H) = \infty$ for $H = 3K_2$. One can verify that the only scheme H such that $k_1(H)$ is not defined by the above results and Claim 1.1 is $2K_3$. In this case the demand problem can be reduced easily to the maximum multiflow problem for the same scheme $2K_3$ (observe that $H = 2K_3$ violates the assumptions of Theorem 1); there are reasons to conjecture that $k_1(2K_3) = 4$.

2) It is known that (1) $k_2(H) = 1$, provided H is a complete bipartite graph (a multi-terminal version of the maximum flow theorem [1]); (2) $k_2(H) = 2$, provided $|\mathscr{A}(H)| > 2$ and $\mathscr{A}(H)$ admits a partition $\{\mathscr{A}_1, \mathscr{A}_2\}$ such that each family $\mathscr{A}_i$ consists of pairwise disjoint anticliques [6] (see also [8] and [4]); and (3) $k_2(H) = 4$, provided the latter condition is violated, yet H does not contain three distinct pairwise intersecting anticliques [6].

Therefore, each scheme H such that $k_2(H)$ is not defined by the above results and Theorem 1 must contain three distinct pairwise intersecting anticliques, and for any such triple A, B, C one must have $A \cap B = B \cap C = C \cap A$. For such schemes H, the values $k_2(H)$ are not established yet (presumably, $k_2(H) \leqslant 4$); recently, the author proved that for such schemes H the fractionality for the problem dual to M(G, H, c) does not exceed 4.

3) It is known that $k_3(H) = 1$, provided H is a complete bipartite graph [1], and that $k_3(H) = 2$, provided $H = K_n$, $n \geqslant 3$ [3].

The latter result can be extended easily to the case of complete multipartite graphs H. Together with Theorem 2, this provides a complete description of the values of $k_3(H)$ for all schemes H.

The solution of fractionality problems for directed multiflows is significantly simpler. Classical results of Ford and Fulkerson [1] imply that $k_1(H) = 1$, provided $H = (T, U)$ is a directed star (that is, either $U = \{(s, t) | t \in T \setminus \{s\}\}$, or $U = \{(t, s) | t \in T \setminus \{s\}\}$ for some $s \in T$), and $k_2(H) = k_3(H) = 1$, provided H is a complete directed bipartite graph (that is, $U = \{(s, t) | s \in S, t \in T \setminus S\}$ for some $S \subset T$). Easy examples show that in all the other cases the fractionalities are infinite.

2. Proof of Theorem 1

The following proposition is true (the proof is left to the reader).

CLAIM 2. *A scheme H satisfies the assumptions of Theorem 1 if and only if H possesses an induced subgraph H' such that $3K_2 \subseteq H' \subseteq H^1$, where H^1 is the graph in Figure 1(b).*

According to Claims 1 and 2, to prove Theorem 1, it suffices to consider only schemes $H = (T, U)$ such that $3K_2 \subseteq H \subseteq H^1$. One must show that for an arbitrary natural k^* there exist a graph $G = (V, E)$, $V \supseteq T$, and a function $c: E \to \mathbb{Z}_+$ such that for any optimal solution f of the problem

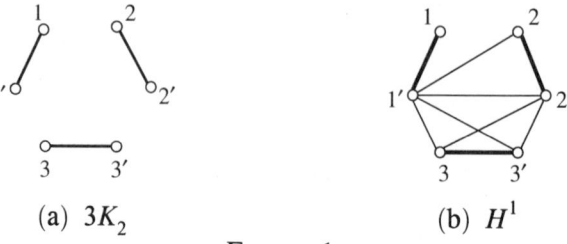

(a) $3K_2$ (b) H^1

FIGURE 1

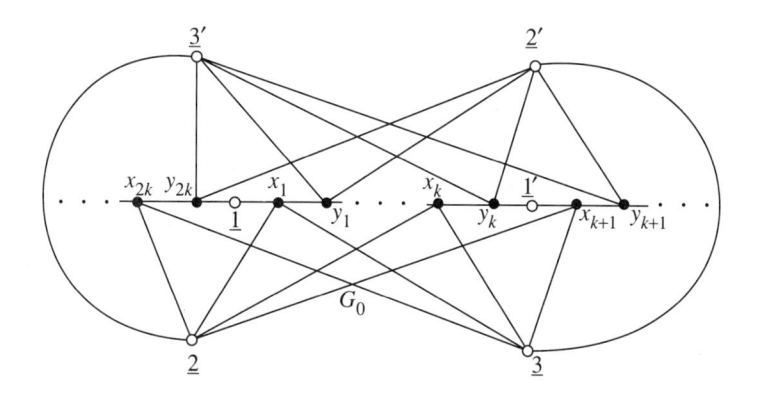

FIGURE 2

D(G, H, c) the vector $k'f$ has at least one noninteger entry for any natural $k' < k^*$. The construction providing such networks (G, c) for the above described schemes H relies on the construction for the "simplest" scheme $H_0 = 3K_2$. For the case of H_0, one can take the counterexample from §12 of [8]; however, below we present a simpler example. Namely, consider the graph $G_0 = (V_0, E_0)$ shown in Figure 2; it consists of the cycle

$$\underline{1}x_1y_1 \cdots x_ky_k\underline{1}'x_{k+1}y_{k+1} \cdots x_{2k}y_{2k}\underline{1}$$

plus vertices $\underline{2}, \underline{2}', \underline{3}, \underline{3}'$, edges $\underline{2}x_i, \underline{3}x_i, \underline{2}'y_i, \underline{3}'y_i$, $i = 1, \ldots, 2k$, and two distinguished edges: $\underline{23}'$ and $\underline{32}'$. The capacities $c_0(e)$ for all the nondistinguished edges $e \in E_0$ equal 1, while those for the distinguished edges equal 2. Put $H_0 = (T_0, U_0)$, where

$$T_0 := \{\underline{1}, \underline{1}', \underline{2}, \underline{2}', \underline{3}, \underline{3}'\}, \qquad U_0 := \{\underline{11}', \underline{22}', \underline{33}'\}.$$

Define in G_0 the following multiflow f_0:

$$f_0(L_i) := f_0(L_i') = (k-1)/k, \qquad i = 1, \ldots, 2k,$$
$$f_0(P_1) := f_0(P_2) = 1/k,$$
$$f_0(Q_i) := f_0(Q_i') = 1/k, \qquad i = 1, \ldots, 2k$$

where

$$L_i = \underline{2}x_iy_i\underline{2}', \qquad L_j' = \underline{3}x_jy_{j-1}\underline{3}' \; (j \neq 1, k+1),$$
$$L_1' = \underline{3}x_1\underline{1}y_{2k}\underline{3}', \qquad L_{k+1}' = \underline{3}x_{k+1}\underline{1}'y_k\underline{3}',$$
$$P_1 = \underline{1}x_1y_1 \cdots x_ky_k\underline{1}', \qquad P_2 = \underline{1}y_{2k}x_{2k} \cdots y_{k+1}x_{k+1}\underline{1}',$$
$$Q_i = \underline{2}x_i\underline{32}', \qquad Q_i' = \underline{23}'y_i\underline{2}';$$

for all the other chains belonging to $\mathscr{L}(G_0, H_0)$ the multiflow f_0 vanishes. A direct calculation shows that all the edges of G_0 are saturated by f_0 (that is, $\zeta^{f_0}(e) = c_0(e)$ for all $e \in E_0$), and that

$$v(f_0, \underline{11}') = 2/k, \qquad v(f_0, \underline{22}') = 2k+2, \qquad v(f_0, \underline{33}') = 2k-2;$$

hence f_0 is c_0-admissible, and its total value $v(f_0)$ equals $4k + 2/k$.

Let us prove that f_0 is an optimal solution for $M(G_0, H_0, c_0)$; this fact would imply that the fractionality for the above problem is at least $k/2$, since the denominator of $v(f_0)$ equals $k/2$. To do this, consider the dual problem $M^*(G_0, H_0, c_0)$. Observe that the admissible solutions for the problem $M^*(G, H, c)$ dual (in the sense of linear programming) to the problem $M(G, H, c)$ are exactly the functions $l: E \to \mathbb{R}_+$ satisfying the restrictions

$$l(EL) \geqslant 1, \quad L \in \mathscr{L}(G, H).$$

The above restrictions can be rewritten as

$$\mathrm{dist}_l(st) \geqslant 1, \quad st \in U,$$

where $\mathrm{dist}_l(xy)$ denotes the distance between the vertices x and y in the graph G with edge lengths $l(e)$; in other words,

$$\mathrm{dist}_l(xy) = \min_{L \in \mathscr{L}(G, xy)} l(EL).$$

By the linear programming duality theorem, admissible solutions f and l are optimal if and only if

$$v(f) = c \cdot l \left(= \sum_{e \in E} c(e)l(e) \right);$$

this is equivalent to the complementary slackness conditions

$$L \in \mathscr{L}(G, H), f(L) > 0 \Rightarrow l(EL) = 1, \tag{C1}$$

$$e \in E, l(e) > 0 \Rightarrow \zeta^f(e) = c(e). \tag{C2}$$

In our case, put

$$l_0(\underline{1}x_1) := l_0(\underline{1}y_{2k}) := l_0(\underline{1}'x_{k+1}) := l_0(\underline{1}'y_k) := 1/4k;$$
$$l_0(\underline{23}') := l_0(\underline{32}') := l_0(x_iy_i) := l_0(y_jx_{j+1}) := 1/2k,$$
$$i = 1, \ldots, 2k, \quad j = 1, \ldots, k-1, k+1, \ldots, 2k-1;$$
$$l_0(\underline{2}x_i) := l_0(\underline{3}x_i) := l_0(\underline{2}'y_i) := l_0(\underline{3}'y_i) := 1/2 - 1/4k, \quad i = 1, \ldots, 2k.$$

It is easy to check that l_0 is an admissible solution for $M^*(G_0, H_0, c_0)$, and that both f_0 and c_0 satisfy (C1) and (C2), as required.

Therefore, we have constructed the proper example for the scheme $H_0 = 3K_2$. To design the required examples for the other schemes H, $H_0 \subset H \subseteq H^1$, we reform G_0, c_0, f_0 in such a way that the values of each of the three flows involved remain integer.

1. Paste together k copies $(G_1, c_1), \ldots, (G_k, c_k)$ of the network (G_0, c_0) by identifying the vertices $\underline{1}_j$ (denote the vertex obtained by $\tilde{1}$) and by identifying the vertices $\underline{1}'_j$ (denote the vertex obtained by $\tilde{1}'$); here w_j stands for the copy of the vertex $w \in V_0$ in the graph G_j.

2. Add to the graph obtained above new vertices $1, 1', 2, 2', 3, 3'$ and new edges $1\tilde{1}, 1'\tilde{1}', 2\underline{2}_j, 2'\underline{2}'_j, 3\underline{3}_j, 3'\underline{3}'_j$, $j = 1, \ldots, k$, with

$$c(1\tilde{1}) := c(1'\tilde{1}') := 2, \qquad c(2\underline{2}_j) := c(2'\underline{2}'_j) := 2k+2,$$
$$c(3\underline{3}_j) := c(3'\underline{3}'_j) := 2k-2, \qquad j = 1, \ldots, k.$$

Denote the resulting network by $(G = (V, E), c)$, and let $\tilde{H} = (\tilde{T}, \tilde{U})$ be the scheme whose edges are $11', 22', 33'$.

One can naturally "extend" the multiflow f_0 to a multiflow f for G, $\tilde{H}, c$. Namely, given a chain $L = \underline{i}v \cdots w\underline{i}'$, $i \in \{1, 2, 3\}$, take k chains $L_j = i\underline{i}_j v_j \cdots w_j \underline{i}'_j i'$, $j = 1, \ldots, k$, in the graph G, and put $f(L_j) := f_0(L)$. Evidently, all the edges of G are saturated by f, and f is an optimal solution for G, c and $\tilde{H}$ as above.

Let us prove that the fractionality for any maximum flow f' for G, c and $\tilde{H}$ as above equals at least $k/4$. Denote by $E(x)$ the set of edges in G that are incident to $x \in V$; put $E' = \bigcup_{s \in \tilde{T}} E(s)$. Consider the following functions l^1 and l^2 defined on E:

$$l^1(e) = \begin{cases} 0 & \text{for } e \in E', \\ l_0(e') & \text{for } e \in E \setminus E', \end{cases} \qquad l^2(e) = \begin{cases} 1/2 & \text{for } e \in E', \\ 0 & \text{for } e \in E \setminus E', \end{cases}$$

where e' is the edge of G_0 such that e is a copy of e'. It is easy to see that both l^1 and l^2 are optimal solutions for the problem $M^*(G, H, c)$. Assume that $L = i \cdots i'$ is a chain in $\mathscr{L}(G, \tilde{H})$ with $f'(L) > 0$. Relation (C1) for f' and l^2 implies that L passes through no vertices from $\tilde{T}$ except i or i'. Next, if $i \in \{2, 3\}$ and L contains a vertex $s \in \{\tilde{1}, \tilde{1}'\}$, then the relations $l^1(EL) = 1$ and $\text{dist}_{l_1}(is) = \text{dist}_{l^1}(si') = 1/2$ (these are easy to verify) yield $l^1(EL') = l^1(EL'') = 1/2$, where L' and L'' are the parts of the chain L from i to s and from s to i', respectively. Denote by $G'_j = (V'_j, E'_j)$ the subgraph of G obtained by adding vertices $2, 2', 3, 3'$ and edges $2\underline{2}_j, 2'\underline{2}'_j, 3\underline{3}_j, 3'\underline{3}'_j$ to G_j. Let g be the multiflow in the graph G'_j with terminals at $T' = \{\tilde{1}, \tilde{1}', 2, 2', 3, 3'\}$ that is induced by f'; that is, $g(L') = \sum f'(L)$ for any chain L' in G'_j such that both its endpoints belong to T', the sum being taken over all chains $L \in \mathscr{L}(G, \tilde{H})$ containing L' as a part. Denote by $\mathscr{L}'$ the set of all chains in G'_j having both endpoints in T'. Since f' saturates all the edges of G, g must saturate all the edges of G'_j; thus

$$\sum_{L' \in \mathscr{L}'} l^1(EL')g(L') = \sum_{e \in E'_j} c(e)l^1(e) = c_0 \cdot l_0 = 4k + \frac{2}{k}.$$

On the other hand, the above arguments imply that for any chain $L' \in \mathscr{L}'$, the value $l^1(EL')$ equals either 1 or $1/2$, provided $g(L') > 0$. Hence the fractionality of g is at least $k/4$.

(Observe that the above example proves also that the fractionality of the demand problem for the scheme $3K_2$ is unbounded.)

Finally, consider an arbitrary scheme $H = (T, U)$ such that $3K_2 \subset H \subseteq H^1$. It is easy to see that H^1 contains exactly one subgraph isomorphic to $3K_2$. Thus, one can assume that $T = \tilde{T}$, $U \supset \tilde{U} = \{11', 22', 33'\}$, and that the edge set of H^1 is $U^1 = \tilde{U} \cup \{1'2, 1'2', 1'3, 1'3', 2'3, 2'3'\}$. Let us prove that for G and c as above, any optimal solution f^* for the problem $\mathrm{M}(G, H, c)$ satisfies $v(f^*, st) = 0$ for all $st \in U \setminus \tilde{U}$; this would yield that the restriction of f^* on $\mathscr{L}(G, \tilde{H})$ is an optimal solution for the problem $\mathrm{M}(G, \tilde{H}, c)$, and thus the fractionality of f^* is at least $k/4$. To do this, it suffices to present an optimal solution l for the dual problem $\mathrm{M}^*(G, \tilde{H}, c)$ such that $\mathrm{dist}_l(st) > 1$ for all $st \in U \setminus \tilde{U}$ (by (C1); this would imply $v(f^*, st) = 0$). The required solution l can be defined as follows:

$$l(1\tilde{1}) = 0, \qquad l(1'\tilde{1}') = 1,$$
$$l(2\underline{2}_j) = \tfrac{1}{4}, \quad l(2'\underline{2}'_j) = \tfrac{3}{4}, \quad l(3\underline{3}_j) = l(3'\underline{3}'_j) = \tfrac{1}{2}, \quad j = 1, \ldots, k,$$

and $l(e) = 0$ for all the other edges of G. It is easy to verify that

$$\mathrm{dist}_l(st) = 1 \quad \text{for} \quad st \in \tilde{U},$$
$$\mathrm{dist}_l(st) > 1 \quad \text{for} \quad st \in U^1 \setminus \tilde{U};$$

besides, relations (C1) and (C2) are valid for the solution l and the multiflow f defined as above (for G, c and $\tilde{H}$); hence l and f (extended by zero to $\mathscr{L}(G, H - \tilde{H})$) are optimal solutions for $\mathrm{M}^*(G, H, c)$ and $\mathrm{M}(G, H, c)$, as required.

Theorem 1 is proved.

3. Proof of Theorem 2

The following proposition is rather trivial (its verification is left to the reader).

CLAIM 3. *Let H be a graph without isolated vertices. Then H possesses two distinct intersecting anticliques if and only if it contains a four-vertex subset $T' = \{1, 1', 2, 2'\}$ such that the subgraph H' induced by T' satisfies $H_0 \subseteq H' \subseteq H_1$, where $H_0 = (T', U_0)$, $H_1 = (T', U_1)$, $U_0 = \{11', 22'\}$, $U_1 = \{12, 12'\} \cup U_0$.*

By Claim 1.2, it suffices to consider only schemes $H = (T, U)$ satisfying the relation $H_0 \subseteq H \subseteq H_1$. Assume that $T = \{1, 1', 2, 2'\}$, and the edge sets U_0 and U_1 for the graphs H_0 and H_1 are defined as in Claim 3.

Given an arbitrary positive odd k, we construct a simple example of the problem $\mathrm{C}(G, H, c, a)$ such that it has exactly one optimal solution f, and the fractionality of this solution is just k. The graph $G = (V, E)$ is shown in Figure 3. Here all the edges have unit capacities, except the edges $2w$

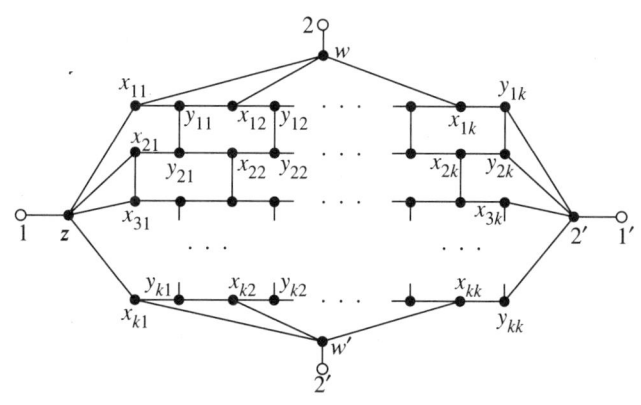

FIGURE 3

and $2'w'$; these two edges are of capacity $k-1$. Edge costs are defined as follows:

$$a(x_{ij}y_{ij}) := 0, \quad a(1z) := 2k, \quad a(1'z') := 0, \quad a(2w) := a(2'w') := k,$$
$$a(y_{ij}x_{i\,j+1}) := a(y_{pq}y_{p+1\,q}) := a(x_{rm}x_{r+1\,m}) := 1,$$
$$a(zx_{i1}) := a(z'y_{ik}) := a(wx_{1j}) := a(w'x_{kj}) := k.$$

Introduce the following chains:

$$L_i = 1zx_{i1}y_{i1}x_{i2}\ldots x_{ik}y_{ik}z'1',$$
$$P_i = 2wx_{1i}y_{1i}y_{2i}x_{2i}x_{3i}y_{3i}\ldots y_{ki}x_{ki}w'2', \quad i = 1, \ldots, k,$$

and define a multiflow f by $f(P_i) = (k-1)/k$, $f(L_i) = 1/k$, $i = 1, \ldots, k$, $f(L) = 0$ for all the other chains in $\mathscr{L}(G, H)$. One can verify the following facts:

(1) $v(f) = k$, and for any c-admissible multiflow $f' : \mathscr{L}(G, H) \to \mathbb{R}_+$ one has $v(f') \leq (c(1z) + c(1'z') + c(2w) + c(2'w'))/2 = k$, whence f is a maximum multiflow.

(2) The costs for the chains L_i and P_i equal $5k-1$, while the cost $a(EL)$ for any other chain L in $\mathscr{L}(G, H)$ is at least $5k$; therefore, the cost of f is minimal among all the multiflows (for G, c and H) of the same total value.

(3) The multiflow f is the unique multiflow of total value k that involves only chains L_i and P_i.

Properties (1)–(3) immediately imply the assertion of the theorem.

References

1. L. R. Ford, Jr., and D. R. Fulkerson, *Flows in networks*, Princeton Univ. Press, Princeton, NJ, 1962.
2. T. C. Hu, *Multi-commodity network flows*, Oper. Res. **11** (1963), 344–360.
3. A. V. Karzanov, *Minimum cost maximum multiflow problem*, Combinatorial Methods in Flow Problems, no. 3 (A. V. Karzanov, editor), Vsesoyuz. Nauchno-Issled. Inst. Sistem. Issled., Moscow, 1979, pp. 138–156. (Russian)

4. _____, *On multicommodity flow problems with integer-valued optimal solutions*, Doklady Akad. Nauk SSSR **280** (1985), 789–792; English transl. in Soviet Math. Dokl. **31** (1985).
5. _____, *Half-integral five-terminus flows*, Discrete Appl. Math. **18** (1987), 263–278.
6. A. V. Karzanov and M. V. Lomonosov, *Flow systems in undirected networks*, Mathematical Programming Etc., Sb. Trudov. Vsesoyuz. Nauchno-Issled. Inst. Sistem. Issled. **1978**, no. 1, 59–66. (Russian)
7. M. V. Lomonosov, *Solutions for two problems on flows in networks*, Unpublished manuscript, 1976. (Russian)
8. _____, *Combinatorial approaches to multiflow problems*, Discrete Appl. Math. **11** (1985), 1–94.
9. P. D. Seymour, *Sums of circuits*, Graph Theory and Related Topics (J. A. Bondy and U. S. R. Murty, eds.), Academic Press, New York, 1978, pp. 341–355.
10. _____, *Four-terminus flows*, Networks **10** (1980), 79–86.

On a Class of Maximum Multicommodity Flow Problems with Integer Optimal Solutions

A. V. KARZANOV

In [4] we announced and briefly outlined a new proof of a theorem, due to Karzanov and Lomonosov, on the existence of an integral optimal solution for a special case of the maximum undirected multicommodity flow problem. In this case the capacity function is inner Eulerian and the commodity graph H possesses the following property: the set of its anticliques has a partition into two subsets, each consisting of pairwise disjoint anticliques (in other words, H is the complement to the line graph of a bipartite graph). Based on the splitting-off techniques (rather than alternating path approaches), this proof looks considerably simpler than the original one and, moreover, it provides a strongly polynomial algorithm to solve the problem.

In this paper we give a detailed description of these results.

1. Introduction

By a *graph* we mean a finite undirected graph without loops and multiple edges; the edge with endpoints x and y is denoted by xy. By a *chain* (or an *xy-chain*) of a graph we mean its subgraph $L = (VL, EL)$ whose vertex-set and edge-set have the following form: $VL = \{x = x_0, x_1, \ldots, x_k = y\}$, $EL = \{x_i x_{i+1} : i = 0, \ldots, k-1\}$; the chain L is denoted by $x_0 x_1 \cdots x_k$.

We deal with the following objects: a graph $G = (V, E)$, *capacities* $c(e) \geq 0$ defined on its edges $e \in E$, and a graph $H = (T, U)$ having no isolated vertices and satisfying the relation $T \subseteq V$. The pair (G, c), the graph H, and its vertex set T are said to be the *network*, the (flow) *scheme*, and the set of *terminals*, respectively.

1991 *Mathematics Subject Classification.* Primary 90B10.

Translation of Modelling and Optimization of Systems of Complex Structures (G. Sh. Fridman, editor), Omsk. Gos. Univ., Omsk, 1987, pp. 103–121.

Consider the well-known problem of finding a maximum multicommodity flow in a network (G, c) with a scheme H. For our purposes, it is convenient to formulate this problem as a chain packing problem. Given $x, y \in V$, denote by $\mathscr{L}(G, xy)$ the set of all xy-chains in G. Put $\mathscr{L} := \mathscr{L}(G, H) := \bigcup_{st \in U} \mathscr{L}(G, st)$. A function $f: \mathscr{L} \to \mathbb{R}_+$ (here $\mathbb{R}_+$ is the set of all nonnegative reals) is said to be a *multicommodity flow*, or, briefly, a *multiflow*; a multiflow f is said to be *c-admissible* if the following capacity restrictions are valid:

$$\zeta^f(e) := \sum_{L \in \mathscr{L},\, e \in EL} f(L) \leqslant c(e), \qquad e \in E. \tag{1}$$

PROBLEM $\mathscr{P} = \mathscr{P}(G, c, H)$. *Find a c-admissible multiflow f having the maximum total value*

$$1 \cdot f = \sum_{L \in \mathscr{L}} f(L).$$

Denote by $v = v(G, c, H)$ the maximum of $1 \cdot f$ over all c-admissible f.

An interesting problem in discrete optimization theory consists in the following: Given a class of linear programs with integer-valued constraint matrices and right-hand side vectors, decide whether each program of this class has an optimal solution with all denominators bounded by a fixed number k. In our case, we have the following situation. Let us classify the problems $\mathscr{P}(G, c, H)$ according to the type of the scheme $H = (T, U)$. We say that a scheme H (and the corresponding class $\mathscr{P}(H)$ of the problems $\mathscr{P}(G, c, H)$ with H fixed) is *solvable in* $\frac{1}{k}\mathbb{Z}_+$, where $\mathbb{Z}_+$ is the set of all nonnegative integers and $k \in \mathbb{Z}_+ \setminus \{0\}$, if for any graph $G = (V, E)$, $V \supseteq T$, and any nonnegative integer-valued function $c: E \to \mathbb{Z}_+$ the problem $\mathscr{P}(G, c, H)$ possesses an optimal solution f such that $kf(L)$ is an integer for all $L \in \mathscr{L}(G, H)$; in other words, if the problem $\mathscr{P}(G, kc, H)$ possesses an integer optimal solution. A scheme H is said to be of *bounded fractionality* if H is solvable in $\frac{1}{k}\mathbb{Z}_+$ for some natural k.

We need also the following notions. Let $X \subseteq V$; the set of all edges of G having one endpoint in X and the other one in $V \setminus X$ is denoted by $\partial X = \partial^G X$ and is called a *cut* of G (the cases $X = \varnothing$ and $X = V$ are allowed). A function c is said to be *inner Eulerian* if it takes integer values and $c(\partial X)$ is even for any $X \subseteq V \setminus T$ (for arbitrary $g: E \to \mathbb{R}$ and $E' \subseteq E$ we denote by $g(E')$ the sum $\sum_{e \in E'} g(e)$). A scheme H is said to be *solvable in* $\frac{1}{k}\mathbb{Z}_+$ *for the inner Eulerian case* if the problem $\mathscr{P}(G, kc, H)$ possesses an integer optimal solution for any graph $G = (V, E)$, $V \supseteq T$, and any inner Eulerian function c on E. Evidently, if H is solvable in $\frac{1}{k}\mathbb{Z}_+$ for the inner Eulerian case, then it is solvable in $\frac{1}{2k}\mathbb{Z}_+$.

By the classical Ford–Fulkerson theorem, H is solvable in $\mathbb{Z}_+ = \frac{1}{1}\mathbb{Z}_+$ for $|U| = 1$ (in this case one has the usual maximum flow problem for an undirected network); this fact is generalized easily to complete bipartite graphs H. One can show that if H is not a complete bipartite graph, then it is not solvable in $\mathbb{Z}_+$. It is well known that the scheme with two edges is solvable in $\frac{1}{2}\mathbb{Z}_+$ (Hu's theorem on half-integer two-commodity flows [3]), as well as schemes that are complete bipartite graphs with an arbitrary number of vertices [7], [9], [1]. These results were sharpened in [10], [9], and [1]; it was proved there that the corresponding schemes are solvable in $\mathbb{Z}_+$ for the inner Eulerian case. In [2], the solvability in $\frac{1}{2}\mathbb{Z}_+$ was proved for certain schemes represented as the union of two complete bipartite graphs. Finally, in [6] large classes of schemes solvable in $\frac{1}{2}\mathbb{Z}_+$ and in $\frac{1}{4}\mathbb{Z}_+$ were provided, which generalize all the previously known schemes with these properties. These classes are defined in terms of the family $\mathscr{A} = \mathscr{A}(H)$ of all anticliques of the graph H (an *anticlique* of a graph is a maximal (with respect to inclusion) independent (i.e., generating the empty subgraph) set of its vertices).

DEFINITIONS. A family $\mathscr{A}$ of anticliques is said to be *bipartite* if it has a partition $\{\mathscr{A}_1, \mathscr{A}_2\}$ such that each $\mathscr{A}_i$ consists of pairwise disjoint anticliques. A family $\mathscr{A}$ is said to be 3-*noncrossing* if it does not contain three pairwise intersecting anticliques. A family is said to be *perfect* if for any three distinct anticliques A, B, C such that $A \cap B \neq \varnothing$, $B \cap C \neq \varnothing$, $C \cap A \neq \varnothing$, one has $A \cap B = B \cap C = C \cap A$.

Each of the above three classes contains the previous one as a proper subset. Here are several instances of these classes:

a) If H contains only two edges, or H is a complete graph, then $\mathscr{A}$ is a bipartite family.

b) If H is the cycle on 5 vertices, then $\mathscr{A}$ is 3-noncrossing, but not bipartite.

c) If H consists of a triangle and an edge not adjacent to the triangle, then $\mathscr{A}$ is perfect, but not 3-noncrossing.

d) If H consists of three pairwise nonadjacent edges, then $\mathscr{A}$ is nonperfect.

According to [6], the following assertions are true:

(1) If $\mathscr{A}$ is bipartite, then H is solvable in $\frac{1}{2}\mathbb{Z}_+$.

(2) If $\mathscr{A}$ is 3-noncrossing, then H is solvable in $\frac{1}{4}\mathbb{Z}_+$.

The proof of the first statement is implied by the existence of a pseudopolynomial algorithm solving the problem $\mathscr{P}(G, c, H)$ for integer c and bipartite $\mathscr{A}(H)$ (the running time for the algorithm is bounded by $c(E)$ times a polynomial in $|V|$ and $|T|$). The second statement follows from a reduction of the problem $\mathscr{P}(G, c, H)$ with a 3-noncrossing $\mathscr{A}(H)$ to the bipartite case; this reduction assigns an optimal solution of the first problem to each optimal solution of the second one; moreover, the fractionality for the former is twice as big as that for the latter. For details of proofs see [4] and [8].

The question of what is the class of H's such that H is solvable in $\frac{1}{k}\mathbb{Z}_+$ for some natural k (depending only on H) is still open, and a conjecture is that this is exactly the set of H's with perfect $\mathscr{A}(H)$ (a strong version of the conjecture suggests that whenever such a k exists for a given H, k can be taken to be 4).[1]

The original proof of the first of the above statements, outlined in a sketched form in [6], provides, in essence, the following stronger result, as was pointed out in [8].

THEOREM 1. *If $\mathscr{A}(H)$ is bipartite and c is inner Eulerian, then the problem $\mathscr{P}(G, c, H)$ has an integer optimal solution (in other words, each scheme H such that $\mathscr{A}(H)$ is bipartite is solvable in $\mathbb{Z}_+$ for the inner Eulerian case).*

This implies that in the inner Eulerian case, H is solvable in $\frac{1}{2}\mathbb{Z}_+$ if $\mathscr{A}(H)$ is 3-noncrossing.

It should be noted that the above-mentioned proof of Theorem 1 used rather intricate augmenting-path techniques. In the present paper, we develop a new proof of this theorem, which seems to be significantly simpler; moreover, this proof provides a strongly polynomial algorithm to solve the problem in question. These proof and algorithm were announced and briefly described in [4].

The proof presented here relies on the following two ideas. First, one can directly establish the existence of a dual optimal solution having a special combinatorial form (in fact, the existence of such dual solutions was stated in [6], but there it appeared as a consequence of the proof of Theorem 1). Second, on the basis of this result, one can apply the splitting-off technique that reduces the initial network step by step to a trivial one, for which an integer optimal solution exists evidently.

The rest of the paper is organized as follows. The special duality theorem is established in §2, and the proof of Theorem 1 is completed in §3. Finally, in §4, we describe a strongly polynomial algorithm for the problem $\mathscr{P}(G, c, H)$ with a bipartite $\mathscr{A}(H)$. It is capable of working with arbitrary "real-valued" capacities c and has complexity $O(n^3 \sigma(tn))$; here $n = |V|$, $t = |T|$, and $\sigma(n')$ is the running time required for finding a maximum flow in a network on n' vertices. Whenever c is inner Eulerian, the algorithm determines an integer-valued optimal solution.

2. Minimal proper families

In what follows we use the abbreviated notation $\mathscr{P}(c)$, $v(c)$, $\mathscr{A}$, $\mathscr{L}$ for $\mathscr{P}(G, c, H)$, $v(G, c, H)$, $\mathscr{A}(H)$, $\mathscr{L}(G, H)$, respectively. Throughout the paper, we assume that $\mathscr{A}$ is bipartite, and $\{\mathscr{A}_1, \mathscr{A}_2\}$ is its partition ($A \cap B = \varnothing$ for any distinct $A, B \in \mathscr{A}_i$, $i = 1, 2$).

---

[1] *Added in translation.* Recently, in [11] it was proved that: (1) if $\mathscr{A}(H)$ is perfect, then the dual problem is solvable in $\frac{1}{4}\mathbb{Z}_+$, and (2) if $\mathscr{A}(H)$ is not perfect, then the primal problem is not solvable in $\frac{1}{k}\mathbb{Z}_+$ for any k.

A family $\mathscr{X} = \{X_A : A \in \mathscr{A}\}$ is said to be *semiproper* if

(1) $X_A \subset V$ and $X_A \cap T \subseteq A$ for any $A \in \mathscr{A}$ (the case $X_A = \varnothing$ is allowed), and

(2) each terminal $s \in T$ is contained exactly in one set $X_A \in \mathscr{X}$.

A semiproper family $\mathscr{X}$ consisting of pairwise disjoint sets is said to be *proper*. Let us define the capacity of a semiproper family $\mathscr{X}$ by $c(\mathscr{X}) := \frac{1}{2}\sum_{A \in \mathscr{A}} c(\partial X_A)$.

In this section our aim is to prove the following theorem establishing a minimax relation between multiflows and proper families.

THEOREM 2.1. *The value $v(c)$ equals the minimum of $c(\mathscr{X})$ over all semiproper families $\mathscr{X}$, and the minimum is attained at a proper family.*

The proof includes several auxiliary facts. Assume that $f: \mathscr{L} \to \mathbb{R}_+$ is an optimal multiflow, and let $\mathscr{L}^+(f)$ denote the set $\{L \in \mathscr{L} : f(L) > 0\}$. A subset $X \subset V$ (as well as the corresponding cut ∂X) is said to be *saturated* by f if $\zeta^f(e) = c(e)$ for any $e \in \partial X$; X is said to be *compatible with* $L \in \mathscr{L}$ if $|\partial X \cap EL| \leq 1$, and *compatible with f* if it is compatible with each chain in $\mathscr{L}^+(f)$. By the definition, for a semiproper $\mathscr{X}$ and an arbitrary $st \in U$ there exists a unique pair $X_A, X_B \in \mathscr{X}$ such that $s \in X_A \not\ni t$ and $s \notin X_B \ni t$. This fact implies easily the following statement, which is often used here and in subsequent sections.

CLAIM 1. *For any semiproper family $\mathscr{X}$ one has $c(\mathscr{X}) \geq v(c)$, and equality holds if and only if each set $X_A \in \mathscr{X}$ is saturated and compatible with f.*

A terminal $s \in T$ is said to be a *1-terminal* (a *2-terminal*) if it belongs to exactly one (respectively, two) anticlique. Given two terminals s and t (not necessarily distinct), we write $s \sim t$ if $s, t \in A \cap B$ for two intersecting anticliques A and B, and $s \not\sim t$ otherwise. In particular, $s \sim s$ if s is a 2-terminal, while $s \not\sim s$ if s is a 1-terminal. One easily obtains the following useful statement.

CLAIM 2. *Let $\mathscr{A}$ be 3-noncrossing, and let $A, B \in \mathscr{A}$.*

1. If $s, t \in A$, $s \not\sim t$, $p \in T \setminus A$, then at least one of the edges sp and tp belongs to U.

2. If $st \in U$ and $s' \sim s$, then $s't \in U$.

3. If A and B are intersecting and either $s \in A \cap B$, $t \in T \setminus (A \cup B)$, or $s \in A \setminus B$, $t \in B \setminus A$, then $st \in U$.

Assume that $l: E \to \mathbb{R}_+$ is an optimal solution for the problem $\mathscr{P}^*(c)$ dual (in sense of linear programming) to $\mathscr{P}(c)$, that is, $l(EL) \geq 1$ for all $L \in \mathscr{L}$ and $\sum_{e \in E} c(e)l(e) = v(c)$. For arbitrary $x, y \in V$ put

$$\mu(xy) := \min_{L \in \mathscr{L}(G, xy)} l(EL);$$

thus, μ is the metric on V induced by the edge lengths l (observe that $\mu(xy) = 0$ if $x = y$, and $\mu(xy) = \infty$ if x and y are from different connected components of G). Then $\mu(st) \geq 1$ for any $st \in U$, and the complementary slackness conditions of linear programming applied to $\mathscr{P}(c)$ and $\mathscr{P}^*(c)$ are specified as

$$e \in E, \quad \zeta^f(e) < c(e) \Rightarrow l(e) = 0; \tag{2}$$
$$L \in \mathscr{L}^+(f) \Rightarrow l(EL) = 1 \ (\Rightarrow \mu(st) = 1), \tag{3}$$

where s and t are the endpoints of L.

Let $A \in \mathscr{A}$ and $x \in V$; put

$$r_A(x) := \min\{\mu(sx): s \in A\},$$
$$d_A(x) := \min\{\mu(sx) + \mu(xt): s, t \in A, s \not\sim t\}.$$

Next, for $A \in \mathscr{A}_1$ put

$$X_A^* := Y_A := \{x \in V: d_A(x) < 1/2\},$$

and for $A \in \mathscr{A}_2$ put

$$Y_A := \{x \in V: d_A(x) \leq 1/2\}, \quad R_A := \{x \in V: r_A(x) = 0\},$$
$$W_A := R_A \cap \{x \in V: d_B(x) \geq 1/2 \ \forall B \in \mathscr{A} \setminus \{A\}\}, \quad X_A^* := Y_A \cup W_A.$$

Finally, put $\mathscr{X}^* = \{X_A^*: A \in \mathscr{A}\}$.

We want to prove that $\mathscr{X}^*$ is a proper family all of whose sets are saturated and compatible with f; by Claim 1, this would yield $c(\mathscr{X}^*) = v(c)$, thus proving the theorem. The proof is divided into several claims. For any two terminals s and t (not necessarily distinct), put

$$\tau(st) = \begin{cases} 1 & \text{if } st \in U, \\ 0 & \text{otherwise.} \end{cases}$$

For any four terminals s, t, p, q (not necessarily distinct), put

$$\tau(s, t, p, q) = \max\{\tau(st) + \tau(pq), \tau(sp) + \tau(tq), \tau(sq) + \tau(tp)\}.$$

CLAIM 3. 1. If $\tau(s, t, p, q) = 0$, then $s, t, p, q \in A$ for some $A \in \mathscr{A}$.

2. If $\tau(s, t, p, q) = 1$, then there is an $A \in \mathscr{A}$ containing exactly three of the terminals s, t, p, q.

PROOF. 1. This follows from the fact that no edge of H can have both of its endpoints in $\{s, t, p, q\}$ (otherwise $\tau(s, t, p, q) \geq 1$).

2. Evidently, each anticlique contains at most three elements out of $\{s, t, p, q\}$. There are two possible cases, up to a permutation:
1) $\tau(st) = \tau(sp) = \tau(tp) = 0$, and
2) $\tau(st) = \tau(sp) = \tau(sq) = 0$.

In the first case, the triple $\{s, t, p\}$ belongs to an anticlique. In the second case, each pair out of $\{s, t\}$, $\{s, p\}$, $\{s, q\}$ belongs to an anticlique, and since $\mathscr{A}$ is 3-noncrossing, two of these anticliques coincide, thus completing the proof.

CLAIM 4. *For any $s, t, p, q \in T$ and $x \in V$, one has*

$$\mu(sx) + \mu(tx) + \mu(px) + \mu(qx) \geq \tau(s, t, p, q).$$

PROOF. Observe that for any $s', t' \in T$ one has $\mu(s't') \geq \tau(s't')$. To be definite, let $\tau(s, t, p, q) = \tau(st) + \tau(pq)$. Then the required inequality follows immediately from the triangle inequalities:

$$\mu(sx) + \mu(xt) \geq \mu(st), \qquad \mu(px) + \mu(xq) \geq \mu(pq).$$

CLAIM 5. *Suppose that $A, B \in \mathscr{A}$ and $x \in V$. Then:*
(1) $d_A(x) + d_B(x) \geq 2$ *if $A \cap B = \varnothing$;*
(2) $d_A(x) + d_B(x) \geq 1$ *if A and B intersect.*

PROOF. Assume that $s, t \in A$ and $p, q \in B$ satisfy the relations

$$s \not\sim t, \qquad p \not\sim q,$$
$$d_A(x) = \mu(sx) + \mu(xt), \qquad d_B(x) = \mu(px) + \mu(xq).$$

By Claim 4, one has $d_A(x) + d_B(x) \geq \tau(s, t, p, q)$. Let $A \cap B = \varnothing$. If $\tau(s, t, p, q) \leq 1$, then by Claim 3 there exists an anticlique A' containing at least three elements out of s, t, p, q, say, $s, t, p \in A'$. Then $A \neq A'$, and $s, t \in A \cap A'$, contrary to $s \not\sim t$. Now let A and B intersect. Then $s \not\sim t$ and $p \not\sim q$ imply that there does not exist any anticlique containing all the four elements s, t, p, q. By Claim 3, this yields $\tau(s, t, p, q) \geq 1$.

CLAIM 6. *Suppose that A and B are two distinct anticliques and $x \in V$. Then:*

(1) If $r_A(x) + d_B(x) < 1$, then A and B intersect and

$$p \in A, \ \mu(px) = r_A(x) \Rightarrow p \in A \cap B.$$

(2) If $A \cap B = \varnothing$ and $r_A(x) + r_B(x) < 1$, then there exists an anticlique C intersecting both A and B and such that $d_C(x) \leq r_A(x) + r_B(x)$.

PROOF. 1. Let $s, t \in B$, $s \not\sim t$, and $d_B(x) = \mu(sx) + \mu(xt)$. Since $\mu(sp) \leq r_A(x) + d_B(x) < 1$, one has $sp \notin U$. Similarly, $tp \notin U$. Assume that B' is an anticlique containing s, t, p. Since $s \not\sim t$ and $\mathscr{A}$ is 3-noncrossing, one obtains $B' = B$.

2. Let $s \in A$, $t \in B$, $r_A(x) = \mu(sx)$, $r_B(x) = \mu(tx)$. Then $\mu(st) < 1$; hence $st \notin U$. Let C be an anticlique containing s and t. Since $\mathscr{A}$ is 3-noncrossing, the anticlique C is defined uniquely; hence $s \not\sim t$ and $d_C(x) \leq \mu(sx) + \mu(xt)$.

Claim 5 yields $Y_A \cap Y_B = \varnothing$ for any two distinct $A, B \in \mathscr{A}$. Next, let $A \in \mathscr{A}_2$ and $x \in R_A$. The definition of W_A implies that for any $B' \in \mathscr{A}_1$ the relations $x \in W_A$ and $x \in Y_{B'}$ are never valid simultaneously, while, by Claim 6(1), $d_{B''}(x) \geq 1$ for any $B'' \in \mathscr{A}_2 \setminus \{A\}$. Hence, $W_A \cap Y_B = \varnothing$ for any $B \in \mathscr{A} \setminus \{A\}$. Next, if $r_B(x) = 0$ for some $B \in \mathscr{A}_2 \setminus \{A\}$, then by Claim 6(2) there exists an anticlique $C \in \mathscr{A}$ such that $d_C(x) = 0$, thus implying $x \notin W_A, W_B$. Therefore, $\mathscr{X}^*$ consists of pairwise disjoint sets.

Consider an arbitrary terminal s. If s is a 1-terminal and $s \in A$, then $s \not\sim s$ and $d_A(s) = \mu(ss) + \mu(ss) = 0$, hence $s \in Y_A$. Assume that s is a 2-terminal and $s \in A \in \mathscr{A}_2$. Then $r_A(s) = \mu(ss) = 0$, and $s \notin W_A$ would imply $d_B(s) < 1/2$ for some $B \in \mathscr{A} \setminus \{A\}$, whence $s \in Y_B$; furthermore, taking Claim 6(1) into account, one obtains $s \in A \cap B$. Therefore each terminal s is contained in exactly one set X_C^*, and $s \in C$. Hence $\mathscr{X}^*$ is a proper family.

It remains to prove that each set in $\mathscr{X}^*$ is saturated and compatible with f. Let us make use of relations (2) and (3) for f and l. The following statement is trivial.

CLAIM 7. *Suppose that $x, y \in V$, $\mu(xy) = 0$, and $A \in \mathscr{A}$. Then:*
(1) *If $x \in Y_A$, then $y \in Y_A$.*
(2) *If $x \in W_A$, then $y \in W_A$.*

Given $A \in \mathscr{A}$ and $xy \in \partial X_A^*$, one applies Claim 7 to obtain $\mu(xy) > 0$, and thus $l(xy) > 0$. Therefore, by (2), the edge xy is saturated by f. Hence X_A^* is saturated by f.

To prove that X_A^* is compatible with f, consider an arbitrary pq-chain $L \in \mathscr{L}^+(f)$ and suppose that there exists a vertex $x \in VL$ contained in X_A^*. To be definite, assume that $q \notin A$, and let L' and L'' be the parts of L from p to x and from x to q, respectively. By (3), one has

$$l(EL) = l(EL') + l(EL'') = \mu(pq) = 1.$$

Let us show that $VL' \subseteq X_A^*$.

1. Suppose that $x \in Y_A$. Choose $s, t \in A$, $s \not\sim t$, such that $d_A(x) = \mu(sx) + \mu(xt)$. Then $\mu(sx) + \mu(tx) + \mu(px) + \mu(qx) \leq 3/2$; hence, by Claims 4 and 3, $\tau(s, t, p, q) = 1$ and there exists an anticlique B containing s, t and $p' \in \{p, q\}$. Evidently, $B = A$ (otherwise $s \sim t$); hence $p' = p$. According to Claim 2(1), one has $\{sq, tq\} \cap U \neq \emptyset$. If $s' \sim p$ for some $s' \in \{s, t\}$, then $s'q \in U$ (by Claim 2(2)), and $t' \not\sim p$, where $\{s', t'\} = \{s, t\}$. Therefore, one can assume that $sq \in U$ and $t \not\sim p$. The relations $\mu(sx) + l(EL'') \geq \mu(sq) \geq 1$ and $l(EL) = 1$ yield $\mu(tx) + l(EL') \leq d_A(x)$. Hence, for each vertex $y \in VL'$ one has $\mu(ty) + \mu(yp) \leq d_A(x)$, and thus $y \in Y_A$.

2. Suppose now that $A \in \mathscr{A}_2$ and $x \in W_A$. Choose $s \in A$ such that $\mu(sx) = r_A(x) = 0$; then $\mu(sx) + l(EL') + l(EL'') = 1$. If $sq \in U$, the above equality together with $\mu(sq) \geq 1$ implies $\mu(sx) + l(EL') = 0$. Hence, for an arbitrary $y \in VL'$ one has $\mu(sy) = 0$; therefore, by Claim 7, $y \in W_A$. Suppose now that $sq \notin U$, and let B be an anticlique containing s and q; evidently, $B \neq A$ and $s \not\sim q$. Since $x \in W_A$, one has $d_B(x) \geq 1/2$. Hence $l(EL'') + \mu(sx) \geq 1/2$, and thus $\mu(sp) \leq \mu(sx) + l(EL') \leq 1/2$. Therefore, $sp \notin U$; since $\mathscr{A}$ is 3-noncrossing, this yields $p \in A$. Now $pq \in U$ and $sq \notin U$ show that $s \not\sim p$. Then an arbitrary $y \in VL'$ satisfies $d_A(y) \leq \mu(sy) + \mu(yp) \leq 1/2$, i.e., $y \in Y_A$.

The proof of Theorem 2.1 is completed.

COROLLARY 2.2. *If c is an inner Eulerian function then $c(\mathscr{X})$ is an integer for any semiproper family $\mathscr{X}$ (hence $v(c)$ is an integer as well).*

PROOF. Indeed, assume that $A \in \mathscr{A}$ and $\mathscr{X}'$ is the proper family consisting of sets $X'_A = X_A \cup (V \setminus T)$ and $X'_B = X_B \cap T$ ($B \in \mathscr{A} \setminus \{A\}$). Since c is inner Eulerian, one obtains $c(\partial X'_C) \equiv c(\partial X_C) \bmod 2$ for any $C \in \mathscr{A}$ (taking into account that $X'_C \cap T = X_C \cap T$); hence $c(\mathscr{X}') - c(\mathscr{X})$ is an integer. Since $\mathscr{X}'$ is a partition of the set V, each edge $e \in E$ occurs in an even number of cuts $\partial X'_C$, $C \in \mathscr{A}$; therefore $c(\mathscr{X}')$ is an integer.

3. Splitting a network

In this section we complete the proof of Theorem 1. It will be convenient to assume that G is a complete graph, that is, $xy \in E$ for any distinct $x, y \in V$ (if the initial graph G lacks an edge xy, add the edge and put $c(xy) = 0$). Let c be inner Eulerian.

A semiproper family $\mathscr{X}$ such that $c(\mathscr{X}) = v(c)$ is called *c-minimal*. Denote by $\mathscr{M}(c)$ the set of all c-minimal proper families. The proof of the theorem proceeds by induction. More precisely, suppose that the assertion of the theorem is true for some fixed G and H and for all inner Eulerian functions c' such that either $|\mathscr{M}(c')| > |\mathscr{M}(c)|$, or $|\mathscr{M}(c')| = |\mathscr{M}(c)|$ and $c'(E) < c(E)$. The assertion is obvious for $c = 0$ (observe that in this case each proper family is c-minimal, hence $|\mathscr{M}(c)|$ takes the maximal possible value).

Denote by f an optimal solution for the problem $\mathscr{P}(G, c, U)$, as above. A triple xyz of vertices such that $y \ne x, z$ (x and z may coincide) is said to be a *fork* for the function c if $c(xy) > 0$ and $c(yz) > 0$. Given a fork xyz, define a function $\theta = \theta_{xyz}$ on E by the following relations:

$$\theta(e) = \begin{cases} 2 & \text{for } e = xy, \\ 0 & \text{otherwise}, \end{cases} \quad \text{if } x = z;$$

$$\theta(e) = \begin{cases} 1 & \text{for } e = xy, yz, \\ -1 & \text{for } e = xz, \\ 0 & \text{otherwise}, \end{cases} \quad \text{if } x \ne z.$$

A fork xyz is said to be *essential* (with respect to f) if $x \ne z$ and the set $\mathscr{L}^+(f)$ contains a chain passing through both xy and yz.

STATEMENT 3.1. *Suppose that xyz is a fork, and $c' = c - \theta_{xyz}$. Then:*
(1) *c' is inner Eulerian.*
(2) *$v(c) \geqslant v(c') \geqslant v(c) - 2$.*
(3) *If $v(c') = v(c)$, then $\mathscr{M}(c') \supseteq \mathscr{M}(c)$.*
(4) *If xyz is essential and $\mathscr{X} = \{X_A : A \in \mathscr{A}\}$ is a proper family such that $c'(\mathscr{X}) < v(c)$, then $c'(\mathscr{X}) = v(c) - 1$, $c(\mathscr{X}) = v(c) + 1$, and there*

exist $X_A, X_B \in \mathscr{X}$ such that $y \notin X_A \ni x, z$ and $x, z \notin X_B \ni y$ (this fact implies in particular that if xyz is essential, then either $v(c') = v(c)$ or $v(c') = v(c) - 1$).

PROOF. The first assertion is obvious. Let $\mathscr{X}$ be an arbitrary proper family. By Corollary 2.2, both $c(\mathscr{X})$ and $c'(\mathscr{X})$ are integers. For any $X_C \in \mathscr{X}$ one has

$$c'(\partial X_C) = \begin{cases} c(\partial X_C) - 2 & \text{if } xy, yz \in \partial X_C, \\ c(\partial X_C) & \text{otherwise.} \end{cases}$$

Since the members of $\mathscr{X}$ are pairwise disjoint, $\mathscr{X}$ contains at most two sets X_C such that $xy, yz \in \partial X_C$; hence $c(\mathscr{X}) \geq c'(\mathscr{X}) \geq c(\mathscr{X}) - 2$. Thus, the second and the third assertions are proved. Now let xyz and $\mathscr{X}$ be defined as in the last assertion, and assume that $X_C \in \mathscr{X}$ is a set such that $xy, yz \in \partial X_C$. Then X_C is not compatible with f; therefore, by Claim 1 of §2, $\mathscr{X}$ fails to be c-minimal. Now the required assertion follows from $c(\mathscr{X}) - c'(\mathscr{X}) \leq 2$.

Denote by $K(f)$ the set of all essential forks for f. If $K(f) = \varnothing$ (that is, $|EL| = 1$ for all $L \in \mathscr{L}^+(f)$), then f is evidently an integer multiflow. Hence, one may assume that $K(f) \neq \varnothing$. A fork xyz is said to be *separable* if $v(c') = v(c)$ with $c' := c - \theta_{xyz}$. We shall prove that there exists at least one separable fork. Once this fact is proved, the proof of Theorem 1 is completed in the following way. Assume that xyz is a separable fork and $c' = c - \theta_{xyz}$. Since $c'(E) < c(E)$ and $\mathscr{M}(c') \supseteq \mathscr{M}(c)$ (by Statement 3.1(3)), the problem $\mathscr{P}(c')$ possesses an integer solution f' by induction. If $x = z$, or $x \neq z$ and $\zeta^{f'}(xz) \leq c(xz)$, then f' is c-admissible, whence it is an optimal solution for $\mathscr{P}(c)$. If $x \neq z$ and $\zeta^{f'}(xz) = c(xz) + 1 = c'(xz)$, put

$$f^*(L) = f'(L) - 1, \qquad f^*(L') = f'(L') + 1,$$
$$f^*(L'') = f'(L''), \qquad L'' \in \mathscr{L} \setminus \{L, L'\},$$

where L is a chain from $\mathscr{L}^+(f')$ passing through the edge xz and L' is a chain from $\mathscr{L}$ such that $EL' \subseteq (EL \setminus \{xz\}) \cup \{xy, yz\}$. Evidently, f^* is c-admissible and $1 \cdot f^* = 1 \cdot f' = v(c)$; therefore f^* is an integer optimal solution for $\mathscr{P}(c)$.

Let us now prove the existence of a separable fork. The proof breaks up into several claims. In what follows we may assume that none of the essential forks is separable. It is then necessary to prove that there exists a separable nonessential fork. Choose an essential fork xyz and put $c' := c - \theta_{xyz}$, $c'' := c' - \frac{1}{2}\theta_{xyz}$, $\tilde{c} := 2c''$. For any proper family $\mathscr{X}$ one has $c''(\mathscr{X}) = (c(\mathscr{X}) + c'(\mathscr{X}))/2$; therefore, by Statement 3.1(4), the following are true:
1) $v(c'') = v(c)$.
2) If $c(\mathscr{X}) = v(c)$, then $c''(\mathscr{X}) = v(c'')$.
3) If $c'(\mathscr{X}) < v(c)$, then $c(\mathscr{X}) > v(c)$ and $c''(\mathscr{X}) = v(c'')$.

Hence $\mathscr{M}(\tilde{c}) = \mathscr{M}(c'') \supset \mathscr{M}(c)$. Since $\tilde{c} = 2c - \theta_{xyz}$ is evidently an inner Eulerian function, the problem $\mathscr{P}(\tilde{c})$ possesses an integer optimal solution $\tilde{f}$ by induction. Furthermore, the fork xyz is separable for the function $2c$; hence, acting as above, one can reform the multiflow $\tilde{f}$ to obtain a $2c$-admissible integer multiflow $\tilde{f}^*$ such that $1 \cdot \tilde{f}^* = 1 \cdot \tilde{f}$. Therefore, the problem $\mathscr{P}(2c)$ possesses an integer optimal solution, and so $\mathscr{P}(c)$ possesses a half-integer optimal solution.

Thus, one may assume that f is *half-integral*, that is, it takes its values in $\frac{1}{2}\mathbb{Z}_+$. Assume as well that the multiflow f has the least possible value of $\zeta^f(E)$ among all the half-integer optimal solutions for $\mathscr{P}(c)$. Given an essential fork xyz, we say that the proper family $\mathscr{X}$ indicated in Statement 3.1(4) is *critical*, while X_A and X_B are said to be the *external* and the *internal* sets (with respect to xyz), respectively.

CLAIM 1. *Suppose that xyz is an essential fork, $\mathscr{X}$ is a critical proper family for xyz, $X_A, X_B \in \mathscr{X}$ are the external and the internal sets, respectively, and $L \in \mathscr{L}^+(f)$ is a chain passing through xy and yz. Then*:
(1) *Each set $X_C \in \mathscr{X}$ is saturated by f.*
(2) *Each set $X_C \in \mathscr{X} \setminus \{X_A, X_B\}$ is compatible with f.*
(3) *The sets X_A and X_B are compatible with each chain in $\mathscr{L}^+(f) \setminus \{L\}$.*
(4) $|EL \cap \partial X_A| = 3$.
(5) $f(L) = 1/2$.

PROOF. Put $c' := c - \frac{1}{2}\theta_{xyz}$ and define a multiflow f' by the following:

$$f'(L) := f(L) - 1/2, \qquad f'(L') = f(L') + 1/2,$$
$$f'(L'') := f(L''), \qquad L'' \in \mathscr{L} \setminus \{L, L'\},$$

where L' is the chain in $\mathscr{L}$ such that $EL' = (EL \setminus \{xy, yz\}) \cup \{xz\}$. Then $v(c) = v(c') = c'(\mathscr{X})$, and f' is an optimal solution for $\mathscr{P}(c')$. Now the first three assertions, as well as the last one, follow immediately from Claim 2 of §2 applied to c', f' and $\mathscr{X}$. Finally, $|EL' \cap \partial X_A| = 1$ (since $VL' \cap X_A \neq \varnothing$) implies $|EL \cap \partial X_A| = |EL' \cap \partial X_A| + 2 = 3$.

Let $\tilde{V}$ be the set of vertices $y \in V$ occurring in at least one essential fork xyz. Consider a vertex $y \in \tilde{V}$ and an essential fork xyz. Claim 1(3) shows that $\mathscr{L}^+(f)$ contains exactly one chain L passing through both edges xy and yz. Since $f(L) = \frac{1}{2}$ and the edge xy is saturated by f, at least one of the following is true:
1) There exists an essential fork $x'yx$, $x' \neq z$.
2) The edge xy belongs to a chain $L' \in \mathscr{L}^+(f)$ having an endpoint at y (in particular, y is a terminal).

This implies immediately that at least one of the following two cases holds:

(C1) G contains a vertex y and distinct vertices $x_1, x_2, \ldots, x_k = x_0$, $k \geq 3$, such that $x_i y x_{i+1}$ is an essential fork for any $i = 1, \ldots, k$.

(C2) $\tilde{V} \subseteq T$, and for some $s \in T$ and $x, z \in V$ there exist a chain $L \in \mathscr{L}^+(f)$ passing through the edges xs and sz and an st-chain $L' \in \mathscr{L}^+(f)$ passing through the edge sx.

Let us prove that case (C2) is in fact impossible. To do this, we need the following statement.

CLAIM 2. *Let* $L \in \mathscr{L}^+(f)$ *be a pq-chain and* $T' := T \cap (VL \setminus \{p, q\}) \neq \varnothing$. *Then* $T' \subseteq A \cap B$ *for some distinct* $A, B \in \mathscr{A}$.

PROOF. The minimality of $\zeta^f(E)$ implies that $p'q' \notin U$ for any pair $\{p', q'\} \subset VL \cap T$ distinct from $\{p, q\}$ (otherwise one would shorten the chain L). Choose $p' \subset T'$, and assume that A is an anticlique containing p and p', while B is an anticlique containing p' and q. Let $q' \in T' \setminus \{p'\}$. The relation $p'q' \notin U$ together with Claim 2(3) of §2 implies $q' \in A \cup B$. Since $pq' \notin U$, the relation $q' \notin A$ would imply the existence of an anticlique $C \neq A, B$ containing p and q', contrary to the fact that $\mathscr{A}$ is 3-noncrossing. Hence $q' \in A$, and similarly, $q' \in B$.

Consider s, x, z, L, L' indicated in (C2). Let p and q be the endpoints of L, and let X_A be the external set of the critical proper family for xsz. Observe that $\tilde{V} \subseteq T$ yields $x, z \in T$. Since $|EL| \geq |EL \cap \partial X_A| = 3$, at least one of the terminals x and z, say x, is distinct from p and q. Claim 2 together with the fact that $x \in X_A \cap T \subseteq A$ yields $s \in A$. Let t be the endpoint of the chain L' distinct from s. Since X_A is compatible with L' (by Claim 1(3)) and $s \notin X_A$, one has $t \in X_A$. Hence both endpoints of L' belong to the anticlique A, a contradiction.

Let us now turn to case (C1); consider the vertices $y, x_1, \ldots, x_k$ as in (C1). Let us prove that any fork of the form $x_i y x_{i+2}$ is separable. Assume that $\mathscr{X}^i = \{X_B^i : B \in \mathscr{A}\}$ is the critical proper family for $x_i y x_{i+1}$, $X_{A(i)}^i \in \mathscr{X}^i$ is the external set (with respect to $x_i y x_{i+1}$), T^i denotes $T \cap X_{A(i)}^i$, and $L^i \in \mathscr{L}^+(f)$ is a chain passing through the edges $x_i y$ and $y x_{i+1}$, $i = 1, \ldots, k$. In what follows all the indices are regarded modulo k.

CLAIM 3. *For any* $i = 1, \ldots, k$, $A(i)$ *intersects* $A(i+1)$ *(possibly,* $A(i) = A(i+1)$*)*.

PROOF. Suppose that $A(i) \cap A(i+1) = \varnothing$. Then $T^i \cap T^{i+1} = \varnothing$; hence the families

$$\mathscr{Y} := (\mathscr{X}^i \setminus \{X\}) \cup \{X \setminus Y\}, \qquad \mathscr{Y}' := (\mathscr{X}^{i+1} \setminus \{Y\}) \cup \{Y \setminus X\},$$

are proper (here $X := X_{A(i)}^i$, $Y := X_{A(i+1)}^{i+1}$). Since the internal (with respect to $x_i y x_{i+1}$) set for $\mathscr{X}^i$ is not compatible with f and occurs in $\mathscr{Y}$, one has $c(\mathscr{Y}) \geq v(c) + 1$. Similarly, $c(\mathscr{Y}') \geq v(c) + 1$. Therefore,

$$c(\mathscr{Y}) + c(\mathscr{Y}') \geq c(\mathscr{X}^i) + c(\mathscr{X}^{i+1}).$$

On the other hand, $x_{i+1}y \in \partial X, \partial Y$ and $x_{i+1}y \notin \partial(X \setminus Y), \partial(Y \setminus X)$; hence the strict submodular inequality

$$c(\partial X) + c(\partial Y) > c(\partial(X \setminus Y)) + c(\partial(Y \setminus X))$$

is valid for X and Y. This yields

$$c(\mathscr{X}^i) + c(\mathscr{X}^{i+1}) > c(\mathscr{Y}) + c(\mathscr{Y}'),$$

a contradiction.

Denote by $L[zz']$ the part of a chain L from z to z', where $z, z' \in VL$. Let s_i denote the endpoint of the chain L^i such that x_i is contained in $L^i[s_i y]$, and let t_i be the other endpoint of L^i, $i = 1, \ldots, k$. Claim 1(3,4) immediately implies the following proposition.

CLAIM 4. *The following assertions are true for any* $j = 1, \ldots, k$:
(1) $t_{j-1}, s_{j+1} \in T^j$.
(2) *Exactly one of the terminals* s_j *and* t_j *belongs to* T^j.

CLAIM 5. *At least one of the terminals* s_{i+1} *and* t_i *belongs to* $T^i \cap T^{i+1}$, $i = 1, \ldots, k$.

PROOF. Suppose the contrary. Claim 4(1) (for $j = i+1$ and $j = i$) yields $t_i \in T^{i+1}$ and $s_{i+1} \in T^i$. Then $t_i \notin T^i$ and $s_{i+1} \notin T^{i+1}$; hence, by Claim 4(2), $s_i \in T^i$ and $t_{i+1} \in T^{i+1}$. Therefore, $s_i, s_{i+1} \in A(i)$, $t_i, t_{i+1} \in A(i+1)$, thus implying $A(i) \neq A(i+1)$. By Claim 3, $A(i)$ intersects $A(i+1)$. Besides, Claim 2(3) of §2 yields $s_i t_{i+1}, s_{i+1} t_i \in U$. Form an $s_i t_{i+1}$-chain L and an $s_{i+1} t_i$-chain L' such that $EL \subseteq EL^i[s_i y] \cup EL^{i+1}[y t_{i+1}]$ and $EL' \subseteq EL^{i+1}[s_{i+1} x_{i+1}] \cup EL^i[x_{i+1} t_i]$. Define the multiflow f' by the relations

$$f'(L) := f(L) + 1/2, \quad f'(L') := f(L') + 1/2, \quad f'(L^i) := f'(L^{i+1}) = 0,$$
$$f'(L'') = f(L'') \quad \text{for all the other chains} \quad L'' \in \mathscr{L}.$$

Evidently, $1 \cdot f' = 1 \cdot f = v(c)$, and $\zeta^{f'}(e) \leq \zeta^f(e)$ for any $e \in E$. Moreover, $\zeta^{f'}(x_{i+1}y) < \zeta^f(x_{i+1}y)$ (since the chains L and L' do not contain the edge $x_{i+1}y$), a contradiction to the minimality of $\zeta^f(E)$.

To be definite, let $s_1 \in T^1$. Then Claims 4 and 5 imply that $s_i \in T^{i-1} \cap T^i$ and $t_i \in T^{i+1} \setminus T^i$, $i = 1, \ldots, k$ (see Figure 1). Taking Claim 3 into account, one obtains that the following facts are true for an arbitrary i:
(F1) $s_i \in A(i-1) \cap A(i)$, $t_i \in A(i+1) \setminus A(i)$.
(F2) $A(i)$ intersects (but does not coincide with) $A(i+1)$ (by Claim 3 and the relations $s_i \in T^i$, $t_i \in T^{i+1}$);
(F3) $A(i) \neq A(i+2)$ (since $s_{i+1} \in T_i$ and $t_{i+1} \in T^{i+2}$).
(F4) $A(i) \neq A(i+3)$ (otherwise $A(i), A(i+1), A(i+2)$ would be pairwise intersecting).

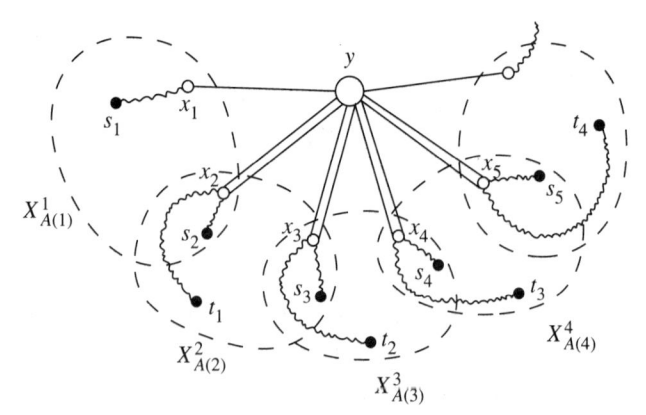

FIGURE 1

In particular, (F4) yields that $k \geq 4$. Besides, by (F2), (F3) and the fact that $\mathscr{A}$ is 3-noncrossing, one has

(F5) $A(i) \cap A(i+2) = \emptyset$.

Now we come to the final claim of our proof.

CLAIM 6. *The fork $x_2 y x_4$ is separable.*

PROOF. By (F1) and (F5) one has $s_1 \notin A(2), A(3)$. Then, taking Claim 2(3) of §2 into account, one obtains $s_1 s_3 \in U$; denote by P_1 an $s_1 s_3$-chain such that $EP_1 \subseteq EL^1[s_1 y] \cup EL^3[y s_3]$. Similarly, $s_2 s_4 \in U$. There are two possible cases: 1) $t_1 t_3 \in U$, and 2) $t_1 t_3 \notin U$. In the first case, choose an $s_2 s_4$-chain P_2 and a $t_1 t_3$-chain P_3 such that

$$EP_2 \subseteq EL^2[s_2 x_2] \cup \{x_2 x_4\} \cup EL^4[x_4 s_4],$$
$$EP_3 \subseteq EL^1[t_1 x_2] \cup \{x_2 x_4\} \cup EL^3[x_4 t_3].$$

In the second case, let B be an anticlique containing t_1 and t_3. By (F1) and (F5), B is distinct from $A(1), A(2), A(3), A(4)$. Then t_1 and t_3 are 2-terminals, $t_3 \notin A(1), A(2)$, $t_1 \notin A(3), A(4)$, and Claim 2(3) of §2 yields $s_2 t_3 \in U$, $t_1 s_4 \in U$. Choose an $s_2 t_3$-chain P_2 and a $t_1 s_4$-chain P_3 such that

$$EP_2 \subseteq EL^2[s_2 x_2] \cup \{x_2 x_4\} \cup EL^3[x_4 t_3],$$
$$EP_3 \subseteq EL^1[t_1 x_2] \cup \{x_2 x_4\} \cup EL^4[x_4 s_4].$$

In both cases, define a multiflow f' by the relations

$$f'(P_j) := f(P_j) + 1/2, \quad j = 1, 2, 3,$$
$$f'(L^m) := f(L^m) - 1/2 \, (= 0), \quad m = 1, 2, 3, 4,$$
$$f'(L) = f(L) \quad \text{for all the other chains } L \in \mathscr{L}.$$

Evidently, f' is c'-admissible with $c' := c - \theta_{x_2 y x_4}$, and $v(c') \geq 1 \cdot f' = 1 \cdot f - 1/2 = v(c) - 1/2$. Now the fact that both $v(c)$ and $v(c')$ are integers implies $v(c') = v(c)$. Therefore, $x_2 y x_4$ is a separable fork.

The proof of Theorem 1 is completed.

4. Algorithm

The algorithm is based on the same splitting-off idea as the proof of Theorem 1. In order to obtain an algorithm having strongly polynomial running time, we determine the maximum possible "weight" that can be assigned to a current fork so that splitting it off according to such a weight still preserves the maximum total value of a multiflow. To do this, one applies a subroutine that finds a c-minimal semiproper family $\mathscr{H}$ (for the current c).

It should be noted that the method of finding such a family presented in §2 is not efficient enough, since it involves the solution of the dual problem $\mathscr{P}^*(c)$. Let us show that finding a c-minimal family can be executed directly, via the reduction to an ordinary minimal cut problem for a certain extended network.

Let $c\colon E \to \mathbb{R}_+$. For each $A \in \mathscr{A}$, take a copy $G_A = (V_A, E_A)$ of the graph G and denote by x_A the copy of a vertex $x \in V$ in the graph G_A. Paste together graphs G_A, $A \in \mathscr{A}$, by identifying the following edges and vertices: for each pair of intersecting anticliques $A, B \in \mathscr{A}$, identify vertices x_A and x_B if $x \in A \cap B$, and identify edges $x_A y_A$ and $x_B y_B$ if $x, y \in A \cap B$ and $xy \in E$. Denote the resulting graph by $\mathscr{G}'$. Retain the same notation G_A for the corresponding subgraph of $\mathscr{G}'$. Let $\tilde{A}$ be the set of 1-terminals for an anticlique A, that is, $\tilde{A} = A - \bigcup_{B \in \mathscr{A} \setminus \{A\}} B$. Construct a graph $\mathscr{G} = (\mathscr{V}, \mathscr{E})$ by adding to $\mathscr{G}'$ the vertices s^0 (the *source*), t^0 (the *sink*) and the following edges:

$$s^0 s_A \quad \text{for } s \in \tilde{A},\ A \in \mathscr{A}_1;$$
$$s^0 t_A \quad \text{for } t \in T \setminus A,\ A \in \mathscr{A}_2;$$
$$t^0 t_A \quad \text{for } t \in T \setminus A,\ A \in \mathscr{A}_1;$$
$$t^0 s_A \quad \text{for } s \in \tilde{A},\ A \in \mathscr{A}_2$$

(see Figure 2 for an example of G, $\mathscr{A}$ and $\mathscr{G}$; here $\mathscr{A}_1 = \{A = \{s, t\},\ C = \{p, q, r\}\}$ and $\mathscr{A}_2 = \{B = \{t, p, q\}\}$). For $A \in \mathscr{A}$, denote by φ_A the natural inclusion map of G into $\mathscr{G}$ that takes G to the subgraph G_A. Define the edge capacities $d = d^c$ for the graph $\mathscr{G}$ in the following way. For $xy \in E$, $A \in \mathscr{A}$, $e = \varphi_A(xy) \in \mathscr{E}$, put

$$d(e) := \begin{cases} 2c(xy) & \text{if } x, y \in A \cap B \text{ for some } B \in \mathscr{A} \setminus \{A\}, \\ c(xy) & \text{otherwise}, \end{cases}$$

and put $d(e) := \infty$ for each edge $e \in \mathscr{E}$ incident to s^0 or t^0 (one can see that d is well-defined).

A cut ∂Z of the graph $\mathscr{G}$ is said to be an (s^0, t^0)-*cut* if $s^0 \in Z \subset \mathscr{V}$ and $t^0 \notin Z$. Let Q denote the set of all $Z \subset \mathscr{V}$ such that ∂Z is an (s^0, t^0)-cut containing no edges incident to s^0 or t^0. Given $Z \in Q$, define a family

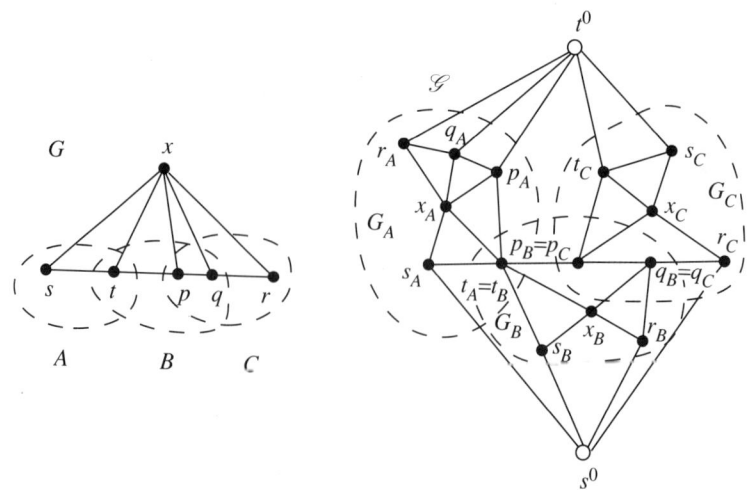

FIGURE 2

$\omega(Z) = \{X_A : A \in \mathscr{A}\}$ of subsets of V by the following rule:

$$X_A := \begin{cases} \varphi_A^{-1}(Z \cap V_A) & \text{for } A \in \mathscr{A}_1, \\ \varphi_A^{-1}(V_A \setminus Z) & \text{for } A \in \mathscr{A}_2. \end{cases}$$

The construction of the graph $\mathscr{G}$ immediately implies the following statement (details of the proof are left to the reader).

STATEMENT 4.1. *The mapping ω is a bijective correspondence between Q and the set of all semiproper families for G and $\mathscr{A}$. Moreover, for $Z \in Q$ and $\mathscr{X} = \omega(Z)$ one has $d^c(\partial Z) = 2c(\mathscr{X})$.*

Thus, to find a c-minimal semiproper family $\mathscr{X}$ and the value $v(c) = c(\mathscr{X})$ one has to construct an (s^0, t^0)-cut ∂Z in the graph $\mathscr{G}$ having the least possible capacity $d^c(\partial Z)$ (we have used the following fact: if $s^0 \in Z' \not\ni t$ and $Z' \notin Q$, then the cut $\partial Z'$ contains an edge having infinite capacity, hence it fails to be a minimum (s^0, t^0)-cut). Observe that one can transform a c-minimal semiproper family $\{X_A : A \in \mathscr{A}\}$ to a c-minimal proper family $\{X'_A : A \in \mathscr{A}\}$ by $X'_A = X_A \setminus (\bigcup_{B \in \mathscr{A} \setminus \{A\}} X_B)$ (this simple fact is left to the reader).

On the basis of on the above arguments, we now design an algorithm for solving the problem $\mathscr{P}(c)$. There are two cases to be distinguished:
1) c takes values in $\mathbb{R}_+$.
2) c is an inner Eulerian function.

In the first case, one has to find a "real-valued" optimal multiflow $f : \mathscr{L} \to \mathbb{R}_+$, and in the second case an integer-valued optimal multiflow $f : \mathscr{L} \to \mathbb{Z}_+$. As before, it is convenient to assume that G is a complete graph.

Let us start with the first case. Given a function $c' : E \to \mathbb{R}_+$ and a triple of vertices xyz ($y \neq x, z$), denote by $b(c', xyz)$ the maximal number $a \in \mathbb{R}_+$

such that $a \leqslant c'(xy)$, $a \leqslant c'(yz)$ and $v(c' - a\theta_{xyz}) = v(c)$. The algorithm consists of the main and the final stages. At the beginning, the number $v(c)$ is determined. At the main stage, the vertices of G are processed one by one in an arbitrary order, and for each $y \in V$ the triples xyz, $x, z \in V \setminus \{y\}$ (including $x = z$), are processed one by one in an arbitrary order. The processing of a current vertex y forms an *iteration* of the stage, while the processing of a current triple xyz forms a *step* of the iteration (therefore, the main stage consists of $|V|$ iterations, while each iteration consists of $|V|(|V|-1)/2$ steps). The processing of a triple xyz consists in finding the number $b = b(c, xyz)$ for the current function c and in changing the function: $c := c - b\theta_{xyz}$. Let $\tilde{c}$ be the current function c at the end of the main stage. Then an optimal solution f for the problem $\mathscr{P}(\tilde{c})$ is defined by $f(L_{st}) = \tilde{c}(st)$, $st \in U$, where L_{st} is the chain with $EL_{st} = \{st\}$ (f is assumed to be extended by zero to all the other chains in $\mathscr{L}$).

The number $b(c, xyz)$ can be computed in the following way. If $a_0 := \min\{c(xy), c(yz)\} = 0$, then, obviously, $b(c, xyz) := 0$; otherwise put $c_0 := c - a_0 \theta_{xyz}$ and find $v(c_0)$ (to do this, according to Statement 4.1, one has to find the value q of the maximum flow from s^0 to t^0 in the network $(\mathscr{G}, d^{c_0})$; then $v(c_0) := q/2$). If $v(c_0) = v(c)$, then, clearly, $b(c, xyz) := a_0$. If $h_0 := v(c) - v(c_0) > 0$, put $a_1 := a_0 - h_0/2$, $c_1 := c - a_1 \theta_{xyz}$ and find $v(c_1)$. The required number $b(c, xyz)$ is equal to $a_1 - h_1$, where $h_1 := v(c) - v(c_1)$.

Let us lay a foundation for the main stage. First of all, observe that during the stage the value of $v(c)$ (for current c) does not change. Let us prove that the numbers $b(c, xyz)$ are computed correctly, and that for any triple xyz one obtains $b(\tilde{c}, xyz) = 0$ for the final function $\tilde{c}$. For arbitrary $y \in V$, $x, z \in V \setminus \{y\}$, $a \in \mathbb{R}_+$ and a proper family $\mathscr{X}$ one has $c'(\mathscr{X}) = c(\mathscr{X}) - ak$, where $c' = c - a\theta_{xyz}$, while $k = k(\mathscr{X}, xyz)$ is the number of $X_A \in \mathscr{X}$ such that $xy, yz \in X_A$; so k can take only one of the following three values: 0, 1, 2. Hence

$$b(c, xyz) = \min\left\{c(xy), c(yz), \min_{\mathscr{X} \in \mathscr{K}} \Delta(\mathscr{X})\right\}, \qquad (4)$$

where $\mathscr{K}$ is the set of all proper families, and

$$\Delta(\mathscr{X}) := \Delta(\mathscr{X}, xyz, c) := \begin{cases} \dfrac{c(\mathscr{X}) - v(c)}{k} & \text{for } k = k(\mathscr{X}, xyz) > 0, \\ \infty & \text{for } k = 0. \end{cases}$$

This fact easily implies that the numbers $b(c, xyz)$ calculated as above have correct values. Next, consider an arbitrary step where a triple xyz is processed. Let us prove the following fact: if relations $b(c, x'y'z') = 0$ are valid for all the previously processed triples $x'y'z'$ before executing the step, they remain valid after finishing it. Denote by c_1 and c_2 the function c at the beginning and at the end of the step, respectively, and let $x'y'z'$ be a previously processed triple. One may assume that both $c_2(x'y')$ and $c_2(y'z')$ are strictly positive. If $c_2(x'y') \leq c_1(x'y')$ and $c_2(y'z') \leq c_1(y'z')$, then the relation $b(c_2, x'y'z') = 0$ is implied by (4) (for $c = c_1$ and $c = c_2$), since evidently $\Delta(\mathcal{X}, x'y'z', c_2) \leq \Delta(\mathcal{X}, x'y'z', c_1)$ for all $\mathcal{X} \in \mathcal{H}$. Suppose now that $c_2(x'y') \neq 0 \neq c_2(y'z')$ and $c_2(x'y') > c_1(x'y')$ (the case $c_2(y'z') > c_1(y'z')$ is treated similarly). Then one has evidently $b(c_1, xyz) > 0$ and $x'y' = xz$, thus implying $y' \neq y$; hence y' is a previously processed vertex. To be definite, let $y' = z$. Since $yy'z'$ is a previously processed triple, one has $b(c_1, yy'z') = 0$, by the hypothesis. However, $c_1(yy') \geq b(c_1, xyz) > 0$ and $c_1(y'z') \geq c_2(y'z') > 0$; therefore there exists $\mathcal{X} \in \mathcal{H}$ such that $\Delta(\mathcal{X}, yy'z', c_1) = 0$. Choose $X_A \in \mathcal{X}$ satisfying $yy', y'z' \in \partial X_A$. The relations $c_1(\mathcal{X}) = v(c)$ and $b(c_1, xyz) > 0$ yield $k(\mathcal{X}, xyz) = 0$; hence, by $yz = yy' \in \partial X_A$, one obtains $xy \notin \partial X_A$ and $x'y' = xz \in \partial X_A$. Therefore, $k(\mathcal{X}, x'y'z') > 0$ and $\Delta(\mathcal{X}, x'y'z', c_2) = 0$ (since $c_2(\mathcal{X}) \leq c_1(\mathcal{X})$ and $c_1(\mathcal{X}) = v(c)$), whence $b(c_2, x'y'z') = 0$ as required. Observe that $b(c_2, xyz) = 0$ as well. Therefore, by induction, $b(\tilde{c}, xyz) = 0$ holds for the final function $\tilde{c}$ and arbitrary $y \in V$, $x, z \in V \setminus \{y\}$.

Assume now that f' is an optimal solution for the problem $\mathcal{P}(\tilde{c})$. If there were a chain $L \in \mathcal{L}^+(f')$ such that $|EL| \geq 2$, then L would contain two distinct edges xy and yz, thus implying $b(\tilde{c}, xyz) \geq f'(L) > 0$. Hence $|EL| = 1$ for all $L \in \mathcal{L}^+(f')$; therefore f' coincides with f.

At the final stage, f is used to determine in a natural way an optimal solution f^* for the initial problem $\mathcal{P}(c)$. Namely, let $c = c^0, c^1, \ldots, c^N = \tilde{c}$ be the sequence of functions at the steps of the main stage. Starting from f, find successively multiflows $f = f^N, f^{N-1}, \ldots, f^0 = f^*$, where f^i is an optimal solution for the problem $\mathcal{P}(c^i)$. How to find f^i via f^{i+1} is obvious.

Let $n = |V|$ and $t = |T|$. Since $|\mathcal{V}| \leq n|\mathcal{A}| + 2$ and $|\mathcal{A}| \leq t$, one has $|\mathcal{V}| \leq tn + 2$. It is easy to see that the running time for the main stage, as well as for the entire algorithm, is $O(n^3 \sigma(tn))$, where $\sigma(n')$ is the running time of the procedure for finding a maximum flow in a network with n' vertices (the final stage can be completed within $O(n^5)$ operations). In fact, one can organize the processing of triples at an iteration of the main stage in such a way that the procedure of finding a maximum flow in the network $(\mathcal{G}, d^c)$ is used only $O(n)$ times per iteration (we omit details here); this enables us to obtain an algorithm with running time $O(n^2 \sigma(tn))$.

The algorithm for the case of an inner Eulerian function g differs from the one described above only in one point: we must put $c := c - \lfloor b(c, xyz) \rfloor \theta_{xyz}$ for each triple xyz at the corresponding processing step (here $\lfloor a \rfloor$ stands for the greatest integer not exceeding a). The verification of the algorithm follows mostly the same lines as above: one proves that at the end of the main stage $b(\tilde{c}, xyz) < 1$ for the final function $\tilde{c}$ and arbitrary $y \in V$, $x, z \in V \setminus \{y\}$ (the proof goes by induction, and is similar to that described above; it relies on Statement 3.1(4) and relation (4) with $b(c, xyz)$ replaced by $\lfloor b(c, xyz) \rfloor$ and $\Delta(\mathscr{X})$ replaced by $\lfloor \Delta(\mathscr{X}) \rfloor$). This implies $|EL| = 1$ for all $L \in \mathscr{L}^+(f')$, provided f' is an optimal solution for the problem $\mathscr{P}(\tilde{c})$; hence $f' = f$.

REFERENCES

1. B. V. Cherkasskiĭ, *Solution for a problem on multicommodity flows in a network*, Èkonom. i Mat. Metody **13** (1977), 143–151. (Russian)
2. _____, *Multi-terminal two-commodity problems*, Studies in Discrete Optimization (A. Fridman, ed.), "Nauka", Moscow, 1976, pp. 261–289. (Russian)
3. T. C. Hu, *Multi-commodity network flows*, Oper. Res. **11** (1963), 344–360.
4. A. V. Karzanov, *Combinatorial methods to solve cut-determined multiflow problems*, Combinatorial Methods in Flow Problems, no. 3 (A. V. Karzanov, ed.), Vsesoyuz. Nauchno-Issled. Inst. Sistem. Issled., Moscow, 1979, pp. 6–69. (Russian)
5. _____, *On multicommodity flow problems with integer-valued optimal solutions*, Dokl. Akad. Nauk SSSR **280** (1985), 789–792; English transl. in Soviet Math. Dokl. **31** (1985).
6. A. V. Karzanov and M. V. Lomonosov, *Flow systems in undirected networks*, Mathematical Programming Etc., Sb. Trudov Vsesoyuz. Nauchno-Issled Inst. Sistem. Issled. **1978**, no.1, 59–66. (Russian)
7. V. L. Kupershtokh, *On a generalization of the Ford–Fulkerson theorem for multiterminal networks*, Kibernetika (Kiev) **1971**, no. 3, 87–93; English transl. in Cybernetics **7** (1971).
8. M. V. Lomonosov, *Combinatorial approaches to multiflow problems*, Discrete Appl. Math. **11** (1985), 1–94.
9. L. Lovász, *On some connectivity properties of Eulerian graphs*, Acta Math. Acad. Sci. Hungar. **28** (1976), 129-138.
10. B. Rothschild and A. Whinston, *On two-commodity network flows*, Oper. Res. **14** (1966), 377–387.
11. A. V. Karzanov, *Polyhedra related to undirected multicommodity flows*, Linear Algebra Appl. **114/115** (1989), 293–328.

On Edge Mappings of Graphs Preserving Subgraphs of a Given Type

A. K. KELMANS

§1. Introduction

In a series of reports to the Seminar on Discrete Mathematics at the Institute of Control Sciences (Moscow, Winter 1976–1977), the author considered (among other things) special decompositions of graphs of various classes into paths [1] (see also §3.) These decompositions were shown to be very useful in establishing many facts in graph theory. Namely, they help to obtain the following results: (1) a simple inductive proof of the Kuratowski theorem on graph planarity (and the corresponding algorithm) [1], (2) a direct proof of a new planarity criterion in terms of nonseparating circuits [2], (3) a simple proof of the Whitney theorem on the 2-isomorphism of circuit isomorphic graphs (see §4), (4) a proof of the fact that a mapping of the edges of 3-connected graphs taking circuits into 2-regular graphs is induced by an isomorphism of these graphs (see§6), and (5) a proof of the fact that an edge mapping of 3-connected graphs with at least 5 vertices preserving (in both directions) 3-skeins is induced by an isomorphism of these graphs (it is a particular case of Theorem 8.1 from [7]; two different proofs of this theorem are given in [4] and [5]). Later (in a report to the Seminar at the Institute of Control Sciences, April 1982), this approach was applied to prove the following strengthening of the Kuratowski theorem: a 3-connected graph distinct from K_5 is nonplanar if and only if it contains a circuit with three crossing chord-edges [6].

This paper presents not only the results (3)–(5) mentioned above, but also some essential strengthenings and extensions of the Whitney theorems on circuit graph isomorphisms. The results of §§ 5 and 6 were reported to

---

1991 *Mathematics Subject Classification.* Primary 05C38.
Translation of Models and Algorithms of Operations Research and Their Application to the Organization of Work in Computational Systems, Yaroslav. Gos. Univ., Yaroslavl, 1984, pp. 19–30.

the Seminar at the Institute of Control Sciences (November, 1981) and at the Georgian Republican Workshop "Methods of Optimization on Networks and Graphs" (Batumi, June 1982); some of them were briefly described in [7].

§2. Basic notions and notation

2.1. Consider undirected *graphs* without loops or isolated vertices (parallel edges are admitted only in graphs of connectivity ≤ 2) [8]. Let VG, EG, and V^*G denote the sets of the *vertices*, *edges*, and *vertices of degree* $\neq 2$ of a graph G, respectively. Let $c(G)$ denote the *cyclomatic number* of a graph G. Let $E(x, G) = E(x)$ and $\mathcal{T}(x, G) = \mathcal{T}(x)$ denote the sets of all edges and all threads ending at a vertex x, respectively. Let K_n denote the complete graph with n vertices, $K_{m,n}$ the complete bipartite graph with parts of size m and n, S^n the graph with two vertices and n parallel edges, N^1 the graph-skeleton of the three-dimensional cube, and N^2 the graph obtained from the circuit with eight vertices by adding the (four) edges that are the main diagonals of the circuit. For $U \subseteq EG$, denote by $G\langle U \rangle$, or just $\langle U \rangle$, the subgraph in G with the edge set U and with the set of vertices incident to U.

2.2. Let xPy denote a path with the end-vertices x and y. We shall write $G(-)P$ instead of $G \setminus (P \setminus \{x, y\})$. A *thread* T in G is a path (or a circuit) maximal under inclusion such that each inner vertex of T is of degree 2 in G. Let $\mathcal{T}G$ denote the set of all threads in G.

2.3. Let $e_G(x, y)$ and $t_G(x, y)$ denote, respectively, the number of edges and threads in G with the end-vertices x and y. A bijection $v: VG \to V\bar{G}$ ($v^*: V^*G \to V^*\bar{G}$) is said to be an *isomorphism* (respectively, a *homeomorphism*) of G onto $\bar{G}$ if $e_G(x, y) = e_{\bar{G}}(v(x), v(y))$ for any $x, y \in VG$ (respectively, $t_G(x, y) = t_{\bar{G}}(v^*(x), v^*(y))$ for any $x, y \in V^*G$). The abbreviation "hom" below stands for "homeomorphic". For example, hom G is a graph homeomorphic to G, while hom 3-connected means "homeomorphic to a 3-connected graph". For a set of graphs $\mathcal{H}$, we have hom $\mathcal{H} = \{\text{hom } G : G \in \mathcal{H}\}$. Graphs G and F are *homotopic* if $V^*G = V^*F \neq \varnothing$ and the identity mapping $v^*: V^*G \to V^*F$ is a homeomorphism. A graph G is an *n-skein between x and y* if $G = \text{hom } S^n$ and $V^*G = \{x, y\}$.

2.4. Let $\mathcal{H}$ be a set of graphs. A bijection $m: EG \to E\bar{G}$ is said to be an $\mathcal{H}$*-isomorphism* ($\mathcal{H}$*-semi-isomorphism*) of G onto $\bar{G}$ if $A \in \text{hom } \mathcal{H} \iff m(A) \in \text{hom } \mathcal{H}$ (respectively, $A \in \text{hom } \mathcal{H} \Rightarrow m(A) \in \text{hom } \mathcal{H}$) for any $A \subseteq G$. In particular, if H is a graph, then an H-isomorphism is an $\mathcal{H}$-isomorphism with $\mathcal{H} = \{H\}$. If C is a circuit, then a C-isomorphism is said to be a *circuit isomorphism*.

2.5. Let $G \cap F = \varnothing$, $g_i \in VG$, $f_i \in VF$, $i = 1, 2$, and let H' (H'') be obtained from G and F by identifying the vertices g_i and f_j, $i, j \in \{1, 2\}$, with a new vertex h_i if $i = j$ (respectively, $i \neq j$). Then G and F are called (h_1, h_2)-*hammocks* of the graphs H' and H'', and we say that

H'' is *obtained from* H' *by the hammock switching* of G (or F), or by an (h_1, h_2)-*switching*. Let $t(G)$ denote the graph obtained from G by a sequence t of hammock switchings in G, and let $s(t)$ denote the number of switchings in t. If t is a sequence of hammock switchings in G and v is an isomorphism of $t(G)$ onto $\bar{G}$, then $g = vt$ is called a 2-*isomorphism* of G onto $\bar{G}$. If each of hammocks switched in t is a path, then the 2-isomorphism vt is called a hom-isomorphism.

2.6. A graph G is called (x, y)-*hammock-like*, $x, y \in VG$, if
(1) $G = G_1 \cup \cdots \cup G_n$, $G_i \cap G_j = \{x, y\}$ for $i \neq j$, and
(2) if G_i is neither a 3-connected graph nor a path, then $G_i \cup u$, where $u = (x, y)$, is 3-connected.

2.7. An edge set $X \subseteq EG$ is called a *blockade* of a connected graph G if $G \setminus X$ has at least two components containing circuits. A graph G is called a *p-blockade graph* if G has blockades and each blockade has at least p edges. A graph G is called *cubic* if all its vertices are of degree 3. It is easy to show that N^1 and N^2 are the minimum cubic 4-blockade graphs. A *cocircuit* of a graph G is a minimal (under inclusion) edge subset X of G with the property that $G \setminus X$ has fewer components than G. A cocircuit X in G is *nontrivial* if the edges of X do not have a common end-vertex; a *k-cocircuit* is a cocircuit of cardinality k.

§3. Preliminary results on graph decomposition

3.1. Following [1], consider a sequence $\mathscr{D} = (G_0, P_1, \ldots, P_k)$ of a graph G consisting of a subgraph G_0 and a sequence of different paths P_i in G. The sequence $\mathscr{D}$ induces a sequence of subgraphs $G_i = G_{i-1} \cup P_i$, $i = 1, \ldots, k$, in G. The last subgraph of this sequence is denoted by $L(\mathscr{D})$. For a graph F, denote by $B(F)$ the set of paths xPy such that $F \cap P = \{x, y\}$, by $W(F)$ the set of paths xPy such that $F \cap P = \{x, y\}$ and $\{x, y\} \not\subseteq VT$ for any thread T in F, and by $H(F)$ the set of paths xPy such that $F \cap P = \{x, y\}$ and x, y are inner vertices of nonadjacent threads in F. A sequence $\mathscr{D}$ of a graph G is called an *A-decomposition* of the graph G, $A \in \{B, W, H\}$, if $P_i \in A(G_{i-1})$ for $i = 1, \ldots, k$ and $L(\mathscr{D}) = G$.

3.2. The following statement is obvious:

PROPOSITION. *Let G be a 2-connected graph and G_0 a 2-connected subgraph of G. Then there exists a B-decomposition $(G_0, P_1, \ldots, P_k)$ of G (implying that $G_i = G_{i-1} \cup P_i$ is 2-connected).*

3.3. THEOREM. *Let G be* hom 3-*connected*, $L \subseteq M \subseteq G$, $M = \hom F$, *where F is 3-connected and L is triangle-free. Then there exists a W-decomposition $(G_0, P_1, \ldots, P_k)$ of G such that*

(a1) $G_0 = \hom F$, *and, moreover*
(a2) G_0 *is homotopic to* M *and* $L \subseteq G_0$.

This theorem represents a useful strengthening of Theorem 3.2 from [1], and the proof is a specialization of the proof of Theorem 3.2 given in [1].

3.4. THEOREM. *Let G be a* $\hom(x, y)$-*hammock-like graph. Then there exists a W-decomposition* $(G_0, P_1, \ldots, P_k)$ *of G such that G_0 is a k-skein between x and y, and $k = m(x, y)$ is the maximum number of inner disjoint paths between x and y (implying that $G_i = G_{i-1} \cup P_i$ is a $\hom(x, y)$-hammock-like graph and G_i contains an k-skein G_0 between x and y).*

The proof of this theorem is similar to that of Theorem 3.2 in [1].

3.5. THEOREM [1]. *Let G and F be homeomorphic to cubic 4-blockade graphs and let G contain a graph homeomorphic to F. Then there exists an H-decomposition $(\hom F, P_1, \ldots, P_k)$ of G. Moreover there exists an H-decomposition $(\hom F, P_1, \ldots, P_k)$ of G with $F \in \{N^1, N^2\}$).*

In the above theorem the last statement follows from the first one because every cubic 4-blockade graph contains $\hom N^1$ or $\hom N^2$.

§4. A simple proof of the Whitney theorem on circuit isomorphisms of 2-connected graphs

4.1. The following classical theorems of Whitney are well known:

4.1.1. THEOREM [9]. *Any circuit isomorphism of a 3-connected graph G onto a graph G' is induced by an isomorphism of G onto G'.*

A simple proof of this theorem was given in [2].

4.1.2. THEOREM [10]. *Any circuit isomorphism of a 2-connected graph G onto a graph G' is induced by a 2-isomorphism of G onto G'.*

The original proof of this theorem is rather complicated and tedious. Truemper [11] has found a shorter proof, based on Whitney's theorem (see 4.1.1) and Tutte's theorem on the decomposition of 2-connected graphs. Below, we give a proof found earlier, which seems quite simple and short and which does not employ the Whitney and Tutte theorems, providing therefore another proof of Whitney's theorem (4.1.1). Moreover, we shall see that the approach developed in this simple proof allows us to obtain essential strengthenings and extensions of the Whitney theorems on circuit graph isomorphisms.

4.2. THEOREM. *Let* (h1) *G be a 2-connected graph, and* (h2) $e: EG \to E\bar{G}$ *be a circuit isomorphism of G onto $\bar{G}$.*

Then there exist (a1) *a sequence t_1 consisting of at most $|V^*G| - 2$ hammock switchings in G such that $t_1(G)$ and $\bar{G}$ are isomorphic, and* (a2) *a sequence t_2 consisting of at most $|VG| - 2$ hammock switchings in G, such that $t_2(G)$ and $\bar{G}$ are isomorphic, and moreover, e is induced by an isomorphism of $t_2(G)$ onto $\bar{G}$.*

REMARK. The above theorem contains Theorem 4.1.2 (Whitney) along with an upper bound on the number $s(t_2)$ of switchings found in [11]. This bound (equal to $|VG|-2$) is treated in [11] as being surprisingly small. The proof below shows that it is not a surprise.

PROOF. (p1) Consider paths xTy and $\bar{x}\bar{T}\bar{y}$ and a bijection $m\colon ET \to E\bar{T}$. One can easily find a sequence t of $|ET|-1 = |VT|-2$ hammock switchings in T such that there exists an isomorphism $v\colon V(t(T)) \to V\bar{T}$ with $v(x) = \bar{x}$ and $v(y) = \bar{y}$ inducing m. It is clear that the same holds true for circuits C and $\bar{C}$.

(p2) Let us assume that G is a 3-skein S. Let T_i, $i \in \{1, 2, 3\}$, be a thread of S and let $S_i = S(-)T_i$. Since S has exactly three circuits and S_i is a circuit, it follows from (h2) that $\bar{S}$ has exactly three circuits as well and $\bar{S}_i$ is also a circuit. Hence, $\bar{T}_i$ is a thread in $\bar{S}$ and a chord of the circuit $\bar{S}_i$. Therefore $\bar{S}$ is a 3-skein. Since S and $\bar{S}$ are isomorphic, we have $s(t_1) = 0 = |V^*G|-2$. From (p1), we have $s(t_2) \leq |VG|-2$.

(p3) It follows from (p1) and (p2) that the theorem holds for graphs F with $c(F) = 1$ and 2. Let us assume that the theorem holds for graphs with $c(F) \leq n$, $n \geq 2$, and let G be a graph with $c(G) = n+1$. It is easy to see (Proposition 3.2) that there exists a thread P in G such that $G' = G(-)P$ is 2-connected. Obviously, $e' = e|_{EG'}$ is a circuit isomorphism of G' onto $\bar{G}'$. Thus, G', $\bar{G}'$, e' satisfy (h1) and (h2), and $c(G') = c(G)-1 = n$. Therefore by the inductive hypothesis, there exist sequences t'_1 and t'_2 of hammock switchings in G' satisfying (a1) and (a2), respectively. In fact, it suffices to prove that $\bar{P}$ is a thread in $\bar{G}$ and that the positions of its end-vertices are specified uniquely in $\bar{G}'$.

(p4) Let us prove that $\bar{P}$ is a path of $\bar{G}$. Since G' is 2-connected, there exists a circuit Z' in G' containing the end-vertices of P. Then $Z = Z' \cup P$ is a 3-skein and P is one of the threads T_1, T_2, T_3 of the 3-skein z, say, $P = T_1$. According to (p2), $\bar{Z}$ is a 3-skein in $\bar{G}$, while $\bar{T}_1 = \bar{P}$, $\bar{T}_2$, and $\bar{T}_3$ are its threads. Thus, $\bar{P}$ is a path in $\bar{G}$, and its end-vertices are the end-vertices of the paths $\bar{T}_2$ and $\bar{T}_3$ as well.

(p5) Let us prove that $\bar{P}$ is a thread of $\bar{G}$. Assume that $\bar{P}$ is not a thread of $\bar{G}$. Then there exists an edge $\bar{q}$ in $E\bar{G} \setminus E\bar{P}$ incident to an inner vertex of the path $\bar{P}$. By (h1), G is 2-connected. Therefore G has a circuit C containing the thread P and the edge $q = m^{-1}(\bar{q})$. By (h2), $\bar{C}$ is a circuit in $\bar{G}$, and $\bar{C}$ contains the path $\bar{P}$ and the edge $\bar{q}$ incident to an inner vertex of $\bar{P}$, a contradiction.

(p6) Put $t'_2 = t'$. According to (p3), the sequence t' of hammock switchings satisfies (a2). Therefore there exists an isomorphism v' of $t'(G)$ onto $\bar{G}$ that induces $e' = e|_{EG'}$, and so $v't'$ is a 2-isomorphism of G onto $\bar{G}$. We may assume that t' has the property:

(pr1) The number $s(t')$ of switchings in t' is minimum.

Let us prove that in this case a switching transformation t' can also be done in G. By (h1), there exists a circuit C of G containing P. Then $C' = C(-)P$ is a path of G'. Suppose that $t' = (s_n, \ldots, s_1)$ can not be done in G. Then there exists $i \in \{1, \ldots, n\}$ such that $t^i(C')$ is not a path of $t^i(G')$; here $t^i = (s_i, \ldots, s_1)$, and $t^i(C')$ is the subgraph of $t^i(G')$ with the edge set EC'. Let $k(t') = k$ denote the maximum i such that $t^{i-1}(C')$ is a path but $t^i(C')$ is not a path. We may assume that t' has the property:

(pr2) $k(t'') \leq k(t')$ for every switching transformation t'' of G' satisfying (a2) and having property (pr1).

By (h2), $\bar{C}$ is a circuit of $\bar{G}$. By (p5), $\bar{P}$ is a path of $\bar{G}$. Therefore $\bar{C}'$ is a path of $\bar{G}'$. Since $v't'$ is a 2-isomorphism of G' onto $\bar{G}'$ inducing e', we have that $t'(C')$ is a path of $t'(G')$. Put $t_k = (s_1, \ldots, s_k)$. Since $k = k(t')$ and t' has property (pr2), one of the following two cases holds:

(c1) $t_k t^{k-1}(G') = t^{k-1}(G')$.

(c2) There exists a switching s' of $t^{k-1}(G')$ such that $t_k t^{k-1}(G') = s' t^{k-1}(G')$.

Let t'' be obtained from t' by deleting t_i in case (c1), and by replacing t_i by s' in case (c2). Then $v't''$ is a 2-isomorphism of G' onto $\bar{G}'$ inducing e', and t'' has fewer switchings than t', which contradicts property (pr1) of t'.

(p7) If both threads T_2 and T_3 in Z (see (p4)) are subpaths of threads from G' or both of them are not subpaths of threads from G', then one can put $t_1 = t_1'$. Therefore let one of T_2, T_3 belong to a thread in G' and the other not. Then, obviously, one can obtain t_1 from t_1' by making at most one switching of a hammock in G containing P. In this case, $|V^*G| \geq |V^*G'| + 1$, and by the inductive hypothesis we have $s(t_1') \leq |V^*G'| - 2$, so that $s(t_1) \leq |V^*G| - 2$. Let t_2^* be the sequence of switchings from (p1), where $T := P$, $\bar{T} := \bar{P}$, and $m := e|_P$. Then $s(t_2^*) \leq |VP| - 2$. Obviously, $t_2 = t_2' t_2^*$. By the inductive hypothesis, $s(t_2') \leq |VG'| - 2$. Therefore $s(t_2) = s(t_2') + s(t_2^*) \leq |VG'| + |VP| - 4 = |VG| - 2$. □

4.3. COROLLARY. *Let G, $\bar{G}$, and e satisfy assumptions* (h1) *and* (h2) *of Theorem 4.2, and let G contain a graph homeomorphic to a 3-connected graph H. Then there exist* (a1) *a sequence t_1 consisting of at most $|V^*G| - |VH|$ hammock switchings in G such that $t_1(G)$ and $\bar{G}$ are isomorphic, and* (a2) *a sequence t_2 consisting of at most $|VG| - |VH|$ hammock switchings in G such that $t_2(G)$ and $\bar{G}$ are isomorphic, and moreover, e is induced by an isomorphism of $t_2(G)$ onto $\bar{G}$.*

§5. On graph $\mathscr{H}$-semi-isomorphisms

The idea of the proof of Whitney's theorem in §4 allows us to establish some similar, but stronger results. In this section, we obtain a sufficient condition that an $\mathscr{H}$-semi-isomorphism of 2-connected (3-connected) graphs is induced by a 2-isomorphism (respectively, isomorphism) of these graphs. Here $\mathscr{H}$ is a set of 2-connected (respectively, 3-connected) graphs. Then we apply this condition to study mappings of graph edges for some special $\mathscr{H}$.

5.1 LEMMA. *Let* (h1) *G be 2-connected (hom 3-connected),* (h2) $c(G) \geq k \geq 2$, *and* (h3) $m: EG \to E\bar{G}$ *be a bijection such that for any 2-connected (respectively, hom 3-connected) subgraph H with* $c(H) = k$, *the mapping* $m|_H: EH \to E\bar{H}$ *is induced by a 2-isomorphism (respectively, hom-isomorphism) of H onto $\bar{H}$.*

Then m is induced by a 2-isomorphism (respectively, hom-isomorphism) of G onto $\bar{G}$.

PROOF. (p1) We prove the lemma by induction on $c(G)$. For $c(G) = k$, the statement is obvious. Let $c(G) > k \geq 2$. By Proposition 3.2 (Theorem 3.3), there exists a thread P in G such that $G' = G(-)P$ is 2-connected (respectively, hom 3-connected). Obviously, $c(G') = c(G) - 1 \geq k$, and $m' = m|_{G'}$ satisfies condition (h3). By the inductive hypothesis, $m': EG' \to E\bar{G}'$ is induced by a 2-isomorphism $v't'$ of G' onto $\bar{G}'$. Actually, it suffices to show that $\bar{P}$ is a thread in $\bar{G}$ and the positions of its end-vertices are specified uniquely in $\bar{G}'$.

(p2) Let us prove that $\bar{P}$ is a path of $\bar{G}$. Since G' is 2-connected, there exists a circuit Z' in G' containing the end-vertices of P. Then $Z = Z' \cup P$ is a 3-skein in which P is a thread. By Proposition 3.2 (Theorem 3.3), there exists a 2-connected (hom 3-connected) subgraph H in G such that P is a thread in H and $c(H) = k$. By (h3), $m|_H$ is induced by a 2-isomorphism $v^0 t^0$ of H onto $\bar{H}$. Therefore $\bar{P}$ is a thread in $\bar{H}$, and so $\bar{P}$ is a path of $\bar{G}$. Moreover the positions of the end-vertices of $\bar{P}$ are specified uniquely in $\bar{H}' = \bar{H} \setminus E\bar{P}$, and thereby, in $\bar{G}'$.

(p3) Let us prove that $\bar{P}$ is a thread of $\bar{G}$. Assume that the path $\bar{P}$ is not a thread in $\bar{G}$. Then there exists an edge $\bar{u} \in E\bar{G} \setminus E\bar{P}$ incident to an inner vertex of the path $\bar{P}$. Since G is 2-connected and $c(G) \geq 2$, there exists a 3-skein in G containing P and u. By Proposition 3.2 (Theorem 3.3), there exists a 2-connected (hom 3-connected) subgraph H in G such that $P, u \in H$, and $c(H) = k$. According to (h3), there exists a 2-isomorphism of H onto $\bar{H}$ inducing $m|_H$. Therefore, if T is a thread in H, then $\bar{T}$ is a thread in $\bar{H}$. Let T be a thread in H containing P. Then $\bar{T}$ is a thread in $\bar{H}$ containing the path $\bar{P}$. Also $\bar{u} \in \bar{H} \setminus E\bar{P}$ and $\bar{u}$ is incident to an inner vertex of $\bar{P}$. Therefore, $\bar{T}$ is not a thread in $\bar{H}$, a contradiction.

(p4) By using the arguments similar to that in (p6) of the proof of Theorem 4.2, one can prove that a 2-isomorpthism $v't'$ in (p1) can be chosen in such a way that the switching transformation t' of G' can also be done in G. Now let us describe a required 2-isomorphism vt of G onto $\bar{G}$. Let H be the subgraph from (p2), so that P is a thread of H. Since H is 2-connected, $H' = H(-)P$ is connected. If G is S^n (the graph with two vertices and n parallel edges), then the theorem obviously holds. Hence we can assume that G is not S^n, and so H' has at least 3 vertices. Thus G' and H are 2-connected, and t' and t^0 (see (p2)) can be done in G. Therefore $t'(G') \cap t^0(H)$ is connected and has at least 3 vertices. Since $v't'$ induces $m|_{G'}$ and $v^0 t^0$ induces $m|_H$, we have $v(x) = v'(x) = v^0(x)$ for $x \in V(t'(G') \cap t^0(H))$. Let t'' be the sequence of switchings of P from t^0. Then obviously $t = t't''$, and for $x \in V(t'(G'))$ we have $v(x) = v'(x)$ if $x \in V(t'(G'))$, and $v(x) = v^0(x)$ if $x \in V(t^0(H))$. □

5.2. Lemma 5.1 immediately yields:

THEOREM. *Suppose that*:

(h1) *G is 2-connected (hom 3-connected) and $c(G) \geqslant k \geqslant 2$,*

(h2) *$\mathcal{H}$ is a set of 2-connected (respectively, hom 3-connected) graphs,*

(h3) *$e: EG \to E\bar{G}$ is a $\mathcal{H}$-semi-isomorphism of G onto $\bar{G}$, and*

(h4) *for any F with $c(F) = k$, any $\mathcal{H}$-semi-isomorphism $e: EF \to E\bar{F}$ is induced by a 2-isomorphism of F onto $\bar{F}$.*

Then $e: EG \to E\bar{G}$ is induced by a 2-isomorphism of G onto $\bar{G}$.

§6. On edge mappings of graphs taking every circuit to a set of disjoint circuits

In this section, we consider graphs G and $\bar{G}$ and a bijection $m: EG \to E\bar{G}$ satisfying the following condition:

(c) if C is a circuit in G, then $m(C) = \bar{C}$ is a 2-regular subgraph (i.e. a set of disjoint circuits) in $\bar{G}$.

We study the properties of such a mapping. In particular, we prove that for 3-connected graphs, such a mapping is induced by an isomorphism of G onto $\bar{G}$.

We recall that a *cocircuit* of a graph G is a minimal (under inclusion) edge subset X of G with the property that $G \setminus X$ has fewer components than G. A cocircuit X in G is *nontrivial* if X is a matching; a *k-cocircuit* is a cocircuit of cardinality k.

6.1. Using the Menger theorem [8], one can easily prove

LEMMA. *Three edges of a 3-connected graph do not belong to a common circuit if and only if they either form a 3-cocircuit or have a common end-vertex.*

6.2. LEMMA. *If $\bar{K}$ is a cocircuit in $\bar{G}$, then $K = m^{-1}(\bar{K})$ contains a cocircuit of the graph G.*

PROOF. Assume that K does not contain a cocircuit of G. Then $G \setminus K$ contains a spanning forest F of the graph G such that every component of F is a spanning tree of a component of G (i.e. EF is a base of the circuit matroid of G). Let $e \in K$. Then $F \cup e$ contains a unique circuit C, and $C \cap K = \{e\}$. Therefore, $\bar{C} \cap \bar{K} = \{\bar{e}\}$. Since C is a circuit in G, it follows from condition (c) that $\bar{C}$ is a 2-regular graph in $\bar{G}$. Since $\bar{K}$ is a cocircuit in $\bar{G}$, we have $\bar{C} \cap \bar{K} \neq \{\bar{e}\}$, a contradiction. □

6.3 LEMMA. *Let G be 3-connected and $\bar{G}$ be connected. If K is a nontrivial 3-cocircuit of G, then $\bar{K}$ is a nontrivial 3-cocircuit of $\bar{G}$.*

PROOF. Let K be a nontrivial 3-cocircuit in G. Then $G \setminus K$ has exactly two components G_i with $|VG_i| \geq 3$, $i = 1, 2$. Let us assume that $\emptyset \neq V\bar{G}_1 \cap V\bar{G}_2 \ni \bar{x}$. Since G is 3-connected, G_i is 2-connected, and therefore each edge in G_i belongs to a circuit in G_i. Then, by condition (c), each edge in $\bar{G}_i$ belongs to a 2-regular graph in $\bar{G}_i$, and hence the degree of each vertex in $\bar{G}_i$ is at least two. Therefore there exists a set $\bar{A}$ of three edges in $\bar{G}$ incident to $\bar{x}$ such that $\bar{A} \cap \bar{G}_i \neq \emptyset$ for $i = 1, 2$. Since all the three edges in $\bar{A}$ have a common vertex, there is no 2-regular graph in $\bar{G}$ containing $\bar{A}$. By condition (c), there is no circuit in G containing A. Since $A \cap G_i \neq \emptyset$ for $i = 1, 2$, A contains two nonadjacent edges. Therefore, by Lemma 6.1, A is a nontrivial 3-cocircuit in G. Let G'_1 and G'_2 be the components of $G \setminus A$. Then $G'_i \cap K \neq \emptyset$ for $i = 1, 2$. Since $|K| = |A| = 3$, we have $|G_i \cap A| = |G_j \cap K| = 1$ for some i and j, say, $G_i \cap A = \{a\}$ and $G'_j \cap K = \{k\}$. Then $\{a, k\}$ is an edge 2-cut in G, which contradicts the 3-connectivity of G. (In other words, the 3-cocircuits K and A are transversal in G, which contradicts the edge 3-connectivity of G). Thus, $\bar{G}_1 \cap \bar{G}_2 = \emptyset$. Since $\bar{G}$ is connected, there exists an edge in $\bar{K}$ connecting $\bar{G}_1$ with $\bar{G}_2$. Since G has no 1- and no 2-cocircuits, neither does $\bar{G}$, by Lemma 6.2. Therefore, each edge in $\bar{K}$ connects $\bar{G}_1$ with $\bar{G}_2$. Let us assume that edges $\bar{p}$ and $\bar{q}$ in $\bar{K}$ have a common vertex $\bar{z}$ in $\bar{G}_i$. Then there exists an edge $\bar{r}$ in $\bar{G}_i$ with the end-vertex $\bar{z}$. Thus, the edges $\bar{p}$, $\bar{q}$ and $\bar{r}$ do not belong to a common 2-regular graph in $\bar{G}$. By condition (c), the edges p, q, and r do not belong to a common circuit in G. Also, $p, q \in K$, while $r \in G_i$. By Lemma 6.1, the edges p, q, and r belong to a common circuit in G, a contradiction. Thus, $\bar{K}$ is a nontrivial 3-cocircuit in $\bar{G}$. □

6.4. LEMMA. *Let G be 3-connected and $\bar{G}$ be connected. If the edges $\bar{u}_1, \bar{u}_2, \bar{u}_3$ have a common vertex in $\bar{G}$, then the edges u_1, u_2, u_3 have a common vertex in G.*

PROOF. Since $\bar{u}_1, \bar{u}_2, \bar{u}_3$ have a common vertex in $\bar{G}$, there is no 2-regular graph in $\bar{G}$ containing $\{\bar{u}_1, \bar{u}_2, \bar{u}_3\}$. Therefore, by (c), u_1, u_2, u_3 do not belong to a common circuit in G. Since $\bar{u}_1$ and $\bar{u}_2$ are adjacent in $\bar{G}$, $\{\bar{u}_1, \bar{u}_2, \bar{u}_3\}$ is not a nontrivial 3-cocircuit in $\bar{G}$ by Lemma 6.3. Hence, by Lemma 6.1, u_1, u_2, u_3 have a common vertex. □

6.4'. By using 6.3, one can easily prove

LEMMA. *Let G be 3-connected and $m: EG \to E\bar{G}$ be a circuit semi-isomorphism of G onto $\bar{G}$. If edges $\bar{u}_1, \bar{u}_2, \bar{u}_3$ have a common vertex in $\bar{G}$, then the edges u_1, u_2, u_3 have a common vertex in G.*

6.5. THEOREM. *Suppose that* (h1) *G is 3-connected and $\bar{G}$ is connected, and* (h2) *$m: EG \to E\bar{G}$ is a bijection such that if C is a circuit of G, then $\bar{C} = m(C)$ is a 2-regular subgraph of $\bar{G}$.*

Then there exists an isomorphism $v: VG \to V\bar{G}$ inducing m.

PROOF. (p1) Let us show first that for every $x \in VG$, there exists $\bar{x} \in V\bar{G}$ such that x is the common end-vertex of all edges of $m^{-1}(E(\bar{x}, \bar{G}))$. Let $u = (x, y) \in EG$ and $\bar{u} = (\bar{a}_1, \bar{a}_2) \in E\bar{G}$. By Lemma 6.4, for every $\bar{a}_i$ in $\bar{G}$, there exists a vertex a_i in G that is the common end-vertex of all edges of $m^{-1}(E(\bar{a}_i, \bar{G}))$, implying $a_i \in \{x, y\}$, $i = 1, 2$. Let us assume that $x \notin \{a_1, a_2\}$, and so $a_1 = a_2 = y$. Then $E(y, G) \supseteq m^{-1}(E(\bar{a}_1, \bar{G}) \cup E(\bar{a}_2, \bar{G}))$. Since $E(\bar{a}_i, \bar{G})$ contains a cocircuit of $\bar{G}$, Lemma 6.2 implies that $m^{-1}(E(\bar{a}_i, \bar{G}))$ contains a cocircuit of G. Since G is 3-connected, $E(y, G)$ is a cocircuit in G. Therefore, $E(y, G) = m^{-1}(E(\bar{a}_1, \bar{G})) = m^{-1}(E(\bar{a}_2, \bar{G}))$. Then $E(\bar{a}_1, G) = E(\bar{a}_2, G)$, and hence $\bar{G}$ is disconnected. This contradicts (h1). Thus, for any $x \in VG$, there exists a unique vertex $\bar{x} \in V\bar{G}$ such that x is the common end-vertex of all edges of $m^{-1}(E(\bar{x}, \bar{G}))$. Put $v(x) = \bar{x}$. Obviously if $v(x) \ne v(y)$, then $x \ne y$.

(p2) If a subset $\bar{U}$ of edges of $\bar{G}$ forms a forest, then, by condition (c), the subset U of edges of G forms a forest as well. Therefore since $\bar{G}$ is connected, we have $|VG| \ge |V\bar{G}|$. Since G is 3-connected, by condition (c) the degree of each vertex in $\bar{G}$ is at least two. Therefore, if $x \ne y$, then $v(x) \ne v(y)$. Thus, $v: VG \to V\bar{G}$ is a bijection, $(x, y) \in EG$ if and only if $(v(x), v(y)) \in E\bar{G}$, and v induces m. □

6.5'. By using Lemmas 6.2 and 6.4 and the arguments in 6.5, we obtain an analogue of Theorem 6.5 (which is, in a sense, weaker):

THEOREM. *Let* (h1) *G be 3-connected, and* (h2) *$m: EG \to E\bar{G}$ be a circuit semi-isomorphism of G onto $\bar{G}$.*

Then there exists an isomorphism $v: VG \to V\bar{G}$ inducing m.

This theorem was proved in a different way in [3].

6.6. REMARK. If one replaces in Theorem 6.5 the condition "$\bar{C}$ is a 2-regular subgraph (i.e. the union of disjoint circuits) in $\bar{G}$" by "$\bar{C}$ is the union of circuits in $\bar{G}$" in Theorem 6.5, or drops the hypothesis "$\bar{G}$ is connected", then the statement of the theorem becomes obviously false.

6.7. With the help of Theorem 3.3 and by induction on the number of edges, or with the help of Theorems 6.5 and 6.5', one can prove the following strengthening of Theorem 6.5':

THEOREM. *Let* (h1) *G be hom 3-connected and have at most two threads that are not edges, and* (h2) *$m: EG \to E\bar{G}$ be a bijection such that either* (h2.1) *m is a circuit semi-isomorphism of G onto $\bar{G}$, or* (h2.2) *$\bar{G}$ is connected and m is a mapping such that if C is a circuit in G, then $\bar{C} = m(C)$*

is a set of disjoint circuits in $\bar{G}$. Then there exists a hom-*isomorphism of G onto $\bar{G}$ inducing m.*

REFERENCES

1. A. K. Kelmans, *Graph expansion and reduction*, Algebraic Methods in Graph Theory (Szeged, 1978), Vol. I, Colloq. Math. Soc. János Bolyai, vol. 25, Akad. Kiadó, Budapest, and North-Holland, Amsterdam, 1981, pp. 317–343.
2. _____, *The concept of a vertex in a matroid, the nonseparating circuits of a graph, and a new criterion for graph planarity*, Algebraic Methods in Graph Theory (Szeged, 1978), Vol. I, Colloq. Math. Soc. János Bolyai, vol. 25, Akademiai Kiadó, Budapest, and North-Holland, Amsterdam, 1981, pp. 345–388.
3. Y. H. Sanders and D. Sanders, *Circuit preserving edge maps*, J. Combin. Theory Ser. B **22** (1977), 91–96.
4. A. K. Kelmans, 3-*skeins in* 3-*connected graphs*, All-Union Conf. Statistical and Discrete Analysis of Nondigital Information, Expert Estimates, and Discrete Optimization, Moscow and Kazakh. Gos. Univ., Alma-Ata, 1981. (Russian)
5. R. L. Hemminger, H. A. Jung, and A. K. Kelmans, *On* 3-*skein isomorphism of graphs*, Combinatorica **2** (1983), 373–376.
6. A. K. Kelmans, *On the existence of subgraphs of a given type in graphs*, Algorithms of Discrete Optimization in Computing Systems, Izdat. Yaroslav. Univ., Yaroslavl, 1983, pp. 3–20. (Russian)
7. _____, *Homeomorphic embeddings of graphs with given properties*, Dokl. Akad. Nauk SSSR **274** (1984), 1298–1303; English transl. in Soviet Math. Dokl. **29** (1984).
8. Frank Harary, *Graph theory*, Addison–Wesley, Reading, MA, 1969.
9. H. Whitney, *Congruent graphs and the connectivity of graphs*, Amer. Math. J. **54** (1932), 150–168.
10. _____, 2-*isomorphic graphs*, Amer. Math. J. **55** (1933), 245–254.
11. K. Truemper, *On Whitney's* 2-*isomorphism theorem for graphs*, J. Graph Theory **4** (1980), 43–49.
12. R. L. Hemminger and H. A. Jung, *On* n-*skein isomorphism of graphs*, J. Combin. Theory Ser. B **32** (1982), 103–111.

On Edge Semi-isomorphisms of Graphs Induced by Their Isomorphisms

A. K. KELMANS

The present paper is a continuation of [1]. It strengthens and generalizes the well-known Whitney theorems on circuit isomorphisms of graphs [2], [3].

§1. Basic notions and notation

1.1. We consider *undirected graphs* without loops or isolated vertices (parallel edges are admitted only in graphs of connectivity $\leqslant 2$) [4]. Let VG, EG, and V^*G denote the sets of the *vertices*, *edges*, and *vertices of degree* $\neq 2$ of a graph G, respectively. Let $c(G)$ denote the *cyclomatic number of* a graph G. Let $E(x, G) = E(x)$ and $\mathcal{T}(x, G) = \mathcal{T}(x)$ denote, respectively, the sets of all edges and all threads ending at a vertex x. Let K_n be the complete graph with n vertices, $K_{m,n}$ be the complete bipartite graph with parts of size m and n, and S^n be the graph with two vertices and n parallel edges. For $U \subseteq EG$, denote by $G\langle U \rangle$, or just $\langle U \rangle$, the subgraph in G with the edge set U.

1.2. Let xPy denote a path with the end-vertices x and y. We write $G(-)P$ instead of $G \setminus (P \setminus \{x, y\})$. A *thread* T in G is a path (or a circuit), maximal under inclusion, such that each inner vertex of T has degree 2 in G. Let $\mathcal{T}G$ be the set of all threads in G.

1.3. Let $e_G(x, y)$ and $t_G(x, y)$ denote, respectively, the number of edges and threads in G having the end-vertices x and y.

A bijection $v: VG \to V\overline{G}$ ($v^*: V^*G \to V^*\overline{G}$) is said to be an *isomorphism* (respectively, a *homeomorphism*) of G onto $\overline{G}$ if $e_G(x, y) = e_{\overline{G}}(v(x), v(y))$ for any $x, y \in VG$ (respectively, $t_G(x, y) = t_{\overline{G}}(v^*(x), v^*(y))$ for any $x, y \in V^*G$). The abbreviation "hom" below stands for "homeomorphic".

1991 *Mathematics Subject Classification*. Primary 05C38.
Translation of Models of Operations Research in Computational Systems, Yaroslav. Gos. Univ., Yaroslavl, 1985, pp. 80–95.

For example, hom G is a graph homeomorphic to G, while hom 3-connected means "homeomorphic to a 3-connected graph". For a collection of graphs $\mathscr{H}$, we have hom $\mathscr{H} = \{\text{hom } G : G \in \mathscr{H}\}$. G is an *n-skein between* x and y if $G = \text{hom } S^n$ and $V^*G = \{x, y\}$.

1.4. Let $\mathscr{H}$ be a collection of graphs. A bijection $m: EG \to E\overline{G}$ is said to be an $\mathscr{H}$-*isomorphism* ($\mathscr{H}$-*semi-isomorphism*) of G onto $\overline{G}$ if $A \in \text{hom } \mathscr{H} \iff m(A) \in \text{hom } \mathscr{H}$ (respectively, $A \in \text{hom } \mathscr{H} \Rightarrow m(A) \in \text{hom } \mathscr{H}$) for any $A \subseteq G$. In particular, if H is a graph, then an H-isomorphism is an $\mathscr{H}$-isomorphism with $\mathscr{H} = \{H\}$. If C is a circuit, then a C-isomorphism is said to be a *circuit isomorphism*.

1.5. Let $G \cap F = \varnothing$, $g_i \in VG$, $f_i \in VF$, $i = 1, 2$, and let H' (H'') be obtained from G and F by identifying the vertices g_i and f_j, $i, j \in \{1, 2\}$, with a new vertex h_i if $i = j$ (respectively, $i \neq j$). Then G and F are the (h_1, h_2)-*hammocks* of the graphs H' and H'', and we say that H'' is obtained from H' by the *hammock switching* of G (or F), or by the (h_1, h_2)-*switching*. Let $t(G)$ denote the graph obtained from G by a sequence t of hammock switchings in G, and let $s(t)$ denote the number of switchings in t. If v is an isomorphism of $t(G)$ onto $\overline{G}$, then $g = vt$ is called a 2-*isomorphism* of G onto $\overline{G}$. If each of hammocks switched in t is a path, then the 2-isomorphism vt is said to be a hom-*isomorphism*.

1.6. A set of edges $X \subseteq EG$ is a *blockade* of a connected graph G if $G \setminus X$ has at least two components with circuits. A graph G is called a *p-blockade graph* if G has blockades and each blockade has at least p edges. A graph G is called *cubic* if all its vertices are of degree 3.

§2. On S^3-semi-isomorphisms of graphs

In this section we investigate when an S^3-semi-isomorphism is induced by a 2-isomorphism, hom-isomorphism, or isomorphism of 2-connected, hom 3-connected, or 3-connected graphs, respectively.

According to Theorem 5.2 from [1], if there exists k such that for any 2-connected (hom 3-connected) graph G with cyclomatic number $c(G) = k$, any $\mathscr{H}$-semi-isomorphism of G onto $\overline{G}$ is induced by a 2-isomorphism (hom-isomorphism, respectively), then the same holds for graphs with cyclomatic number $c(G) \geq k$. We establish below that for $\mathscr{H} = \{S^3\}$, such a k does exist, and for 2-connected, hom 3-connected, and 3-connected graphs, the minimum k equals 4. Therefore, we obtain an essential strengthening of the Whitney theorem on circuit isomorphisms of 2-connected graphs [3].

2.1. LEMMA. *Let* $EG = E_1 \cup \cdots \cup E_6$, $E_i \cap E_j = \varnothing$ *for* $i \neq j$, *and let each subgraph* $G^i = \langle EG \setminus E_i \rangle$ *be a 3-skein in* G (*i.e.*, $G^i = \text{hom } S^3$). *Then* $G = \text{hom } K_4$ *and each* $\langle E_i \rangle$ *is a thread of* G.

PROOF. Obviously, G is 2-connected, and therefore every vertex of G is of degree at least 2. Any four sets E_i belong to a 3-skein, while any 3-skein has no vertices of degree ≥ 4. Therefore, G has no vertices of degree ≥ 4 as well, and either

(a1) exactly one subgraph among the $\langle E_i\rangle$'s has vertices of degree 3, and there are exactly two such vertices, or

(a2) each subgraph $\langle E_i\rangle$ has at most one vertex of degree 3, and at most two subgraphs in $\{\langle E_i\rangle, i = 1, \ldots, 6\}$ have exactly one vertex of degree 3.

Since G^i is a 3-skein, we have $|E_i \cap E(x, G)| = 0$ for $x \in V^*G^i$ and $|E_i \cap E(x, G)| = 1$ or 3 for $x \in V^*G \setminus V^*G^i$. Therefore, in case (a1), we have
$$3|V^*G| = 5(|V^*G| - 2) + (|V^*G| - 4 + 6),$$
and hence $|V^*G| = 8/3$, a contradiction. In case (a2), we get
$$3|V^*G| = (6 - n)(|V^*G| - 2) + n(|V^*G| - 3 + 3),$$
where n is the number of the subgraphs $\langle E_i\rangle$ having a vertex of degree 3, and hence $n \leq 2$. Therefore, $|V^*G| = 4 - 2n/3$, which means that $n = 0$ and $|V^*G| = 4$. Since G is 2-connected, we have $G = \hom Q$, where Q is one of the graphs shown in Figure 1. Obviously, G satisfies the hypothesis of the lemma only if $Q = K_4$ and each of the $\langle E_i\rangle$'s is a thread in G. □

2.2. LEMMA. *Let*

(h1) *G be homeomorphic to one of the graphs F_i, $i = 1, 2, 3$, shown in Figure 2, and*

(h2) *$m: EG \to E\overline{G}$ be a S^3-semi-isomorphism of G onto $\overline{G}$.*

Then there exists a hom-*isomorphism of G onto $\overline{G}$ inducing m.*

PROOF. Let $G(T) = G(-)T$ for $T \in \mathscr{T}G$.

(p1) Obviously, for any $T \in \mathscr{T}G$, there exist $P_1, P_2 \in \mathscr{T}G \setminus \{T\}$ such that $G(P_i) = \hom K_4$ and $T \in \mathscr{T}G(P_i)$, $i = 1, 2$. By (h2) and Lemma 2.1, we

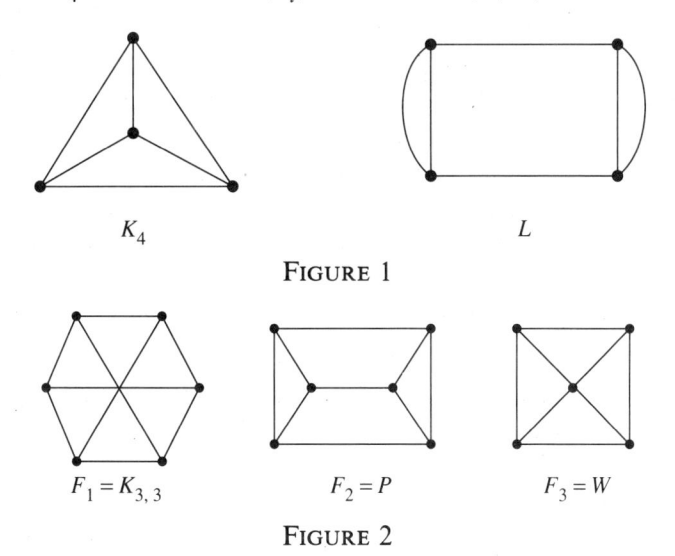

$K_4 \qquad\qquad L$

FIGURE 1

$F_1 = K_{3,3} \qquad F_2 = P \qquad F_3 = W$

FIGURE 2

have $\overline{G}(P_i) = \text{hom } K_4$ and $\overline{T} \in \mathscr{T}\overline{G}(P_i)$ for $i = 1, 2$. Since $E\overline{P}_1 \cap E\overline{P}_2 = \varnothing$, we get $\overline{T} \in \mathscr{T}\overline{G}$.

(p2) Let $x \in V^*G$ and $\mathscr{T}(x, G) = \mathscr{T}(x) = \{T_x^1, T_x^2, T_x^3\}$. Obviously, there exist $A_x^1, A_x^2 \in \mathscr{T}G$ such that $G(A_x^i) = \text{hom } K_4$, $i = 1, 2$, and $A_x^2 \notin \mathscr{T}(x)$, while $A_x^1 \in \mathscr{T}(x)$, say, $A_x^1 = T_x^1$. Then $T_x^2 \cup T_x^3 \in \mathscr{T}G(A_x^1)$ and $T_x^2, T_x^3 \in \mathscr{T}G(A_x^2)$. By (h2) and Lemma 2.1, we have $\overline{G}(A_x^i) = \text{hom } K_4$, $i = 1, 2$, and $\overline{T}_x^2 \cup \overline{T}_x^3 \in \mathscr{T}\overline{G}(A_x^1)$, $\overline{T}_x^2, \overline{T}_x^3 \in \mathscr{T}\overline{G}(A_x^2)$. Hence, the threads $\overline{T}_x^1, \overline{T}_x^2, \overline{T}_x^3$ have a common vertex, say, $\overline{x}$. Therefore, if $G = \text{hom } F_1$ or $\text{hom } F_2$, then a mapping $v^*: V^*G \to V^*\overline{G}$, where $v^*(x) = \overline{x}$, is obviously a bijection and can be extended to a hom-isomorphism of G onto $\overline{G}$ inducing m. Now let $G = \text{hom } F_3$. For the above reasons, $T \in \mathscr{T}G$ if and only if $\overline{T} \in \mathscr{T}\overline{G}$ and $|V^*\overline{G}| \geq 5$. From (h2), we derive $|VG| \geq |V\overline{G}|$, and hence $5 = |V^*G| \geq |V^*\overline{G}| \geq 5$, i.e. $|V^*\overline{G}| = 5$. Thus, a mapping $v^*: V^*G \to V^*\overline{G}$ such that $v^*(x) = \overline{x}$ for any vertex of degree 3 in G is clearly a bijection and can be extended to a hom-isomorphism of G onto $\overline{G}$ inducing m. □

2.3. Lemma 2.2 considers all homeomorphisms of 3-connected graphs with cyclomatic number 4 and states that their S^3-semi-isomorphisms are induced by hom-isomorphisms. Therefore, Theorem 5.2 from [1] along with Lemma 2.2 immediately yields

THEOREM. *Let*

(h1) *G be 3-connected and $G \neq \text{hom } K_4$, and*

(h2) *$m: EG \to E\overline{G}$ be a S^3-semi-isomorphism of G onto $\overline{G}$.*

Then there exists a hom-*isomorphism of G onto $\overline{G}$ inducing m.*

Obviously, this theorem does not hold for $G = K_4$ even if m is an S^3-isomorphism of G onto $\overline{G}$.

At the end of this section, we prove the more general Theorem 2.8.

2.4. LEMMA. *Let $EG = E_1 \cup \cdots \cup E_5$, $E_i \cap E_j = \varnothing$ for $i \neq j$, $i, j \in \{1, \ldots, 5\}$. Let $E_k \neq \varnothing$, and let $G_k = \langle EG \setminus E_k \rangle$ be a 3-skein for any $k \in \{1, \ldots, 4\}$ (G can be regarded as a graph whose edges are colored with colors $1, \ldots, 4$ if $E_5 = \varnothing$, or with colors $1, \ldots, 5$ if $E_5 \neq \varnothing$, such that upon deleting edges of any given color of the first four one obtains a 3-skein up to isolated vertices). Then G is homeomorphic to one of the colored graphs of Figure* 3.

PROOF. Let $E(x, G) = E(x)$.

(p1) Obviously, G is 2-connected, implying that the degree $|E(x)|$ of any vertex x in G is at least 2, and if $|E(y)| = 2$ then $E(y) \subseteq E_k$ for some $k \in \{1, \ldots, 4\}$. Since G_k is a 3-skein, we have $E_k \cap E(x) \neq \varnothing$ for any $x \in VG$ such that $|E(x)| \geq 4$, $k \in \{1, \ldots, 4\}$. Therefore:

(1) $|E(x)| \leq 4$ for $x \in VG$, and
(2) $|E(x, G_k)| = 3$ for any $k = 1, \ldots, 4$, provided $|E(x, G)| = 4$.

ON EDGE SEMI-ISOMORPHISMS OF GRAPHS 117

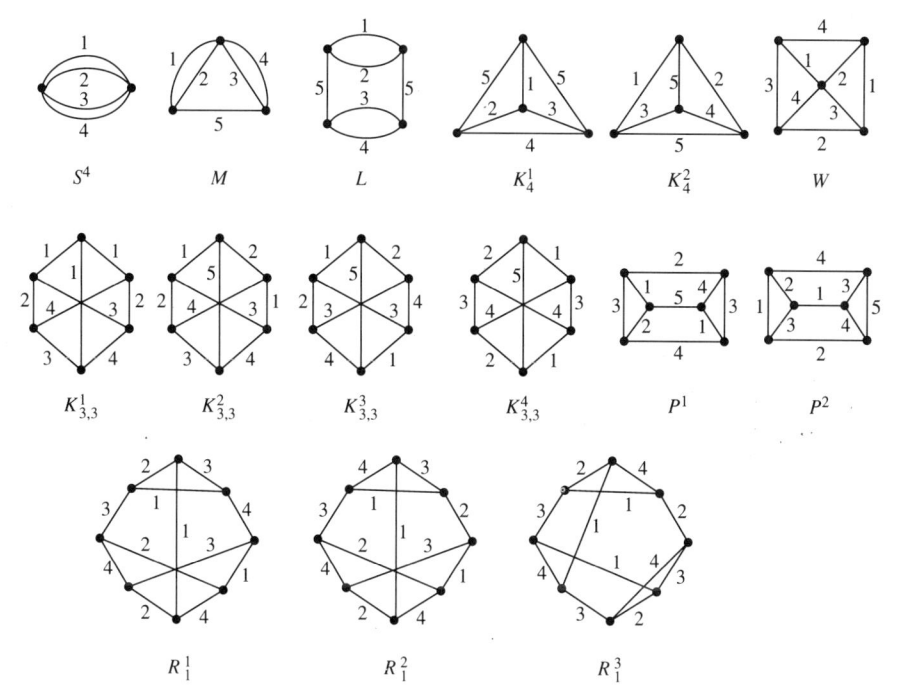

FIGURE 3. $K_{3,3}^4$ contains a two-color square, $K_{3,3}^3$ contains a three-color square with color 5 and no two-color square, and $K_{3,3}^2$ contains neither a two-color square nor a three-color square with color 5.

Thus, G has at most two vertices of degree 4.

(p2) Let us assume that G has exactly two vertices x_1, x_2 of degree 4; hence $E(x_i) \cap E_k \neq \emptyset$ for any $k = \{1, \ldots, 4\}$ and $i = \{1, 2\}$. Let us assume that G has a vertex y of degree 3. Then $E(y) \cap E_j = \emptyset$ for some $j \in \{1, \ldots, 4\}$, and hence G_j is not a 3-skein, a contradiction. Therefore, G has no vertices of degree 3. Let us assume that $E_5 \neq \emptyset$. Since $E_5 \cap E(x_i) = \emptyset$ for $i = 1, 2$ and G has no vertices of degree 3, there exists a vertex z of degree 2 in G such that $|E_5 \cap E(z)| = 1$. Let $E(z) \setminus E_5 \subseteq E_r$, $r \neq 5$. Then $|E(z, G_r)| = 1$, and hence G_r is not a 3-skein, a contradiction. Therefore $E_5 = \emptyset$. Then, obviously, it is exactly the 4-skein between the vertices x_1 and x_2 that satisfies the hypothesis of the lemma, and E_k, $k = 1, \ldots, 4$, is the edge set of one of its threads. Thus, G is homeomorphic to the colored graph S^4 in Figure 3.

(p3) Let us assume that G has exactly one vertex x of degree 4, and hence $E(x) \cap E_k \neq \emptyset$ for any $k \in \{1, \ldots, 4\}$. Then for each E_k there exists exactly one vertex z_k of degree 3 in G such that $E_k \cap E(z_k) = \emptyset$. Therefore, one can distinguish five cases:

(c1) G has exactly four vertices $z_1, \ldots, z_4$ of degree 3 and $E_k \cap E(z_k) = \emptyset$, and hence $E_5 \cap E(z_k) = \emptyset$, $k = 1, \ldots, 4$;

(c2) G has exactly three vertices of degree 3;

(c3) G has exactly two vertices z^1, z^2 of degree 3 and $E_i \cap E(z^1) \neq \emptyset$ only for $i = 2, 3, 5$ and $E_i \cap E(z^2) \neq \emptyset$ only for $i = 1, 4, 5$;

(c4) G has exactly two vertices z^1 and z^2 of degree 3, $E_i \cap E(z^1) \neq \emptyset$ only for $i = 2, 3, 4$ and $E_i \cap E(z^2) \neq \emptyset$ only for $i = 1, 5$, and hence $|E_5 \cap E(z^2)| = 2$.

(c5) G has exactly one vertex of degree 3.

(p3.1) Consider case (c1). Following the reasoning of (p2), we obtain $E_5 = \emptyset$, and therefore each $\langle E_k \rangle$, $k = 1, \ldots, 4$, consists of two disjoint threads $xT_k^1 y_k^1$ and $t_k^2 T_k^2 y_k^2$ in G. We can take $y_1^1 = z_4$, $t_1^2 = z_3$, and $y_1^2 = z_2$. Let us assume that $y_2^1 = z_4$ (the case $y_3^1 = z_4$ is similar). Then the threads T_1^1 and T_2^1 are parallel and G_3 is not a 3-skein, a contradiction. Therefore, z_4 is not an end-vertex of the threads T_2^1 and T_3^1: $y_2^1 \neq z_4$ and $y_3^1 \neq z_4$, and moreover, the threads T_i^1 and T_j^1 with $i \neq j$ are not parallel. Obviously, G_4 is a 3-skein between x and z_4. If $y_2^2 = y_3^2 = z_1$, or if $y_2^2 = z_3$, $y_3^2 = z_2$, then G_4 has a circuit that does not contain z_4, a contradiction. Hence, we can put $y_2^2 = z_3$ and $y_3^2 = z_1$. If $y_4^1 = z_1$, and hence $\{t_4^2, y_4^2\} = \{z_2, z_3\}$, then G_2 is not a 3-skein, a contradiction. Therefore, either $y_4^1 = z_3$, and then $\{t_4^2, y_4^2\} = \{z_1, z_2\}$, or $y_4^1 = z_2$, and then $\{t_4^2, y_4^2\} = \{z_1, z_2\}$. In both cases, G is homeomorphic to the colored graph W in Figure 3.

(p3.2) In case (c2), G has an odd number of vertices of odd degree, a contradiction.

(p3.3) In case (c3), each $\langle E_i \rangle$, $i = 1, \ldots, 5$, is a thread T_i in G. Thus, G is homeomorphic to the colored graph M of Figure 3.

(p3.4) In case (c4), $\langle E_5 \rangle$ is a circuit, and hence each G_k is not a 3-skein, a contradiction.

(p3.5) In case (c5), G has an odd number of vertices of odd degree, a contradiction.

(p4) Let us assume now that each vertex in G is of degree ≤ 3 (and ≥ 2, as indicated above). Since G_k is a 3-skein, there exist exactly two vertices x_k^1 and x_k^2 of degree 3 in G such that $E_k \cap E(x_k^i) = \emptyset$ for $i = 1, 2$, $k = 1, \ldots, 4$. Therefore $|V^*G| = 4, 6$ or 8.

(p4.1) Let us assume that $|V^*G| = 4$. Since G is 2-connected, G is homeomorphic to one of the colored graphs L, K_4^1, K_4^2 of Figure 3.

(p4.2) Let us assume that $|V^*G| = 6$. Since each E_i, $i = 1, \ldots, 5$, belongs to a 3-skein, $\langle E_i \rangle$ has at most two vertices of degree 3.

(p4.2.1) Let us assume that some $\langle E_k \rangle$, $k \in \{1, \ldots, 4\}$, has exactly two vertices z_1, z_2 of degree 3, say, $\langle E_1 \rangle$. Since G_r is a 3-skein for $r = 2, 3, 4$, we have $E(x) \cap E_r \neq \emptyset$ for any $x \in V^*G \setminus \{z_1, z_2\}$, and thus $E(x) \cap E_1 = \emptyset$. Hence, each vertex in $V^*G \setminus \{z_1, z_2\}$ is of degree 3 in G_1, and therefore G_1 is not a 3-skein, a contradiction.

(p4.2.2) Let us assume that each $\langle E_k \rangle$, $k = 1, \ldots, 4$, has at most one vertex of degree 3, while, say, $\langle E_1 \rangle$ has exactly one vertex z of degree 3. Let $r \in \{2, 3, 4\}$. Since G_r is a 3-skein, there exists exactly one vertex $x_r \neq z$ of degree 3 in G such that $E_r \cap E(x_r) = \emptyset$ (and then $E_r \cap E(x) \neq \emptyset$ for any $x \in V^*G \setminus \{x_r, z\}$). Let $y \in V^*G \setminus \{z, x_2, x_3, x_4\}$. Then $E_r \cap E(y) \neq \emptyset$ for every $r \in \{2, 3, 4\}$. Therefore every vertex in $V^*G \setminus \{z, x_2, x_3, x_4\}$ is of degree 3 in G_1. Since G_1 is a 3-skein, we have $|V^*G \setminus \{z, x_2, x_3, x_4\}| = 2$, and so all three vertices x_2, x_3, x_4 are distinct, and $E_1 \cap E(x_r) \neq \emptyset$ for any $r \in \{2, 3, 4\}$. Then, following the reasoning of (p2), we obtain $E_5 = \emptyset$. Thus, $\langle E_1 \rangle$ consists of three threads in G having exactly one common end-vertex z, while $\langle E_r \rangle$, $r = 2, 3, 4$, consists of two disjoint threads in G. Then G is homeomorphic to the colored graph $K_{3,3}^1$ in Figure 3.

(p4.2.3) Let us assume now that each $\langle E_k \rangle$, $k = 1, \ldots, 4$, has no vertices of degree 3. Then, clearly, $\langle E_k \rangle$ has no circuits, i.e. it consists of disjoint threads of the graph G. Since G_k is a 3-skein, there exist exactly two vertices x_k^1, x_k^2 of degree 3 in G such that $E_k \cap E(x_k^i) = \emptyset$, $i \in \{1, 2\}$. Therefore, each $\langle E_k \rangle$, $k = 1, \ldots, 4$, consists of two disjoint threads T_k^1, T_k^2, while $\langle E_5 \rangle$ is a thread $z_1 T_5 z_2$ in G, and hence $E_5 \neq \emptyset$. Let $V^*G \setminus \{z_1, z_2\} = \{y_1, \ldots, y_4\} = \mathscr{Y}$. We can assume that $E(z_1) \cap E_i \neq \emptyset$ only for $i = 1, 2, 5$. The following cases are distinguished:

(c1) $E(z_2) \cap E_i \neq \emptyset$ for $i = 3, 4, 5$, and hence $E(y_k) \cap E_i = \emptyset$ if and only if $i = k, 5$ for $k \in \{1, \ldots, 4\}$ with an appropriate numbering of the vertices in $\mathscr{Y}$;

(c2) $E(z_2) \cap E_i \neq \emptyset$ for $i = 2, 3, 5$, and hence $E(y_k) \cap E_i = \emptyset$ if and only if $i = k, 5$ for $k \in \{1, 3\}$ and $E(y_k) \cap E_i = \emptyset$ if and only if $i = 2, 5$ for $k \in \{2, 4\}$;

(c3) $E(z_2) \cap E_i \neq \emptyset$ for $i = 1, 2, 5$, and hence $E(y_k) \cap E_i = \emptyset$ if and only if $i = 1, 5$ for $k \in \{1, 3\}$ and $E(y_k) \cap E_i = \emptyset$ if and only if $i = 2, 5$ for $k \in \{2, 4\}$.

Now we can verify by simple exhaustion that G is homeomorphic to one of the colored graphs $K_{3,3}^2$, $K_{3,3}^3$, $K_{3,3}^4$, P^1, P^2 of Figure 3. Case (c1) refers to $K_{3,3}^2$, case (c2) to one of the graphs $K_{3,3}^3$, P^1, P^2, and case (c3) to $K_{3,3}^4$.

(p4.3) Let us assume now that $|V^*G| = 8$. Then $V^*G = \{x_k^1, x_k^2 : k = 1, \ldots, 4\}$, and $E_k \cap E(x_r^i) = \emptyset \iff r = k$ for $k \in \{1, \ldots, 4\}$ and $i \in \{1, 2\}$. As in (p2), we obtain $E_5 = \emptyset$. Now we can verify by exhaustion that G is homeomorphic to one of the colored graphs R_1^1, R_1^2, R_2 of Figure 3. □

Lemma 2.4 yields the following theorem, which is interesting by itself.

2.4′. THEOREM. (1) *Each of the 4-color graphs in Figure* 3 *has the property* (s) *the edges of three arbitrary colors form a 3-skein.*

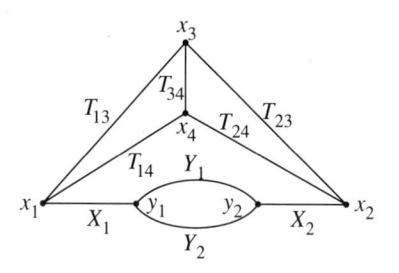

FIGURE 4

(2) *Every edge 4-colored graph with the property* (s) *is homeomorphic to one of the edge 4-color graphs in Figure 3.*

2.5. LEMMA. *Let*

(h1) G *be the graph shown in Figure 4, where* T_{ij}, $i, j \in \{1, \ldots, 4\}$, *and* Y^k, $k \in \{1, 2\}$, *are threads in* G, *while* X_k, $k \in \{1, 2\}$, *is either a thread or a "degenerate" thread consisting of one vertex* $x_k = y_k$, *and*

(h2) $m: EG \to E\overline{G}$ *be an* S^3-*semi-isomorphism of* G *onto* $\overline{G}$.

Then $\overline{T}_{ij}$, $i, j \in \{1, \ldots, 4\}$, *and* $\overline{Y}^k$, $k \in \{1, 2\}$, *are threads in* $\overline{G}$, *and the graph* $\overline{G}$ *is obtained from* G *by replacing* T_{ij} *by* $\overline{T}_{ij}$, Y^k *by* $\overline{Y}^k$, $k \in \{1, 2\}$, *and* X_k *by a thread* $\overline{Z}_k$, $k \in \{1, 2\}$, *such that* $E\overline{Z}_1 \cup E\overline{Z}_2 = E\overline{X}_1 \cup E\overline{X}_2$ (*and hence there exists a 2-isomorphism of* G *onto* $\overline{G}$ *inducing* m).

PROOF. Obviously, $G^i = G(-)Y^i = \hom K_4$, $i = 1, 2$. By Lemma 2.1, $\overline{G}^i = \hom K_4$ and if $T \in \mathcal{T}\overline{G}^i$, then $\overline{T} \in \mathcal{T}\overline{G}^i$. Put $T_{12}^i = X_1 \cup X_2 \cup Y^i$. Since $T_{12}^i, T_{p,q} \in \mathcal{T}G^i$ for $i \in \{1, 2\}$ and $\{p, q\} \neq \{1, 2\}$, we have $\overline{T}_{12}^i, \overline{T}_{p,q} \in \mathcal{T}\overline{G}^i$. Then each component of $\overline{T}_{12}^1 \cap \overline{T}_{12}^2$ is a path. Since $\overline{T}_{pq} \in \mathcal{T}\overline{G}^i \cap \mathcal{T}\overline{G}_2$, we see that $\overline{T}_{12}^1$ and $\overline{T}_{12}^2$ have the same pair of end-vertices, say, $\overline{x}_1$ and $\overline{x}_2$. Consider the 3-skein $S_k = T_{1k} \cup T_{2k} \cup T_{12}^1 \cup T_{12}^2$, $k \in \{3, 4\}$. By (h2), $\overline{S}_k$ is a 3-skein in $\overline{G}$. Hence $\overline{T}_{12}^1 \cup \overline{T}_{12}^2$ has at most one circuit, $\overline{T}_{12}^1$ and $\overline{T}_{12}^2$ are paths with the same end-vertices, and $\overline{T}_{12}^1 \cap \overline{T}_{12}^2 = \overline{X}_1 \cup \overline{X}_2$ consists of disjoint paths. Therefore, $\overline{Y}_1$ and $\overline{Y}_2$ are parallel threads in $\overline{G}$ (let $\overline{y}_1$ and $\overline{y}_2$ be the end vertices of the threads $\overline{Y}_1$ and $\overline{Y}_2$), $\overline{T}_{12}^i = \overline{Z}_1 \cup \overline{Z}_2 \cup \overline{Y}_i$, and so $\overline{T}_{12}^1 \cap \overline{T}_{12}^2 = \overline{Z}_1 \cup \overline{Z}_2$, where $\overline{Z}_1$ and $\overline{Z}_2$ are disjoint paths. Since $\overline{S}_k$ is a 3-skein, $k = 3, 4$, the threads $\overline{x}_1^k \overline{T}_{1k} \overline{x}_k^1$ and $\overline{x}_2^k \overline{T}_{2k} \overline{x}_k^2$ have a common end-vertex, say, $\overline{x}_k^1 = \overline{x}_k^2 = \overline{x}_k$, and $\{\overline{x}_1^3, \overline{x}_2^3\} = \{\overline{x}_1^4, \overline{x}_2^4\} = \{\overline{x}_1, \overline{x}_2\}$, say, $\overline{x}_i^3 = \overline{x}_i^4 = \overline{x}_i$, $i = 1, 2$. Since $\overline{G}^1 = \hom K_4$ and $\overline{T}_{pq} \in \mathcal{T}\overline{G}_1$, $\{p, q\} \neq \{1, 2\}$, we see that $\overline{x}_3$ and $\overline{x}_4$ are the end-vertices of the thread $\overline{T}_{34}$. □

ON EDGE SEMI-ISOMORPHISMS OF GRAPHS 121

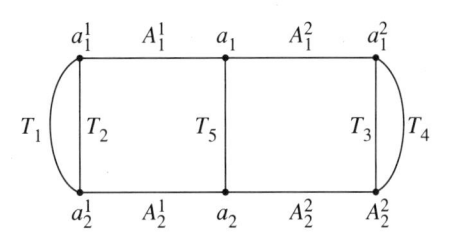

FIGURE 5

2.6. LEMMA. *Let*

(h1) *G be the graph in Figure 5, where* T_k, $k = 1, \ldots, 5$, *is a thread in G, while* A_j^i, $i, j \in \{1, 2\}$, *is either a thread or a "degenerate" thread consisting of one vertex* $a_j^i = a_j$, *and*

(h2) $m: EG \to E\overline{G}$ *be an* S^3-*semi-isomorphism of G onto* $\overline{G}$.

Then $\overline{T}_k$, $k = 1, \ldots, 5$, *is a thread in* $\overline{G}$, *and* $\overline{G}$ *is obtained from G by replacing* T_k *by* $\overline{T}_k$, $k = 1, \ldots, 5$, *and* A_j^i *by a thread* $\overline{B}_j^i$, $i, j \in \{1, 2\}$, $E\overline{B}_j^1 \cup E\overline{B}_2^i = E\overline{A}_1^i \cup E\overline{A}_2^i$, $i, j \in \{1, 2\}$ (*hence there exists a 2-isomorphism of G onto* $\overline{G}$ *inducing m*).

PROOF. (p1) Let $G_s = G(-)T_s$, $s = 1, \ldots, 5$, $A^r = A_1^r \cup A_2^r$, $r = 1, 2$, and $A = A^1 \cup A^2$. Obviously, $G_s = \hom\{S^4, L, M\}$ (Figure 1). We can assume that all threads in G_s are colored with different colors and the edge $\overline{e} = m(e)$ in $\overline{G}_s$ is colored with the same color as the edge e in G_s. Then, by (h2), $\overline{G}_s$ satisfies the hypothesis of Lemma 2.4. Therefore by Lemma 2.4, $\overline{G}_s$ is homeomorphic to one of the colored graphs of Figure 3. Denote by $R(G_s)$ the set of threads T in G_s such that $G_s(-)T$ is a 3-skein and put $\overline{R}(G_s) = \{\overline{T} : T \in R(G_s)\}$. Then for $s \neq 5$, we have

$$R(G_s) = \{T_5; T^s = A^1 \cup T_r, \text{ if } \{s, r\} = \{1, 2\};$$
$$T^s = A^2 \cup T_r \text{ if } \{s, r\} = \{3, 4\}; T_i \text{ for } i \in \{1, \ldots, 4\} - \{s, r\}\},$$

while $R(G_5) = \{T_k : k = 1, \ldots, 4\}$.

(p2) Let us assume that $\overline{G}_5$ is homeomorphic to K_4^1 or K_4^2: $\overline{G}_5 \in \hom\{K_4^1, K_4^2\}$, implying that $\overline{A} \neq \varnothing$ (see Figure 3). Then $\overline{T}_k$ is a thread in $\overline{G}_s$, $k = 1, \ldots, 4$, and $\overline{A}$ is the union of the two threads $x_1\overline{P}_1y_1$ and $x_2\overline{P}_2y_2$ in G_5: $\overline{A} = \overline{P}_1 \cup \overline{P}_2$. Put $\mathscr{D}_1 = \{L, M, K_4^1, K_4^2\}$ (Figure 3). Since each $\overline{T}_k$ is a path in $\overline{G}$ and $\overline{A} \neq \varnothing$, we have $\overline{G}_k \in \hom \mathscr{D}_1$ for $k = 1, \ldots, 4$ and $\overline{R}(G_k) \subseteq \mathscr{T}\overline{G}_k$.

(p2.1) Let us assume that $\overline{G}_5 = \hom K_4^1$, i.e. $\overline{P}_1$ and $\overline{P}_2$ have a common end-vertex, say, $x_1 = x_2 = x$. Let $\overline{T}_1$ be the third thread in $\overline{G}_5$ ending at x. As indicated above, $\overline{G}_1 \in \hom \mathscr{D}_1$ and $\overline{T}_1$ is a thread in $\overline{G}_1$. Obviously, $\overline{G}_1$ is obtained from $\overline{G}_5(-)\overline{T}_1$ by adding a new thread $\overline{T}_5$. However, the graph obtained in this way cannot both belong to $\hom \mathscr{D}_1$ and contain the thread $\overline{T}^1 = \overline{A}^1 \cup \overline{T}_2$, where $\overline{A}^1 \subseteq \overline{A} = \overline{P}_1 \cup \overline{P}_2$, a contradiction.

(p2.2) Let us assume that $\overline{G}_5 = \hom K_4^2$, i.e. $x_1 \overline{P}_1 y_1$ and $x_2 \overline{P}_2 y_2$ have no common end-vertices. Let x_1 and x_2 be the end-vertices of the thread $\overline{T}_1$ in $\overline{G}_5$. As indicated above, $\overline{G}_i \in \hom \mathscr{D}_1$ and for $i \in \{1, 2\}$, $\overline{T}_3, \overline{T}_4, \overline{T}_5$, and $\overline{T}^i = \overline{A}^1 \cup \overline{T}_j$ with $\{i, j\} = \{1, 2\}$ are threads in $\overline{G}_i$. Therefore, $\overline{T}_5$ is a thread in $\overline{G}$ with the end-vertices $x = x_i$ and $y \in \overline{P}_j$, while x_j and y_i are the end-vertices of the thread $\overline{T}_2$. Then either $\overline{T}_3$ or $\overline{T}_4$ is not a thread in $\overline{G}_2$, a contradiction.

(p3) Let us assume that $\overline{G}_5 \in \hom\{S^1, L, M\}$ (Figure 3). Then $\overline{R}(G_5) \subseteq \mathscr{T}(\overline{G}_5)$, and $\overline{A} = \overline{B}_1 \cap \overline{B}_2$, where $\overline{B}_1$ and $\overline{B}_2$ are paths in $\overline{G}_5$. Then each $\overline{G}_k \in \hom \mathscr{D}_1$ and $\overline{R}(G_k) \subseteq \mathscr{T}(\overline{G}_k)$. Obviously, $\overline{G}_k$ is obtained from $\overline{G}_5(-)\overline{T}_k$ by adding a new thread $\overline{T}_5$. Since for $\{i, j\} = \{1, 2\}$, $\overline{T}^i = \overline{A}^1 \cup \overline{T}_j$ is a thread in $\overline{G}_i$, and for $\{s, r\} = \{3, 4\}$, $\overline{T}^s = \overline{A}^2 \cup \overline{T}_r$ is a thread in $\overline{G}_s$, we see that $\overline{G}$ is obtained from G in the way described in the lemma.

(p4) Let us assume that $\overline{G}_5 \in \hom \mathscr{D}_2$, where $\mathscr{D}_2 = \{W, K_{3,3}^1\}$ (Figure 3). Then $E\overline{A} = \varnothing$ (implying $G = \hom S^5$) and $\overline{G}_k \in \hom \mathscr{D}_2$ for $k = 1, \ldots, 4$. Hence, $\overline{T}_5$ in $\overline{G}$ consists either of two nonadjacent threads, or of three threads with a common end-vertex. It is easy to verify that in any case, $\overline{G} = \overline{G}_5 \cup \overline{T}_5$ does not have the following property: $\overline{G}_k = \langle E\overline{G} \setminus E\overline{T}_k \rangle \in \hom \mathscr{D}_2$ for any $k \in \{1, \ldots, 4\}$, a contradiction.

(p5) Assume that $\overline{G}_5 \in \mathscr{D}_3$, where $\mathscr{D}_3 = \{K_{3,3}^2, K_{3,3}^3, K_{3,3}^4, P^1, P^2\}$ (Figure 3). Then $\overline{A}$ is a thread in $\overline{G}_5$ (hence $E\overline{A} \neq \varnothing$), and each $\overline{T}_k$, $k = 1, \ldots, 4$, consists of two nonadjacent threads in $\overline{G}_5$. Therefore, $\overline{G}_k \in \hom \mathscr{D}_3$ for $k = 1, \ldots, 4$. Then $\overline{T}_5$ consists of two nonadjacent threads in each $\overline{G}_k$, $k = 1, \ldots, 4$, and thus in G as well. Now, it is easy to verify that in any case the graph $\overline{G}$ obtained from $\overline{G}_5 \in \hom \mathscr{D}_3$ by adding two nonadjacent threads does not have the property $\overline{G}_k = \overline{G}(-)\overline{T}_k \in \hom \mathscr{D}_3$ for any $k = 1, \ldots, 4$, a contradiction.

(p6) Let us assume that $\overline{G}_5 \in \hom \mathscr{D}_4$, where $\mathscr{D}_4 = \{R_1^1, R_1^2, R_2\}$ (Figure 3). Then $E\overline{A} = \varnothing$, and each $\overline{T}_k$, $k = 1, \ldots, 4$, consists of three nonadjacent threads in $\overline{G}_5$. Therefore, $\overline{G}_k \in \hom \mathscr{D}_4$, $k = 1, \ldots, 4$, and hence $\overline{T}_5$ consists of three nonadjacent threads in $\overline{G}$. Since each of the $\overline{G}_k$'s belongs to $\hom \mathscr{D}_4$, we see that for any $k = 1, \ldots, 4$, $\overline{T}_k$ and $\overline{T}_5$ have the same set of vertices of degree 1. However, this is impossible, a contradiction. $\square$

2.7. LEMMA. *Let*

(h1) *G be the graph of Figure 6, where T_j^i is a thread in G, while A_j is either a thread or a "degenerate thread" consisting of one vertex, $i \in \{1, 2\}$, $j \in \{1, 2, 3\}$, and*

(h2) *$m: EG \to E\overline{G}$ be an S^3-semi-isomorphism of G onto $\overline{G}$.*

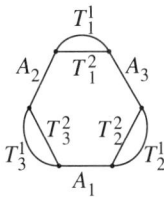

FIGURE 6

Then each of the $\overline{T}_j^i$'s is a thread in $\overline{G}$, and $\overline{G}$ is obtained from G by replacing each T_j^i by $\overline{T}_j^i$ and each A_i by a thread $\overline{B}_j$ such that $E(\overline{A}_1 \cup \overline{A}_2 \cup \overline{A}_3) = E(\overline{B}_1 \cup \overline{B}_2 \cup \overline{B}_3)$ (hence we obtain a 2-isomorphism of G onto $\overline{G}$ inducing m).

The proof is similar to that of Lemma 2.6.

PROOF. Let $G_j^i = G(-)T_j^i$, $i \in \{1, 2\}$, $j \in \{1, 2, 3\}$, and let $A = A_1 \cup A_2 \cup A_3$. We can regard all threads in G_j^i colored with different colors so that the edge $\overline{e} = m(e)$ in $\overline{G}_j^i$ is of the same color as the edge e in G_j^i. Obviously, $G_j^i \in \hom\{L, M\}$. Then, by (h2), $\overline{G}_s$ satisfies the hypothesis of Lemma 2.4. Therefore by Lemma 2.4, $\overline{G}_j^i$ is homeomorphic to one of the graphs of Figure 3 colored with 5 colors. Denote by $R(G_j^i)$ the set of threads T in G_j^i such that $G_j^i(-)T$ is a 3-skein, and let $\overline{R}(G_j^i) = \{\overline{T} : T \in R(G_j^i)\}$. Then $R(G_j^i) = \{T_r^k : k \in \{1, 2\}, r \in \{1, 2, 3\} - j\}$.

(p1) Let us assume that $\overline{G}_1^1 \in \hom \mathscr{D}^1$, where $\mathscr{D}^1 = \{L, M, K_4^1, K_4^2\}$. Then, by Lemma 2.4, $\overline{R}(G_1^1) \subseteq \mathscr{T}(\overline{G}_1^1)$ (and hence $\overline{R}(G_1^1) = R(\overline{G}_1^1)$), while $\overline{A} \cup \overline{T}_1^2$ is the union of at most two threads $x_1\overline{P}_1y_1$ and $x_2\overline{P}_2y_2$ in $\overline{G}_1^1$: $\overline{A} \cup \overline{T}_1^2 = \overline{P}_1 \cup \overline{P}_2$. Since each $\overline{T}$ in $\overline{R}(G_1^1)$ is a path, we have $\overline{G}_j^i \in \hom \mathscr{D}^1$ and $\overline{R}(G_j^i) \subseteq \mathscr{T}(\overline{G}_j^i)$ (so that $\overline{R}(G_j^i) = R(\overline{G}_j^i)$) for any $i \in \{1, 2\}$, $j \in \{1, 2, 3\}$. Therefore, each $\overline{T}_j^i$ is a thread in $\overline{G}$.

(p1.1) Let us assume that $\overline{G}_1^1 \in \hom\{L, M\}$. Let $\overline{T}_1^2 \subseteq \overline{P}_1$ (possibly, $E\overline{P}_2 = \emptyset$). Since $\overline{T}_2^1$ and $\overline{T}_1^2$ are threads in $\overline{G}_2^1 = \overline{G}_1^1(-)\overline{T}_2^1 \cup \overline{T}_1^1$, we see that $\overline{T}_1^1$ and $\overline{T}_1^2$ are parallel threads in $\overline{G}$. Then $\overline{G}_j^i \in \hom\{L, M\}$ for $i \in \{1, 2\}$ and $j \in \{1, 2, 3\}$, and, for the above reasons, $\overline{T}_j^1$ and $\overline{T}_j^2$ are parallel threads in $\overline{G}$. Then $\overline{G}$ is obtained from G by replacing each T_j^i by $\overline{T}_j^i$ and each A_j by a thread $\overline{B}_j$, where $\overline{B}_1 \cup \overline{B}_3 = \overline{P}_1(-)\overline{T}_1^2$ and $\overline{B}_2 = \overline{P}_2$, so that $E(\overline{A}_1 \cup \overline{A}_2 \cup \overline{A}_3) = E(\overline{B}_1 \cup \overline{B}_2 \cup \overline{B}_3)$.

(p1.2) Let us assume that $\overline{G}_1^1 = \hom K_4^1$, i.e. $E\overline{P}_i = \emptyset$, $i = 1, 2$, and $\overline{P}_1$ and $\overline{P}_2$ have a common end, say, $x_1 = x_2 = x$. Let, say, $x\overline{T}_2^1y$ be the third thread in $\overline{G}_1^1$ with the end x. Since $\overline{T}_r^i$, $i \in \{1, 2\}$, $r \in \{1, 3\}$, are threads in $\overline{G}_2^1 = \overline{G}_1^1(-)\overline{T}_2^1 \cup \overline{T}_1^1$, the end-vertices of the thread $\overline{T}_1^1$ are the vertex y and some vertex $z \in \overline{P}_1 \cup \overline{P}_2$, say, $z \in \overline{P}_1$. Then $\overline{T}_1^2 = z\overline{P}_1 y_1$. Let $\overline{P}$ be a thread in $\overline{G}$ with the end-vertices y_1 and y_2. Then $\overline{P} \in R(\overline{G}_1^1) = R(\overline{G}_1^2) \subseteq \mathcal{T}(\overline{G}_1^1) \cap \mathcal{T}(\overline{G}_1^2)$. However, $\overline{P}$ is not a thread in $\overline{G}_1^2 = \overline{G}(-)\overline{T}_1^2$, a contradiction.

(p1.3) Let us assume that $\overline{G}_1^1 = \hom K_4^2$, i.e. $EP_i \neq \emptyset$, $i = 1, 2$, and $\overline{P}_1$ and $\overline{P}_2$ have no common end-vertices. We can take x_1 and x_2 as the end-vertices of the thread $\overline{T}_2^1$. Let $\overline{F}_i$ be a thread in $\overline{G}_1^1$ with the end-vertices x_i and y_j, where $\{i, j\} = \{1, 2\}$. Obviously, at most one of $\overline{F}^1$ and $\overline{F}^2$ can fail to be a thread in $\overline{G}_2^1$. Therefore, $\overline{T}_1^1$ has its end-vertices at x_r and $y \in \overline{P}_s$, where $\{r, s\} = \{1, 2\}$. Then $\overline{T}_1^2 = y\overline{P}_s y_s$. Obviously, $\overline{F}_r \in R(\overline{G}_1^1) = R(\overline{G}_1^2) \subseteq \mathcal{T}(\overline{G}_1^1) \cap \mathcal{T}(\overline{G}_1^2)$. However, $\overline{F}_r$ is not a thread in $\overline{G}_1^r = \overline{G}(-)\overline{T}_1^2$, a contradiction.

(p2) Let $\overline{G}_1^1 \in \hom \mathscr{D}^2$, where $\mathscr{D}^2 = \{K_{3,3}^2, K_{3,3}^3, K_{3,3}^4; P^1, P^2\}$. According to Lemma 2.4, each $\overline{T}$ in $\overline{R}(G_1^1)$ consists of two nonadjacent threads in $\overline{G}_1^1$. Therefore, $\overline{G}_j^i \in \hom \mathscr{D}^2$ and $\overline{T}_j^i$ is a thread in $\overline{G}$ for $i \in \{1, 2\}$ and $j \in \{1, 2, 3\}$. Then $\overline{G}_2^1$ is obtained from $\overline{G}_1^1$ by deleting the two nonadjacent threads that form $\overline{T}_1^2$ and by adding the two nonadjacent threads that form $\overline{T}_1^1$. It is easy to see that the colored graph obtained in this way does not belong to $\hom \mathscr{D}^2$, a contradiction.

(p3) The remaining two cases $\overline{G}_1^1 \in \hom \mathscr{D}^3 = \{W, K_{3,3}^1\}$ and $\overline{G}_1^1 \in \hom \mathscr{D}^4 = \{R_1^1, R_1^2, R_2\}$ are similar, respectively, to (p4) and (p5) in the proof of Lemma 2.6. □

2.8. Lemmas 2.2, 2.5, 2.6, 2.7 consider all graphs homeomorphic to 2-connected graphs and having cyclomatic number 4, and claim that their S^3-semi-isomorphisms are induced by 2-isomorphisms. Therefore, Theorem 5.2 from [1] and Lemmas 2.2, 2.5, 2.6, and 2.7 immediately yield the following strengthening of the Whitney theorem on graph 2-isomorphisms [3]:

THEOREM. *Suppose that*

(h1) *G is 2-connected,*

(h2) *$c(G) \geq 4$ (i.e. G is neither a circuit nor a 3-skein and G is not a member of $\hom\{S^4, L, M, K_4\}$, Figure 3), and*

(h3) *$m: EG \to E\overline{G}$ is an S_1^3-semi-isomorphism of G onto $\overline{G}$.*

Then there exist a sequence t of hammock switchings in G and an isomorphism $v: V(t(G)) \to V\overline{G}$ such that v induces m (i.e. there exists a 2-isomorphism vt of the graph G onto $\overline{G}$ inducing m).

2.8.1. REMARK. Lemma 2.4 implies that the assertion of this theorem does not hold for any graph with $c(G) = 3$. If we replace hypothesis (h3) of the theorem by

(h3)' $m: EG \to E\overline{G}$ is an S^3-isomorphism of G onto $\overline{G}$,

then the assertion of the theorem holds for $G \in \hom\{S^4, L, M\}$, but is still false for $G = \hom K_4$.

2.8.2. REMARK. Since a graph circuit isomorphic to a 3-skein is a 3-skein (see (p2) in the proof of Theorem 4.2 in [1]), the Whitney theorem [3] on circuit isomorphisms of 2-connected graphs follows immediately from this theorem.

§3. On K_4-semi-isomorphisms of 3-connected graphs

3.1. Following the scheme of the proof of Theorem 2.3 and making use of Theorem 3.5 from [1], one can easily prove the following statement:

THEOREM. *Let G be a* hom-*cubic and 4-blockade (i.e. cyclically 4-connected) graph and let $m: EG \to E\overline{G}$ be a K_4-semi-isomorphism of G onto $\overline{G}$. Then there exists a* hom-*isomorphism of G onto $\overline{G}$ inducing m.*

3.2. THEOREM. *Let G be homeomorphic to a cubic 3-connected graph and let $\overline{G}$ be a planar graph. Then any K_4-semi-isomorphism of G to $\overline{G}$ is induced by a* hom-*isomorphism of G onto $\overline{G}$ if and only if either $c(G) \geq 6$ (and G is planar) or G is homeomorphic to the skeleton K of the cube ($c(K) = 5$); all the hypotheses of the theorem are essential.*

Using Theorem 5.2 from [1], the proof of this theorem is reduced to the case when $c(G) \leq 6$.

§4. On S^n-isomorphisms of hammock-like graphs

The present section provides some sufficient conditions for graph S^n-isomorphisms to be induced by their isomorphism.

4.1. LEMMA. *Let the edges of a graph G be colored with $k + 1$ colors, $k \geq 4$, such that the edges of any k colors form a k-skein, i.e. $EG = \bigcup\{E_i, i = 1, \ldots, k+1\}$, $E_i \cap E_j = \varnothing$ for $i \neq j$, and $S_i = \langle EG \setminus E_i \rangle$ is a k-skein for $i = 1, \ldots, k+1$. Then G is a $(k+1)$-skein and each thread in G is colored with its own color (i.e. $\langle E_i \rangle$ is a thread in G).*

PROOF. Let us prove that $\langle E_j \rangle$ is a thread in S_i, $i \neq j$, i.e. that all edges of one thread in S_i are colored with the same color. Let us assume the opposite. Then a thread in S_i contains two edges e_r and e_s of different colors r and s having a common end x: $e_r \in E_r$, $e_s \in E_s$. Then $E(x, G) \setminus \{e_r, e_s\} \subseteq E_i$. If $|E(x, G)| = 2$ or k, then $|E(x, S_r)| = 1$ or $k-1$. Since $k \geq 4$, S_r is not a k-skein, a contradiction. If $3 \leq |E(x, G)| \leq k-1$ or $|E(x, G)| \geq k+1$, then $|E(x, S_j)| = |E(x, G)|$ for $j \neq i, r, s$. Since

$k \geqslant 4$, we observe that S_j is not a k-skein, a contradiction. Thus G is a $(k+1)$-skein and $\langle E_i \rangle$ is a thread in G for $i = 1, \ldots, k+1$. □

Observe that the above lemma naturally supplements Theorem 2.4'.

4.2. Lemma 2.4 yields the following lemma, which can be easily proved directly:

LEMMA. *Let* $m: S^4 \to G$ *be an* S^3-*isomorphism. Then there exists a homisomorphism of* S^4 *onto* G *inducing* m *(and so if* T *is a thread of* S^4, *then* $m(T)$ *is a thread of* G).

4.3. A graph G is called (x, y)-hammock-like, $x, y \in VG$, if

(1) $G = G_1 \cup \cdots \cup G_h$, $G_i \cap G_j = \{x, y\}$ for $i \neq j$, and

(2) if G_i is neither a 3-connected graph nor a path, then $G_i \cup u$ is 3-connected, where $u = (x, y)$.

By using Lemmas 4.1 and 4.2, we prove the following theorem.

THEOREM. *Suppose that*

(h1) G *is a* $\hom(x, y)$-*hammock-like graph*, $x, y \in VG$,

(h2) *the maximum number* $k(x, y; G)$ *of inner disjoint paths connecting* x *and* y *is at least* $n + 1 \geqslant 4$, *and*

(h3) $m: EG \to E\overline{G}$ *is an* S^h-*isomorphism of* G *onto* $\overline{G}$.

Then there exists a sequence t *consisting of at most* $k(x, y; G)$ *hammock* (x, y)-*switchings such that* m *is induced by a* hom-*isomorphism of* $t(G)$ *onto* $\overline{G}$.

The particular case of this theorem when G is a 3-connected graph was proved in a different way (namely, by reducing to the Whitney theorem on circuit isomorphisms of 3-connected graphs) in [5].

The results of the present paper were reported at the Seminar on Discrete Mathematics at the Institute of Control Problems (November, 1981) and at the Georgian Republican Workshop "Methods of Optimization on Networks and Graphs" (Batumi, June, 1982); some of them were briefly described in [6].

REFERENCES

1. A. K. Kelmans, *On edge mappings of graphs preserving subgraphs of a given type*, Models and Algorithms of Operations Research and Their Application to the Organization of Work in Computing Systems, Izdat. Yaroslav. Univ., Yaroslavl, 1984, pp. 19–30; English transl. in this volume.
2. Hassler Whitney, *Congruent graphs and the connectivity of graphs*, Amer. J. Math. **54** (1932), 150–168.
3. _____, *2-isomorphic graphs*, Amer. J. Math. **55** (1933), 245–254.
4. F. Harary, *Graph theory*, Addison–Wesley, Reading, Mass., 1969.
5. R. L. Hemminger and H. A. Jung, *On n-skein isomorphisms of graphs*, J. Combin. Theory Ser. B **32** (1982), 103–111.
6. A. K. Kelmans, *Homeomorphic embeddings of graphs with given properties*, Dokl. Akad. Nauk SSSR **274** (1984), 1298–1303; English transl. in Soviet Math. Dokl. **29** (1984).

Constructions of Cubic Bipartite 3-connected Graphs without Hamiltonian Cycles

A. K. KELMANS

1

Many combinatorial problems are known to be NP-complete. They are the "most complicated" in the large class NP of the so-called polynomially verifiable problems. If a polynomial algorithm is found to solve an NP-complete problem, then a polynomial algorithm can be found to solve any problem in NP. One of the reasons why an NP-complete problem is difficult is that nobody knows a "polynomial" certificate for the nonexistence of a solution of the problem. In this respect, the study of a variety of sufficient "polynomial reasons" for the nonexistence of a solution to an NP-complete problem seems valuable. From this point of view, the present paper treats the NP-complete problem on finding a Hamiltonian circuit in a graph. Some nontrivial sufficient "polynomial reasons" for the nonexistence of a solution are revealed; namely, infinite series of cubic bipartite 3-connected graphs without Hamiltonian circuits are constructed. The constructions given here strengthen earlier results and disprove some related conjectures.

Let $\mathscr{C}$ be the set of all cubic bipartite 3-connected graphs. Tutte [3] proposed the following conjecture: all graphs in $\mathscr{C}$ have a Hamiltonian circuit (such graphs are usually called Hamiltonian). Horton [1] constructed a counterexample to this conjecture. His non-Hamiltonian graph from $\mathscr{C}$ has 96 vertices, and the corresponding construction uses essentially nontrivial 3-cuts (3-blockades) in the graph. The following questions naturally arise:

(1) Do there exist non-Hamiltonian graphs in $\mathscr{C}$ with fewer vertices?
(2) Do there exist non-Hamiltonian graphs in $\mathscr{C}$ having no 3-blockades?

The constructions described below (see 18 and 24) give positive answers to both questions. Namely, they provide an infinite set $\mathscr{H}$ of non-Hamiltonian

---

1991 *Mathematics Subject Classification.* Primary 05C38.

Translation of Analysis of the Problems of Formulation and Choice of Alternatives (The Problem of Choice on Even and Odd Data), vyp. 10, Sb. Trudov Vsesoyuz. Nauchno-Issled. Inst. Sistem. Issled., Moscow, 1986, pp. 64–72.

graphs in $\mathscr{C}$ without 3-blockades. The minimum graph of this construction has 50 vertices, and for any even number $p \geqslant 50$ there exist graphs in $\mathscr{H}$ with p vertices (obviously, any cubic graph has an even number of vertices). Therefore, any graph in $\mathscr{H}$ is a counterexample to the conjecture from [2].

It was shown by computer search [3] that all graphs in $\mathscr{C}$ having 22 or fewer vertices are Hamiltonian. Together with M. V. Lomonosov, we proved, without any computer assistance, the following four facts:

(a1) Any graph G in $\mathscr{C}$ with the number of vertices $|VG| \leqslant 30$ has a Hamiltonian circuit.

(a2) For any pair (G, e) such that $G \in \mathscr{C}$, $e \in EG$, and $|VG| \leqslant 24$ ($|VG| \leqslant 22$), there exists a Hamiltonian circuit in G containing (respectively, avoiding) e.

(a3) For any pair $(G, (x, y))$ such that $G \in \mathscr{C}$, $\{x, y\} \subset EG$, $|VG| \leqslant 18$, and $(G, (x, y)) \neq (M, (x, y))$ (see Figure 8 below), there exists a Hamiltonian circuit in G containing $\{x, y\}$; the graph M has no Hamiltonian circuit containing $\{x, y\}$.

(a4) For any pair $(G, (x, y))$ such that $G \in \mathscr{C}$, $\{x, y\} \subset EG$, $|VG| \leqslant 16$ and $(G, (x, y)) \neq (W, (a, b))$ (see Figure 6 below), there exists a Hamiltonian circuit in G containing x and avoiding y; the graph W has no Hamiltonian circuit containing a and avoiding b (and, by symmetry, containing b and avoiding a).

No non-Hamiltonian planar graphs in $\mathscr{C}$ are known. Barnette [4] proposed the following conjecture,

(B) All planar graphs in $\mathscr{C}$ are Hamiltonian.

In 17 and 18 we show that (B) is equivalent to a stronger statement,

(K) For any planar graph G in $\mathscr{C}$ and for two edges x and y belonging to the same circuit-face in G, there exists a Hamiltonian circuit of the graph G containing x and avoiding y (as well as containing both x and y).

The results of this paper were reported at the Seminar on Discrete Mathematics at the Institute of Control Sciences (December, 1985).

2

We need the following notions and notation.

2.1. We consider *undirected graphs* without loops or isolated vertices (see [1] and [5]). Let VG and EG denote the sets of vertices and edges of a graph G, respectively, and let $V(x, G)$ denote the set of all vertices of G adjacent to a vertex x.

A graph G is *cubic* if all its vertices are of degree 3.

A graph G is *bipartite* if the set of its vertices VG can be split into two parts X and Y (colored with two colors) such that every edge of G connects vertices of different parts.

A *matching* in G is a set of edges of G that are pairwise nonincident. A matching X in G is said to be *perfect* if every vertex of G is an end-vertex of an edge in X.

A graph G is said to be *planar* if it can be embeded in the sphere in such a way that distinct vertices of G become distinct points of the sphere, each edge of G is a segment of a Jordan curve on the sphere connecting the points that correspond to the end-vertices of this edge, and no pair of segment-edges has common points other than the ends.

A graph G is 2-*connected* if $|VG| \geq 2$, the deletion of every vertex from G results in a connected graph, and for $|VG| = 2$ the graph G has parallel edges. A graph G is k-*connected*, $k \geq 3$, if G has no parallel edges, $|VG| \geq k+1$, and the deletion of $k-1$ (or fewer) arbitrary vertices from G results in a connected graph.

As above, $\mathscr{C}$ denotes the set of all cubic, bipartite, and 3-connected graphs.

A *blockade* of a connected graph G is a set of edges $A \subseteq EG$ such that $G \backslash A$ is disconnected and consists of two components having circuits. A graph G is a k-*blockade graph* if G has blockades and each blockade contains at least k edges. Let $\mathscr{C}_k$ denote the set of all cubic, bipartite, and k-blockade graphs.

As usual, K_4 is the complete graph on four vertices, and $K_{3,3}$ the complete bipartite graph with parts of cardinality 3. Obviously, K_4 is the minimum 3-connected graph, and $K_{3,3}$ is the minimum bipartite 3-connected graph. It is also clear that both K_4 and $K_{3,3}$ are cubic and have no blockades, so $K_{3,3} \notin \mathscr{C}_k$ and $\mathscr{C}_3 = \mathscr{C} \backslash \{K_{3,3}\}$.

A subgraph H of a graph G is called an α-*subgraph* if H has a property α. For $x \in EG$, an x^+-*subgraph* (x^--*subgraph*) is a subgraph containing (respectively, avoiding) the edge x.

We focus on properties α that are derived from the properties x^+ and y^-, $x, y \in EG$, by the operations of disjunction $\vee$ and conjunction $\wedge$. In particular, $x^{\pm} \wedge y^{\mp} = (x^+ \wedge y^-) \vee (x^- \wedge y^+)$, and for $A \subseteq EG$, $A^+ = \bigwedge\{x^+ : x \in A\}$. In other words an $x^{\pm} \wedge y^{\mp}$-subgraph is a subgraph containing exactly one edge of x, y, while an A^+-subgraph is a subgraph containing A.

A *circuit* is a connected graph having all vertices of degree 2. A *Hamiltonian circuit* of a graph G is a circuit in G containing all its vertices. A graph G is α-*Hamiltonian* if it contains an α-Hamiltonian circuit, and α-*non-Hamiltonian* otherwise.

2.2. A *four-pole* $H^* = [H, (x_1, x_2, y_1, y_2)]$ is a graph H having four specified and ordered vertices x_1, x_2, y_1, y_2 of degree 1 called *terminals*. If all other vertices in H are of degree 3, then the four-pole H is said to be *cubic*.

A four-pole $[H, (x_1, x_2, y_1, y_2)]$ is said to be *bipartite* if
(b1) H is a bipartite graph, and
(b2) x_1 and x_2, as well as y_1 and y_2, belong to different parts of H.

A four-pole $[H, (x_1, x_2, y_1, y_2)]$ is said to be *planar* if the graph H^S obtained from H by adding the square S of four new edges (x_1, x_2), (x_2, y_1), (y_1, y_2), and (y_2, x_1) is planar and S is a circuit-face in H^S.

A cubic four-pole $H^* = [H, (x_1, x_2, y_1, y_2)]$ is said to be a *k-blockade graph* (a *strongly k-blockade graph*), $k \geq 3$, if the graph H^p obtained from H by identifying the vertices x_1 and x_2 with a new vertex, say x', and the vertices y_1 and y_2 with a new vertex, say y', and adding a new edge $p = (x', y')$ satisfies the following conditions:

(c1) H^p is a $(k-1)$-blockade graph (respectively, a k-blockade graph).
(c2) Any $(k-1)$-blockade in H^p contains the edge p.

A four-pole $[H, (x_1, x_2, y_1, y_2)]$ is *trivial* if it has exactly two nonincident edges, each of them connecting $\{x_1, x_2\}$ with $\{y_1, y_2\}$.

Let $x, y \in EB$. A pair $[B, (x, y)]$ is called *planar* if B is a planar graph and the edges x, y belong to the same circuit-face in B.

A four-pole $[H, (x_1, x_2, y_1, y_2)]$ is said to be $x^+ \alpha y^-$-*nontraceable* if $H \setminus \{y_1, y_2\}$ has no α-path containing $V(H \setminus \{y_1, y_2\})$.

A four-pole H^* is said to be $x^{\pm} \alpha y^{\mp}$-*nontraceable* if it is $x^+ \alpha y^-$-nontraceable and $x^- \alpha y^+$-nontraceable.

A four-pole H^* is said to be $x^+ \alpha y^+$-*nontraceable* if H has no α-subgraph Q with $VQ = VH$ consisting of two disjoint paths, each beginning at $\{x_1, x_2\}$ and ending at $\{y_1, y_2\}$.

A four-pole H^* is said to be *semi-α-nontraceable* if it is $x^{\pm} \alpha y^{\mp}$-nontraceable and $x^+ \alpha y^+$-nontraceable.

A four-pole $[H, (x_1, x_2, y_1, y_2)]$ is said to be *totally α-nontraceable* if

(a1) for every pair $x, y \in \{x_1, x_2, y_1, y_2\}$, the graph $H \setminus \{x, y\}$ has no α-path that contains $V(H \setminus x \setminus y)$, and

(a2) H has no α-subgraph Q with $VQ = VH$ consisting of two disjoint paths.

If $\alpha = \varnothing$, then, instead of α-nontraceable, $(x^+ \alpha y^-)$-nontraceable, etc., we say just *nontraceable*, (x^+, y^-)-*nontraceable*, etc.

2.3. Consider two four-poles $B^i = [H^i, (u_1^i, u_2^i, v_1^i, v_2^i)]$, $i = 1, 2$. Let G^1 be a cubic graph, $H^1 \subset G^1$, and $H^2 \cap G^1 = \varnothing$. We say that G^2 is obtained from G^1 by *replacing the four-pole* B^1 *with the four-pole* B^2 if G^2 is obtained from $H^2 \cup G^1 \setminus (H^1 \setminus \{u_1^1, u_2^1, v_1^1, v_2^1\})$ by identifying the vertices u_j^1 and u_j^2, as well as v_j^1 and v_j^2, $j = 1, 2$. Let e be an edge in G with the end-vertices x, y, and let $\{y, x_1, x_2\}$ and $\{x, y_1, y_2\}$ be the sets of the vertices in G adjacent to x and y, respectively. Let H_e denote the graph with $VH_e = \{x, y, x_1, x_2, y_1, y_2\}$ and $EH_e = \{(x, y), (x_1, x), (x_2, x), (y_1, y), (y_2, y)\}$. The graph $H_e^* = [H_e, (u_1, u_2, v_1, v_2)]$ is said to be a *four-pole of the edge* e in G if either $\{u_1, u_2\} = \{x_1, x_2\}$ and $\{v_1, v_2\} = \{y_1, y_2\}$, or $\{u_1, u_2\} = \{y_1, y_2\}$ and $\{v_1, v_2\} = \{x_1, x_2\}$ (and so there exist eight different four-poles of an edge e). *To replace an edge e in a graph G by a four-pole B* means to replace one of the four-poles of the edge e in G by B.

3

Since the number of vertices of a cubic four-pole is even, the following lemma is true.

LEMMA. *Let $[H, (x_1, x_2, y_1, y_2)]$ be a cubic bipartite four-pole, and let $\{x, y\}$ be a pair of its terminals of the same color. Then the graph $H \setminus \{x, y\}$ has no path containing all of its vertices.*

4

It is easy to prove the following lemma.

LEMMA. *Let G^1 be a graph or a four-pole, and e be an edge in G^1. Let G^2 be obtained from G^1 by replacing the edge e by a four-pole H (and so if G^1 is a four-pole, then e is a nonterminal edge in G^1).*

(a1) *If G^1 and H are cubic, so is G^2.*

(a2) *If G^1 and H are k-blockade graphs with $k \geq 3$, so is G^2.*

(a3) *If G^1 and H are planar, then there exists a replacement of e by H such that the resulting graph G^2 is planar.*

(b1) *If G^1 is α^1-non-Hamiltonian and H is semi-α-nontraceable, then G^2 is α^2-non-Hamiltonian with $\alpha^2 = \alpha^1 \wedge \alpha$.*

(b2) *If the four-terminal graph G^1 is $x^+\alpha^1 y^\pm$-nontraceable and H is semi-α-nontraceable, then the four-pole G^2 is $x^+\alpha^2 y$-nontraceable with $\alpha^2 = \alpha^1 \wedge \alpha$.*

5

It is easy to prove the following lemma.

LEMMA. *Let G^1 be a graph (or a four-pole), and G^2 be obtained from G^1 by replacing H^1 in G^1 with H^2.*

(a1) *If G^1 and H^2 are cubic, so is G^2.*

(a2) *If G^1 is a 3-blockade graph and H^2 is a strongly 3-blockade graph, then G^2 is a 3-blockade graph.*

(a3) *If G^1 and H^2 are 4-blockade graphs, so is G^2.*

(a4) *If G^1, H^1, and H^2 are planar, so is G^2.*

(b) *If H^2 is totally α-nontraceable, then G^2 is α-non-Hamiltonian (totally α-nontraceable, respectively).*

6

Let A and B be disjoint graphs, $a \in VA$, $V(a, A) = \{a_1, a_2, a_3\}$, and $V(b, B) = \{b_1, b_2, b_3\}$. Let $p : V(a, A) \to V(b, B)$ be a bijection. Let $AapbB$ denote the graph obtained from $(A \setminus a) \cup (B \setminus b)$ by adding three new edges (x, px), $x \in V(a, A)$, i.e. $AapbB = (A \setminus a) \cup (B \setminus b) \cup \{x, px : x \in V(a, A)\}$ (Figure 1).

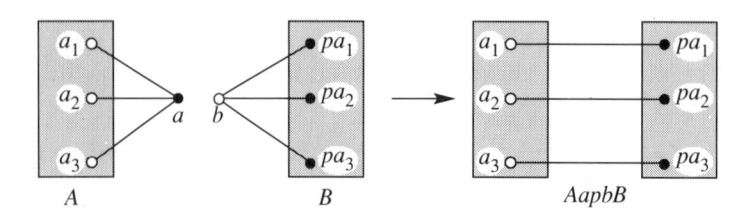

FIGURE 1

It is easy to prove the following lemma.

LEMMA. (a) *Let* Pr *be a property from the list {cubic, 3-connected}. Then AapbB has the property* Pr *if and only if A and B do.*
  (b1) *If A and B are bipartite, so is AapbB.*
  (b2) *If AapbB is bipartite and cubic, so are both A and B.*
  (p1) *If both A and B are planar, so is AapbB.*
  (p2) *Let* $AapbB \setminus \{(x, px) : x \in V(a, A)\}$ *consist of two components. If AapbB is planar, so are A and B.*
  (h1) *AapbB is Hamiltonian if and only if A is* $(x, a)^-$-*Hamiltonian for some* $x \in V(a, A)$, *and B is* $(px, b)^-$-*Hamiltonian.*
  (h2) *Let* $V(a, A) = \{x, y, z\}$. *If A is* $(x, a)^-$-*non-Hamiltonian and B is* $(py, b)^-$-*non-Hamiltonian, then AapbB is* $(z, pz)^+$-*non-Hamiltonian (this follows from* (h1)).

7

Let x and y be distinct edges of a graph H. Denote by $H_i^* = [H_i, (x_1^*, x_2^*, y_1^*, y_2^*)]$, $i = 1, 2$, the four-poles obtained from the pair $[H, (x, y)]$ as shown in Figure 2. Put $H_1^* = I_x(H_0^*) = I(H_0^*)$, $H_0^* = C[H, (x, y)]$, and $H_2^* = O[H, (x, y)]$. It is easy to prove the following lemma.

LEMMA. (a1) *Let* Pr *be a property from the list {cubic, bipartite, planar}. Then* $[H, (x, y)]$ *has the property* Pr $\Leftrightarrow H_0^*$ *does* $\Leftrightarrow H_j^*$, $j = 1, 2$, *do.*
  (a2) *Let* $k \in \{3, 4\}$. *If* H_0^* *is a k-blockade graph, so are* H_j^*, $j = 1, 2$.
  (a3) *If H is a 3-blockade graph, then* H_j^*, $j = 1, 2$, *are strongly 3-blockade graphs. If H is a 4-blockade graph, and the edges x, y are non-incident, then* H_0^* *is a strongly 4-blockade graph, and* H_j^*, $j = 1, 2$, *are 4-blockade graphs.*
  (b1) *H is* $x^+\alpha y^-$-*non-Hamiltonian* $\Leftrightarrow H_0^*$ *is* $\Leftrightarrow H_1^*$ *is.*
  (b2) *H is* $x^+\alpha y^+$-*non-Hamiltonian if and only if* H_0^* *is* $x^+\alpha y^+$-*nontraceable if and only if* H_1^* *is* $x^+\alpha y^+$-*nontraceable* (H_1^* *is* $x^-\alpha y^+$-*nontraceable if and only if* H_2^* *is semi-*α-*nontraceable*).

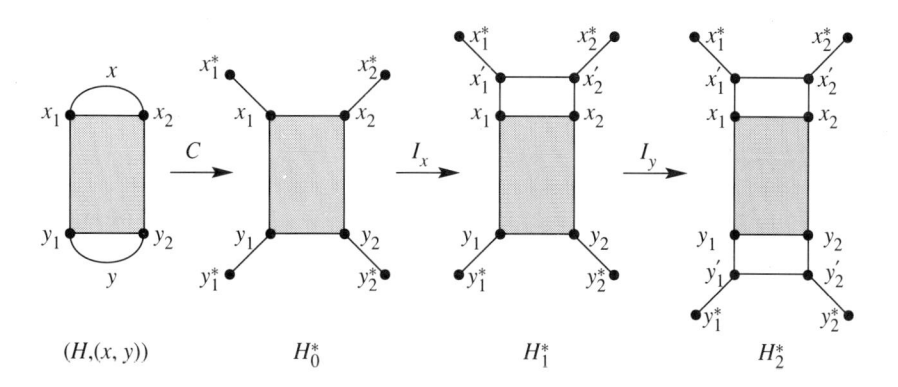

FIGURE 2

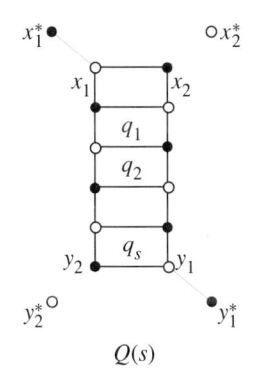

FIGURE 3

8

It is easy to prove directly, or to derive from Lemma 7, the following result.

LEMMA. *The four-pole* $Q^*(s) = [Q(s), (x_1^*, x_2^*, y_1^*, y_2^*)]$ *of Figure* 3 *is a cubic, bipartite, planar, 4-blockade, and semi-* q_i^+-*nontraceable,* $i = 1, \ldots, s$.

9

Let G be a cubic graph and X a matching in G. Let $\{H_x : x \in X\}$ be a set of bipartite 3-blockade four-poles. The matching X is said to be *replaceable* in G if there exists a collection of replacements of each edge by its own four-poles H_x in G such that the final graph is bipartite. Obviously, the property of a matching X to be replaceable in G does not depend on the set $\{H_x : x \in X\}$. It is easy to prove the following lemma.

LEMMA. *Any perfect matching in a cubic graph* G *is replaceable in* G.

10

By the Petersen theorem [5], any cubic 3-connected graph has a perfect matching. Therefore, the following statement is derived from 9.

PROPOSITION. *Any cubic 3-connected graph has a replaceable matching.*

The following problem may be interesting in itself: find a minimum replaceable matching in a cubic 3-connected graph.

11

Any perfect matching in a cubic 3-connected graph G has a common edge with any 3-blockade in G. Therefore, the following statement is derived from 9.

PROPOSITION. *Let G be a cubic 3-connected graph and X be a perfect matching in G. Let $\{H_x : x \in X\}$ be a set of four-poles, and let a graph G' be obtained from G by replacing each edge x in X by its own four-pole H_x. If each of the H_x's is a k-blockade graph, $k \in \{3, 4\}$, then G' is a k-blockade graph as well.*

12

Using the results from §7, one can prove the following statement.

PROPOSITION. *Let G be a cubic planar graph and X be a replaceable matching in G. Let $\{H_x : x \in X\}$ be a set of planar and bipartite four-poles. Then for each edge x in X there exists a four-pole $E_x \in \{H_x, I(H_x)\}$ (see §7) and a replacement of x in G by E_x such that the graph $G\{E_x : x \in X\}$ obtained from G as the result of all these replacements is bipartite and planar.*

13

Let $\mathcal{K}_4(X, Y)$ be the set of all bipartite graphs obtained from K_4 by replacing two of its nonincident edges x and y by the bipartite four-poles $X^* = [X, (x_{11}^*, x_{12}^*, x_{22}^*, x_{21}^*)]$ and $Y^* = [Y, (y_{11}, y_{12}, y_{22}, y_{21})]$, respectively (Figure 4). By the symmetry of K_4, we have $\mathcal{K}_4(X, Y) = \mathcal{K}_4(Y, X)$. It is easy to prove the following lemma.

LEMMA. *Let $G \in \mathcal{K}_4(X, Y)$.*

(a) G is cubic and planar if and only if X^* and Y^* are cubic and planar, respectively.

(b) If X^* is $x_1^+ \alpha x_2^-$-nontraceable and Y^* is $y_1^+ \beta y_2^-$-nontraceable, then G is $(\alpha \wedge \beta \wedge e^-)$-non-Hamiltonian for $e = (x_{22}, y_{22})$. If X^* is $x_1^\pm \alpha x_2^\mp$-nontraceable and Y^* is $y_1^+ \beta y_2^-$-nontraceable, then G is $(\alpha \wedge \beta \wedge e^-)$-non-Hamiltonian for $e \in \{(x_{22}, y_{22}), (x_{12}, y_{21})\}$. If X^* is $x_1^\pm \alpha x_2^\mp$-nontraceable

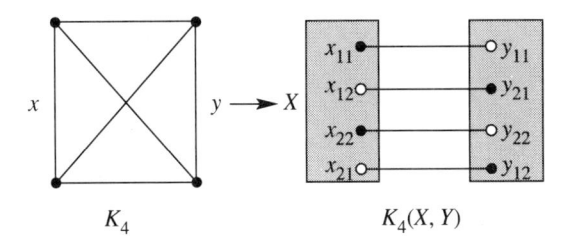

FIGURE 4

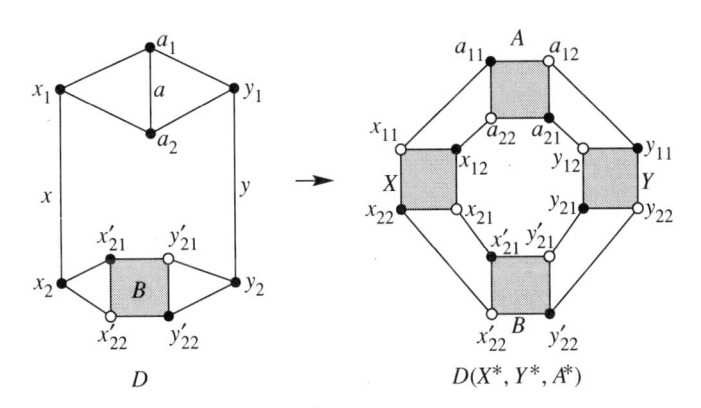

FIGURE 5

and Y^* is $y_1^\pm \beta y_2^\mp$-nontraceable, then G is $(\alpha \wedge \beta \wedge e^-)$-non-Hamiltonian for any edge e from $E(X \cap Y)$ in G.

The proof of (b) is based on results from §3.

14

Consider the cubic graph D in Figure 5. Let $\{x, y, a\}$ be a replaceable matching in D such that either B' is a bipartite subgraph in D, or $VB' = \emptyset$. Let $G = (X^*, Y^*, A^*)$ be the bipartite graph obtained from D by replacing the edges x, y, and a with the bipartite four-poles $X^* = [X, (x_{11}^*, x_{12}^*, x_{21}^*, x_{22}^*)]$, $Y^* = [Y, (y_{11}^*, y_{12}^*, y_{21}^*, y_{22}^*)]$, and $A^* = [A, (a_{11}^*, a_{12}^*, a_{21}^*, a_{22}^*)]$, respectively, as shown in Figure 5. Let B be the subgraph of D obtained from B' by adding the edges (x_{21}, x_{21}'), (x_{22}, x_{22}'), (y_{21}, y_{21}'), and (y_{22}, y_{22}'), so that $B^* = [B, (x_{21}, x_{22}, y_{22}, y_{21})]$ is a cubic bipartite four-pole.

It is easy to prove the following lemma.

LEMMA. (a1) *If X^*, Y^*, A^*, and B^* are planar, so is G.*

(a2) *If X^* and Y^* are strongly 3-blockade graphs, and each of A^* and B^* either has at most one inner edge or is a 3-blockade graph, then G is a 3-blockade graph.*

[Figure 6]

W

FIGURE 6

(a3) *If X^* and Y^* are 4-blockade graphs and each of A^* and B^* either has at most one inner edge or is a 4-blockade graph, then G is a 4-blockade graph.*

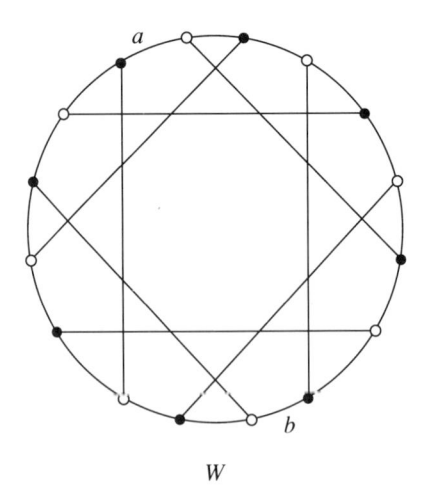

(b1) *If X^* is (x_1^+, x_2^+)-nontraceable, Y^* is (x_1^+, x_2^+)-nontraceable, A^* is $(a_1^\pm, a_2^\mp)$-nontraceable, and B' is connected, then G is non-Hamiltonian.*

(b2) *If X^* and Y^* are semi-nontraceable and A^* is $(a_1^\pm, a_2^\mp)$-nontraceable (or, moreover, semi-nontraceable), then G is non-Hamiltonian.*

Statements (b1) and (b2) of this lemma are proved with the help of results from §3.

15

The following statement on the graph W (see Figure 6) can be easily proved.

PROPOSITION. (a) *W is a cubic bipartite 5-blockade graph.*

(b) *For any Hamiltonian circuit C in W, either $\{a, b\} \subset C$, or $\{a, b\} \cap C = \varnothing$, i.e. W is $(a^\pm \wedge b^\mp)$-non-Hamiltonian.*

16

Let W_x and W_y be copies of the four-pole $W_0^* = [W_0, (a_1^*, a_2^*, b_1^*, b_2^*)] = C[W, (a, b)]$ (see Figures 2, 6). By the symmetry of K_4 and $[W, (a, b)]$, all the graphs of the set $\mathscr{H}_4(W_x, W_y)$ (see §13) are isomorphic to the same graph, to be denoted by S. The following lemma is derived from §§11, 13, and 15.

LEMMA. *S is a cubic, bipartite, 4-blockade, and e^--non-Hamiltonian graph for any e in the unique 4-blockade $E(W_x \cap W_y)$ in S.*

17

Using §§6 and 16, it is easy to construct non-Hamiltonian (but not 4-blockade) graphs in $\mathscr{C}$. The minimum graph in this class has 92 vertices. With the help of §§6, 7, and 11–13, we prove the following theorem.

THEOREM. *Suppose that any planar graph in $\mathscr{C}$ is Hamiltonian (i.e. that Barnette's conjecture holds for 3-connected graphs). Then for any planar graph G in $\mathscr{C}$ and any two edges x and y belonging to the same circuit-face in G, there exists a Hamiltonian circuit in G containing x and avoiding y.*

18

With the help of §§7, 14, 15, and 20, it is easy to construct non-Hamiltonian graphs in $\mathscr{C}_4$, of which the minimal one has 54 vertices. With the help of §§7 and 14, we prove the following theorem.

THEOREM. *Suppose that any planar graph in $\mathscr{C}_k$ is Hamiltonian for some $k \in \{3, 4\}$ (i.e. that Barnette's conjecture holds for k-blockade graphs). Then for any planar graph G in $\mathscr{C}_k$ and any two edges x and y belonging to a common circuit-face in G, there exists a Hamiltonian circuit in G containing both x and y.*

For 3-blockade graphs, this theorem follows directly from §17. It follows also from §§4, 11, 12, Petersen's theorem [5], and the existence of cubic, 3-connected, planar, and non-Hamiltonian graphs [1], [5].

19

PROPOSITION (well-known). *Petersen's graph P (Figure 7) is a cubic, 5-blockade, and non-Hamiltonian graph.*

PROOF (different from exhaustive search). The first two properties of the graph P are evident. Let us prove that P is non-Hamiltonian. There is a matching $\{x, y, z\}$ in P (see Figure 7) such that the distance between any two of its edges equals 2. Let T be the set of 6 vertices that are the end-vertices of the edges x, y, and z. It easy to check that any circuit in P containing T contains x as well. For any edge x in P, there exists an automorphism of P taking x to e. Thus, if P has a Hamiltonian circuit C, then $EP \subseteq EC$, a contradiction. □

20

Let X and Y be bipartite four-poles. Denote by $\mathscr{P}(X, Y)$ the set of all bipartite graphs obtained from P by replacing x and y by the four-poles X and Y, respectively. By the symmetry of P, we have $\mathscr{P}(X, Y) = \mathscr{P}(Y, X)$. Let Q_x and Q_y be the copies of the four-pole $Q(1) = Q$ (Figure 3) with the "middle" edges x and y, respectively. By the symmetry of P and Q,

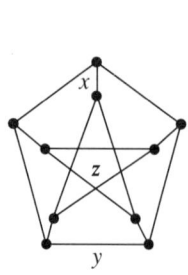

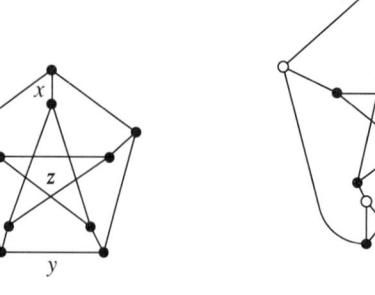

FIGURE 7   FIGURE 8

all graphs in $\mathscr{P}(Q_x, Q_y)$ are isomorphic to the same graph, to be denoted by M (Figure 8). From §§8 and 19 we derive the following lemma.

LEMMA. *M is a cubic, bipartite, 4-blockade, and $\{x, y\}^+$-non-Hamiltonian graph.*

PROOF. According to §§8 and 19, P and Q are cubic 4-blockade graphs; hence, by §4, so is M. Since Q is a bipartite graph, so is M (see Figure 8). Since Q is semi-q^+-nontraceable by §8, and P is non-Hamiltonian by §19, we see from §4 that M is $\{x, y\}^+$-non-Hamiltonian. □

21

Denote by R the four-pole $O[M, (x, y)]$ (see §7 and Figure 2). Let R_x and R_y be copies of R. By the symmetry of the graph P and the four-poles Q and R, all graphs in $\mathscr{P}(Q_x, R_y)$, as well as all graphs in $\mathscr{P}(R_x, R_y)$, are isomorphic; denote these graphs by L and K, respectively (Figures 9 and 10).

LEMMA. *The graphs K and L (Figures 9 and 10) are cubic, bipartite, and 4-blockade graphs; K is non-Hamiltonian, and L is x^+-non-Hamiltonian.*

PROOF. By §§7 and 20, R is a cubic, bipartite, 4-blockade, and semi-nontraceable four-pole. Then, according to §4, L and K are cubic, bipartite, 4-blockade four-poles. Moreover, L is x^+-non-Hamiltonian, and K is non-Hamiltonian. □

22

Denote by $\mathscr{F}(X, Y)$ the set of all bipartite four-poles obtained from the four-pole $F^* = [F, (f_1, f_2, f_3, f_4)]$ (Figure 11) by replacing the edges x and y by the bipartite four-poles $X^* = [X, (x_1^*, x_2^*, x_3^*, x_4^*)]$ and $Y^* = [Y, (y_1^*, y_2^*, y_3^*, y_4^*)]$, respectively (Figure 11). By symmetry, we have $\mathscr{F}(X, Y) = \mathscr{F}(Y, X)$. Using §3, we prove the following lemma.

LEMMA. *Let $H^* \in \mathscr{F}(X, Y)$. If X^* is semi-α-nontraceable and Y^* is semi-β-nontraceable, then H^* is totally $\alpha \wedge \beta$-nontraceable.*

L   K

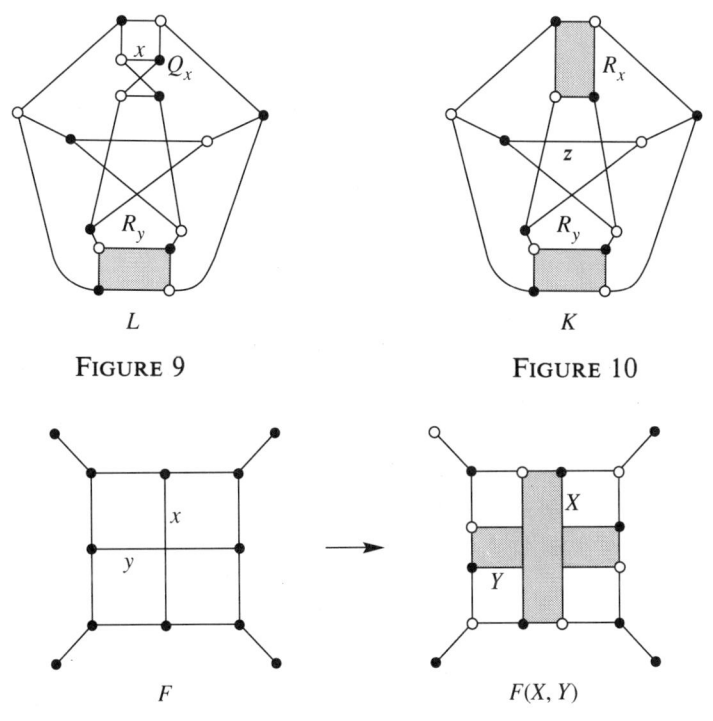

FIGURE 9   FIGURE 10

F   F(X, Y)

FIGURE 11

23

Denote by $\mathscr{K}$ the set of all bipartite graphs obtained from the graph K by replacing the edge z (Figure 10) by the trivial four-pole T that consists of the two edges a and b. By the symmetry of K and T, all graphs in $\mathscr{K}$ are isomorphic to the same graph, to be denoted by N (and so $a, b \in EN$). Since K is non-Hamiltonian, it follows that N is $(a^{\pm}, b^{\mp})$-non-Hamiltonian. From §§7, 20, and 22, we have:

LEMMA. *The graph N is a cubic bipartite 4-blockade graph, and every Hamiltonian circuit in N contains neither a nor b.*

24

With the help of §§4, 15, 16, 20–23, we prove the following theorem.

THEOREM. *Let $\mathscr{C}_4$ be, as above, the set of all cubic bipartite 4-blockade graphs.*

(a) *The graphs K, L, M, N, S, and W belong to $\mathscr{C}_4$.*

(k) *K has no Hamiltonian circuit, $|VK| = 50$ and for every even $k \geq 50$, there exists a non-Hamiltonian graph $K(k)$ in $\mathscr{C}_4$ having k vertices.*

(l) *L has no Hamiltonian circuit containing the edge x, $|VL| = 34$, and* *for every even $l \geq 34$ there exists an x^+-non-Hamiltonian graph $L(l)$ in $\mathscr{C}_4$ having l vertices, where x is an edge of $L(l)$.*

(m) *M has no Hamiltonian circuit containing both edges x and y, $|VM| = 18$, and for every even $m \geq 18$ there exists an $\{x, y\}^+$-non-Hamiltonian graph $M(m)$ in $\mathscr{C}_4$ having m vertices, where x and y are edges of $M(m)$.*

(n) *N has no Hamiltonian circuit containing at least one of the edges a, b, $|VN| = 48$, and for every even $n \geq 48$ there exists an $x^+ \wedge y^+$-non-Hamiltonian graph $N(n)$ in $\mathscr{C}_4$ having n vertices, where x and y are edges of $N(n)$.*

(s) *S has no Hamiltonian circuit containing the edge e, $|VS| = 32$, and for every even $s \geq 32$ there exists an x^--non-Hamiltonian graph $S(s)$ in $\mathscr{C}_4$ having s vertices, where x is an edge of $S(s)$.*

(w) *W has no Hamiltonian circuit containing only one of the two edges a, b, $|VW| = 16$, and for every $w \geq 16$ there exists an $(x^{\pm} \wedge y^{\mp})$-non-Hamiltonian graph $W(w)$ in $\mathscr{C}_4$ having w vertices, where x and y are edges of $W(w)$.*

25

Denote by K^*, L^*, and M^* the four-poles $F(R_x, R_y)$, $F(Q_x, R_y)$, and $F(Q_x, Q_y)$, respectively. From §§3, 4, 7, 8, and 22, we derive the following theorem.

THEOREM. *Let $\mathscr{R}_4$ be the set of all cubic, bipartite, and strongly 4-blockade four-poles.*

(a^*) *The four-poles K^*, L^*, M^* are four-poles from $\mathscr{R}_4$.*

(k^*) *K^* is totally nontraceable, $|VK^*| = 52$, and for every even $k^* \geq 52$ there exists a totally nontraceable four-pole $K^*(k^*)$ in $\mathscr{R}_4$ having k^* vertices.*

(l^*) *L^* is totally x^+-nontraceable, $|VL^*| = 36$, and for any even $l^* \geq 36$ there exists a totally x^+-nontraceable four-pole $L^*(l^*)$ in $\mathscr{R}_4$ having k^* vertices, where x is an edge of $L^*(l^*)$.*

(m^*) *M^* is totally $\{x, y\}^+$-nontraceable, $|VM^*| = 20$, and for every even $m^* \geq 20$ there exists a totally $\{x, y\}^+$-nontraceable four-pole $M^*(m^*)$ in $\mathscr{R}_4$ having m^* vertices, where x and y are edges of $M^*(m^*)$.*

REFERENCES

1. J. A. Bondy and U. S. R. Murty, *Graph theory with applications*, Amer. Elsevier, New York, and Macmillan, London, 1976.
2. J. Csima, *Problem 65*, Discrete Mathematics **55** (1985), no. 3.
3. A. M. Baraev and I. A. Faradzhev, *Construction and computer study of homogeneous bipartite graphs*, Algorithmic Studies in Combinatorics (Internat. Colloq., Orsay, 1976; I. A. Faradzhev, editor), "Nauka", Moscow, 1978, pp. 25-60. (Russian)
4. D. Barnette, *Conjecture 5*, Recent Problems in Combinatorics (W. T. Tutte, editor), Academic Press, New York, 1969, p. 343.
5. Frank Harary, *Graph theory*, Addison–Wesley, Reading, MA, 1969.

Nonseparating Circuits and the Planarity of Graph-Cells

A. K. KELMANS

§1. Introduction

The problem of embedding graphs into the plane [1] (or, briefly, the graph planarity problem) is known to reduce easily to similar problems for 3-connected graphs. Therefore, planarity criteria for 3-connected graphs play an important role in the theory of graph planarity. For 3-connected graphs, [2] strengthens the Kuratowski planarity criterion, while [3] gives a new planarity criterion in terms of nonseparating circuits (see also §4.2, below). It turns out that the planarity problem for 3-connected graphs can be easily reduced (see §3) to similar problems for 3-connected graphs having no essential 3-cuts and no triangles (we call them graph-cells). This explains why planarity criteria for graph-cells are of interest.

This paper provides a criterion for a given edge in a 3-connected graph without essential 3-cuts to belong to exactly two nonseparating circuits (Theorem 6.1). This criterion is used to obtain a planarity criterion for graph-cells in terms of nonseparating circuits (Corollary 6.3) The latter strengthens the corresponding planarity criterion for 3-connected graphs (Theorem 4.2).

Reference [5] gives a planarity criterion for graph-cells in terms of forbidden subgraphs (a strengthening of the Kuratowski theorem for graph-cells).

The results of this paper were reported at the Seminar on Discrete Mathematics at the Institute of Control Problems (Moscow, 1982), and at the Third National Conference on Methods and Algorithms for Solving Optimization Problems on Graphs and Networks (Tashkent, 1984), and are recapitulated in [4] and [5].

1991 *Mathematics Subject Classification.* Primary 05C10.
Translation of Problems of Discrete Optimization and Methods for Their Solution, Tsentral. Èkonom.-Mat. Inst. Akad. Nauk SSSR, Moscow, 1987, pp. 224–232.

§2. Basic notions and notation

We consider *undirected graphs* [1]. Denote by VG and EG the sets of the *vertices* and the *edges* of a graph G, respectively. For $\diamond \in \{\cap, \cup\}$, we write $G \diamond H$ if $V(G \diamond H) = VG \diamond VH$ and $E(G \diamond H) = EG \diamond EH$. A graph H is a *subgraph* of a graph G, $H \subseteq G$, if $VH \subseteq VG$ and $EH \subseteq EG$.

A graph G is *2-connected* if G has no loops and at least two vertices, for $|VG| = 2$ the graph G has parallel edges, and for $|VG| \geq 3$ the deletion of any vertex from G results in a connected graph. A graph G is *k-connected*, $k \geq 3$, if G has no loops or parallel edges, $|VG| \geq k + 1$, and the deletion of $k - 1$ (or fewer) arbitrary vertices from G results in a connected graph. A *block* of a graph G is a maximal (under inclusion) connected subgraph H in G with $EH \neq \varnothing$ such that any two of its vertices belong to a circuit.

A set $X \subseteq VG$ is a *vertex cut* of a connected graph G if $G \setminus X$ is disconnected. A cut X is *nonessential* if $G \setminus X$ consists of two components, one of which is an isolated vertex, and *essential* otherwise. A *k-cut* is a vertex cut of cardinality k. A *vertex cut X of a graph G cuts its subgraph H off* if $H \cap P_1 = \varnothing$ and $H \cap P_2 \neq \varnothing$ for some components P_1 and P_2 of $G \setminus X$.

A *graph-cell* is a 3-connected graph without essential 3-cuts or triangles.

For $H \subseteq G$, denote by G/H the graph obtained from G by contracting each edge of H (to *contract an edge* $e = (x, y)$ in G means to delete e and to identify its end-vertices x and y). A *circuit C does not separate a subgraph A* in a 2-connected graph G if there exists a block in G/C containing $EA \setminus EC$. A circuit C of a 2-connected graph G is *nonseparating* if C does not separate G, i.e. if G/C is 2-connected.

If T is a path, and x, y are vertices in T, then xTy denote the subpath in T with the end-vertices x, y.

§3. Triangle and 3-cut reductions of the graph planarity problem

3.1. Let $G = G_1 \cup G_2$ and $G_1 \cap G_2 = X \subseteq VG$. Put $G_k^+ = G_k \cup \{y_k\} \cup \{(x, y_k) : x \in X\}$, where y_k is a new vertex, $k = 1, 2$.

It is easy to prove

LEMMA. *Let G be 3-connected, $G = G_1 \cup G_2$, $G_1 \cap G_2 = X \subseteq VG$, $VG_1 \setminus X \neq \varnothing$, $VG_2 \setminus X \neq \varnothing$, and $|X| = 3$. Then G is planar if and only if both G_1^+ and G_2^+ are planar.*

3.2. Let T be a triangle in G. Denote by G_T' the graph $G \setminus ET$ if G has a vertex of degree three adjacent to each vertex in T, and the graph $G \setminus ET \cup \{y\} \cup \{(x, y) : x \in VT\}$ otherwise (here y is a new vertex). Denote by G_T the graph obtained from G_T' by replacing each thread in G_T' by the corresponding edge.

It is easy to see that either $|EG_T| > |EG|$, or $|EG_T| = |EG|$, but G_T contains fewer triangles than G. It is easy to prove

LEMMA. *Let G be 3-connected, T be a triangle in G, and VT be not an essential 3-cut in G. Then G is planar if and only if G_T is planar.*

3.3. It follows from Lemmas 3.1 and 3.2 that the planarity problem for 3-connected graphs can be reduced to that for 3-connected graphs without essential 3-cuts or triangles (i.e. graph-cells).

§4. On nonseparating circuits in 3-connected graphs

We need two facts on nonseparating circuits in 3-connected graphs.

4.1. THEOREM [3], [7]. *For every edge $e = (x, y)$ of a 3-connected graph G, there exist two nonseparating circuits C and P such that $C \cap P = xey$.*

This theorem follows also from Lemma 5.1 given below.

§6 provides a criterion for a given edge to belong to exactly two nonseparating circuits in the case of 3-connected graphs without essential 3-cuts.

4.2. THEOREM [3] (a planarity criterion in terms of nonseparating circuits). *A 3-connected graph is planar if and only if each of its edges belongs to exactly two nonseparating circuits.*

A strengthening of this criterion for graph-cells is presented in §6

§5. Some auxiliary statements

5.1. Let $e \in EG$ and $A \subseteq G$. Denote by $\mathscr{C}(G, A, e)$ the set of circuits C of the graph G such that $e \in C \not\subseteq A$ and C does not separate A. If $C \in \mathscr{C}(G, A, e)$, then, clearly, there exists a block $B_A C$ in G/C containing $A \setminus C$.

LEMMA. *Let $e \in EG$ and $A \subseteq G$. Suppose that G is 3-connected and $\mathscr{C}(G, A, e) \neq \varnothing$. Let $P \in \mathscr{C}(G, A, e)$ and $|E(B_A P)| \geq |E(B_A C)|$ for every $C \in \mathscr{C}(G, A, e)$. Then P is a nonseparating circuit in G (and $e \in P$).*

PROOF (similar to the proof of Theorem 7.13 in [3]). Let C be a circuit in G and B be a block in G/C. Denote by $\bar{B}$ the subgraph in G with the edge set $EB \cup EC$, and by HB the set of the vertices of C of degree ≥ 3 in $\bar{B}$.

Suppose the contrary, i.e. suppose that P is not a nonseparating circuit in G. Then there exists a block B in G/P distinct from $B_A P$. Let $pTBc$ be a path in P (with end-vertices p and c) such that $e \in TB$ and $TB \cap HB = \{p, c\}$.

Since B is a block in G/P, there exists a circuit C in $\bar{B}$ such that $C \cap P = TB$. As $P \in \mathscr{C}(G, A, e)$, we also have $C \in \mathscr{C}(G, A, e)$.

Suppose that $H(B_A C) \not\subseteq V(TB)$. Then

$$\varnothing \neq EP \setminus E(TB) \subset E(B_A C) \setminus E(B_A P)$$

and $E(B_A C) \supseteq E(B_A P)$. Therefore $|E(B_A C)| > |E(B_A P)|$, in contradiction to the choice of P (see the hypotheses of the lemma).

Suppose now that $H(B_A P) \subseteq V(TB)$ for any block B in G/P distinct from $B_A P$. Then there exists a unique path xMy in P such that

$H(B_A P) \cap M = \{x, y\}$ and $H(B) \subset M$ for any block B in G/P distinct from $B_A P$. For any two vertices x, y of a graph K, denote by $K/\{x, y\}$ the graph obtained from K by deleting an edge with the end-vertices x, y (if any) and identifying the vertices x and y with a new vertex denoted by $(x \circ y)$. Obviously, if K is 3-connected, then $K/\{x, y\}$ is 2-connected. Let $A' = \bar{B}_A P \setminus (M \setminus \{x, y\})$ and $B' = G \setminus (A' \setminus \{x, y\})$. Then $E(A'/\{x, y\}) \neq \emptyset$, $E(B'/\{x, y\}) \neq \emptyset$, $G/\{x, y\} = (A'/\{x, y\}) \cup (B'/\{x, y\})$, and $(A'/\{x, y\}) \cap (B'/\{x, y\}) = (x \circ y)$, and hence G is not 3-connected, a contradiction. Thus, P is a nonseparating circuit of G. □

5.2. Lemma 5.1 immediately yields

LEMMA. *Let G be 3-connected, $e = (x, y) \in EG$, and C_1, C_2 be nonseparating circuits in G such that $C_1 \cap C_2 = xey$ (hence $C_1 \neq C_2$). Then the following conditions are equivalent:*

(a) *There exists a nonseparating circuit C_3 in G containing e and distinct from C_1 and C_2.*

(b) *There exists a circuit C in G containing e, not belonging to $C_1 \cup C_2$, and not separating $C_1 \cup C_2$.*

5.3. LEMMA. *Let G be 3-connected, $e = (x_1, x_2) \in EG$, and let C_1, C_2 be nonseparating circuits in G such that $C_1 \cap C_2 = x_1 e x_2$. Then the following conditions are equivalent:*

(b) *There exists a circuit C in G containing e, not belonging to $C_1 \cup C_2$, and not separating $C_1 \cup C_2$.*

(c) *There exist two disjoint paths $x_1 X x_2$ and $y_1 Y y_2$ in $G \setminus e$ such that $y_k \in C_k \setminus \{x_1, x_2\}$, $k = 1, 2$.*

PROOF. (p1) Assume that a circuit C does not separate $C_1 \cup C_2$, $e \in C$, $C \not\subseteq C_1 \cup C_2$. Then $EC_k \setminus EC \neq \emptyset$, $k = 1, 2$. Since C does not separate $C_1 \cup C_2$, there exists a path $y_1 Y y_2$ in G/C going from $C_1 \setminus C$ to $C_2 \setminus C$ (i.e. $y_k \in C_k \setminus C$, $k = 1, 2$, and $[(C_1 \setminus C) \cup (C_2 \setminus C)] \cap y = \{y_1, y_2\}$).
Then $Y \cap C = \emptyset$. Put $X = C \setminus e$. Then X and Y are the required paths.

(p2) Assume that there exist two disjoint paths $x_1 X x_2$ and $y_1 Y y_2$ in $G \setminus e$ such that $y_k \in C_k \setminus \{x_1, x_2\}$, $k = 1, 2$. Let $y_1' Y' y_2'$ be the minimal (under inclusion) subpath in Y such that $y_k' \in C_k \setminus \{x_1, x_2\}$, $k = 1, 2$. Denote by T_k, $k = 1, 2$, the path in $C_1 \cup C_2$ having the ends y_1', y_2' and containing the vertex x_k. Let $x_1' X' x_2'$ be the minimal (under inclusion) subpath in X such that $x_k' \in T_k$, $k = 1, 2$ (since $X \cap Y = \emptyset$, we have $x_k' \in T_k \setminus \{y_1', y_2'\}$). Obviously, the paths X' and Y' do exist. Then the circuit $C = x_1' T_1 x_1 e x_2 T_2 x_2' X' x_1'$ surely has the desired properties, namely, $e \in C$, $C \not\subseteq C_1 \cup C_2$, and C does not separate $C_1 \cup C_2$. □

5.4. A path $p_1 P p_2$ is called a *path-chord* of a circuit C if $C \cap P = \{p_1, p_2\}$. Two path-chords $p_1 P p_2$ and $a_1 A a_2$ of a circuit C are called *crossing* if their end-vertices occur in the order $a_1 p_1 a_2 p_2$ when tracing C.

Let $e = (x_1, x_2) \in EG$, let C_1 and C_2 be circuits in G, and $C_1 \cap C_2 = x_1 e x_2$. Denote by G^* the graph obtained from $G \setminus e$ by adding two new vertices c_k, $k = 1, 2$, and the new edges $\{(c_k, y_k): y_k \in C_k\}$, $k = 1, 2$. Denote by K the circuit in G^* with the vertex set $\{c_1, x_1, c_2, x_2\}$ (obviously such a circuit is unique).

It is easy to prove the following result,

LEMMA. *The following conditions are equivalent*:

(c) *There exist two disjoint paths $x_1 X x_2$ and $y_1 Y y_2$ in $G \setminus e$ such that $y_k \in C_k \setminus \{x_1, x_2\}$, $k = 1, 2$.*

(d) *There exist two disjoint crossing path-chords of the circuit K in G^*.*

§6. The main results

6.1. THEOREM. *Suppose that* (c1) *G is a 3-connected graph*, (c2) *G is a nonplanar graph, and* (c3) *$e = (x_1, x_2) \in EG$ and G has no essential 3-cut cutting the subgraph $x_1 e x_2$ off. Then the following conditions are equivalent*:

(a1) *The edge e belongs to exactly two nonseparating circuits in G.*

(a2) *One of the end-vertices of e is of degree 3, and e belongs to two triangles C_1 and C_2 in G (and hence C_1 and C_2 are the two nonseparating circuits containing the edge e).*

PROOF. Let us first prove (a2) $\Longrightarrow$ (a1). Every circuit in G containing e and distinct from the circuit-triangles C_1 and C_2 contains the edge-chord $C_1 \setminus x$ or $C_2 \setminus x$ (where x is the end-vertex of the edge e of degree 3 in G), and hence it cannot be a nonseparating circuit. Therefore, the only circuits in G containing e which may be nonseparating are C_1 and C_2. According to (c1), G is 3-connected, and, by Theorem 4.1, e belongs to at least two nonseparating circuits. This means that exactly two nonseparating circuits in G contain e, namely, C_1 and C_2.

Now we prove now the converse: (a1) $\Longrightarrow$ (a2). Suppose that (G, e) satisfies (a1). According to (c1), G is 3-connected, and, by Theorem 4.1, there are two nonseparating circuits C_1 and C_2 such that $C_1 \cap C_2 = x_1 e x_2$. It follows from Lemmas 5.2, 5.3 and 5.4 that (a1) is equivalent to

(f) The graph G^* (see 5.4) has no pair of disjoint crossing path-chords of the circuit K.

Since G is 3-connected, so is G^*.

Assume first that G^* has no essential 3-cuts cutting K off. Since G^* satisfies (f), G^* is planar by Theorem 3.2 from [6]. But then G is planar as well, which contradicts (c2).

Assume now that G^* has an essential 3-cut, say, $A = \{a_1, a_2, a_3\}$, cutting K off. Then $G^* = G_1(A) \cup G_2(A)$, $G_1(A) \cap G_2(A) = A$, $K \subseteq G_1(A)$, and $K \cap G_2(A) \subseteq A$. We can assume that the subgraph $G_2(A)$ above is maximal (under inclusion); hence exactly one of the following statements holds:

(k1) $|VG_2(A) \setminus A| \geq 2$.

(k2) $|VG_2(A) \setminus A| = 1$, and $G_1(A) \setminus A$ consists of exactly two components, each having a common vertex with K, implying that $|A \cap VK| = 2$.

If $A \cap VK = \varnothing$, then A is an essential 3-cut in G cutting $x_1 e x_2$ off, which contradicts (c3). Therefore, $A \cap VK \neq \varnothing$.

Put $A^- = A \setminus C_1 \setminus C_2$. Since $G^* \setminus A$ is disconnected, the graph $G \setminus A^- \setminus e$ is disconnected in the case $|A^-| = 1$, and the graph $G \setminus A^-$ is disconnected in the case $|A^-| = 2$. Both cases contradict (c1) claiming that G is 3-connected.

Thus, $\varnothing \neq A \cap VK \subseteq \{x_1, x_2\}$.

Assume first that $|A \cap VK| = 1$ with, say, $A = \{x_1, a_1, a_2\}$ and $x_2 \neq a_1, a_2$. We know that $G = G_1 \cup G_2(A)$, $G_1 \cap G_2(A) = A$, and $e \in G_1$, $x_2 \in G_1 \setminus A$, implying that A is a 3-cut in G cutting the subgraph $x_1 e x_2$ off. Then, by (c3), A is a nonessential 3-cut in G. It follows that $|VG_2(A) \setminus A| \geq 2$, and so $|VG_1 \setminus A| = 1$. Therefore, $G^* \setminus G_2(A) = \{x_2, C_1, C_2\}$, $G \setminus G_2(A) = \{x_2\}$, and, because $C_1 \cap C_2 = x_1 e x_2$, we have $VC_k = \{x_1, x_2, p_k\}$, where $\{p_1, p_2\} = \{a_1, a_2\}$ and $k = 1, 2$.

Assume now that $|A \cap VK| = 2$, i.e. $A \cap VK = \{x_1, x_2\}$ for any 3-cut A in G^* separating K. Denote by $G_1^+(A)$ the graph obtained from $G_1(A)$ by adding a new vertex y_1 and three new edges (y_1, a), $a \in A$ (as in §3). Then $K \subseteq G_1^+(A)$. Suppose that A is chosen such that $G_1(A)$ is minimal with respect to inclusion. Then $G_1^+(A)$ has no essential 3-cut cutting K off. Since G^* is 3-connected, $G_1^+(A)$ is also 3-connected up to the vertices of degree 2 in A. Since G^* satisfies (f), $G_1^+(A)$ is planar by Theorem 3.2 from [6]. Then $G_1^+(A) \setminus x_1 \setminus x_2$ is disconnected, and therefore $G^* \setminus x_1 \setminus x_2$ is disconnected as well, which contradicts the fact that G^* is 3-connected. □

6.2. Theorems 4.1, 4.2, and 6.1 directly yield the following strengthening of Theorem 4.2.

THEOREM. *Suppose that* (c1) *G is a 3-connected graph*, (c2) *G has no essential 3-cuts, and* (c3) *every vertex of degree 3 in G belongs to at most one triangle. Then the following assertions are true*:

(a1) *If G is planar, then each edge of G belongs to exactly two nonseparating circuits.*

(a2) *If G is nonplanar, then each edge of G belongs to at least three nonseparating circuits.*

6.3. COROLLARY. *Let G be a 3-connected graph without essential 3-cuts or triangles (i.e. a graph-cell), or a 4-connected graph. Then G satisfies* (a1) *and* (a2) *of Theorem 6.2.*

REFERENCES

1. Frank Harary, *Graph theory*, Addison–Wesley, Reading, MA, 1969.
2. A. K. Kelmans, *On the existence of subgraphs of a given type in graphs*, Algorithms of Discrete Optimization in Computing Systems, Izdat. Yaroslav. Univ., Yaroslavl, 1983, pp. 3–20. (Russian)

3. _____, *The concept of a vertex in a matroid, the nonseparating circuits of a graph, and a new criterion for graph planarity*, Algebraic Methods in Graph Theory (Szeged, 1978), Vol. 1, Colloq. Math. Soc. János Bolyai, vol. 25, Akad. Kiadó, Budapest, and North-Holland, Amsterdam, 1981, pp. 345–388.

4. _____, *On the planarity of 3-connected graphs without essential 3-cuts or triangles*, Third National Conf. on Methods and Programs for Solving Optimization Problems on Graphs and Networks, Abstracts of Reports, vol. 2, Novosibirsk, 1984, pp. 59–61. (Russian)

5. _____, *On 3-connected graphs without essential 3-cuts or triangles*, Dokl. Akad. Nauk SSSR **288** (1986), 531–535; English transl. in Soviet Math. Dokl. **33** (1986).

6. _____, *A criterion for the existence of two crossing path-chords of a given circuit*, Algorithmic Constructions and Their Efficiency, Izdat. Yaroslav. Univ., Yaroslavl, 1983, pp. 50–60. (Russian)

7. W. T. Tutte, *How to draw a graph*, Proc. London Math. Soc. (3) **13** (1963), 743–768.

Extremal Sets and Covering and Packing Problems in Matroids

A. K. KELMANS AND V. P. POLESSKIĬ

§1. Introduction

This paper deals with the study of various aspects of covering and packing problems in matroids [1], including interrelations between these problems, relations between these problems and the problems of constructing bases for the union of matroids, algorithms to solve them, the variety of possible solutions, etc. Basic notions and notation, as well as some information on the matroid theory, are given in §2.

In 1971, Kelmans suggested a constructive description of circuits of the union of matroids and the notion of an augmenting path for augmenting an independent set of the union of matroids [2]. These notions and results are discussed in §3 and play an important role in the subsequent reasoning. In particular, understanding the construction of circuits of the union of matroids leads directly to the well-known theorems about the rank of the union of matroids and the fact that the union of matroids is itself a matroid, and also to an algorithm for finding a base of the union of matroids, and solving thereby the packing and covering problems, with $O(|E|^3)$ calls to the corresponding independence oracles.

Combining these results with the results of Polesskiĭ about algorithms for packing and covering for graphs [3]–[5], we manage to construct algorithms that solve the above problems in the case of identical matroids, requiring the smallest known number $O(|E|^2)$ of calls to the corresponding independence oracle (see §8). For the sake of completeness, §4 describes applications of the rank theorem to the covering, packing, and intersection problems in matroids, §5 deals with the so-called extremal sets of a matroid sequence [6], which are naturally interpreted as obstacles for finding a base of given size in the

---

1991 *Mathematics Subject Classification.* Primary 05B35, 05B40.
Translation of Studies in Applied Graph Theory (A. S. Alekseev, editor), "Nauka", Novosibirsk, 1986, pp. 140-168.

union of the matroids from the sequence. It is established that the family of extremal sets constitutes a lattice. This lattice was discovered independently by many authors (see [7] and [8]). Various descriptions of this lattice are given in §5.

In §6 the notion of a base sequence of a matroid sequence S is introduced (see also [2]). The notion of a regular packing of spanning forests of a graph introduced by Polesskiĭ in 1974 [4] serves as a prototype for the notion of a base sequence of a set of identical matroids. A base sequence of a sequence of matroids on the same ground set is a special partition of a base of the union of matroids from S into independent sets of the corresponding matroids. In particular, a base sequence contains a maximum packing of bases of the matroids, as well as a minimum covering of the ground set by independent sets of the matroids. Thus, construction of a base sequence provides solutions to both the covering and packing problems for given matroids. §7 indicates that a base sequence of a matroid sequence admits a natural decomposition with respect to an extremal set of this sequence, and §8 contains an algorithm for finding a base sequence of a matroid sequence with $O(|E|^3)$ calls to the independence oracles that describe the matroid sequence. In the case of identical matroids, an algorithm is given for constructing a base sequence that requires $O(|E|^2)$ calls to the corresponding independence oracle. In particular, this algorithm allows us to construct a base of the union of a given number of identical matroids, and hence to solve the covering and packing problems for a given matroid. A qualitative comparison of some matroid algorithms is given in §9.

§2. Basic notions and notation. Some concepts and results in matroid theory

2.1. Let E be a finite set. For $X \subseteq E$, put $\overline{X} = E \setminus X$, and let $|X|$ denote the number of elements in X. If $A, B \subseteq E$ and $A \cap B = \varnothing$, then we write $A + B$ instead of $A \cup B$. A set $\mathscr{I} \subseteq 2^E$ of subsets of E is called an *independence set*, and the sets in $\mathscr{I}$ are called *independent*, if
($\mathscr{I}0$) $\varnothing \in \mathscr{I}$, and
($\mathscr{I}1$) $A \subseteq B$ and $B \in \mathscr{I}$ $\Rightarrow$ $A \in \mathscr{I}$.

A set in $\mathscr{I}$ that is maximal under inclusion is called a *base of the set $\mathscr{I}$*. Let $\mathscr{B}(\mathscr{I})$ denote the set of all bases of $\mathscr{I}$. A subset A of E is called *dependent* if $A \notin \mathscr{I}$. A minimal (under inclusion) dependent subset in E is called a *circuit of the set $\mathscr{I}$*. Let $\mathscr{C}(\mathscr{I})$ denote the set of all circuits of $\mathscr{I}$.

An independent set $\mathscr{I}$ is called a *matroid* [1] and is denoted by M if it has the following property:
($\mathscr{I}2$) $A, B \in \mathscr{I}$ and $|A| > |B|$ $\Rightarrow$ $\exists a \in A \setminus B: B + a \in \mathscr{I}$.

We sometimes describe a matroid M by a pair $(E, \mathscr{I})$, where E is the ground set and $\mathscr{I}$ is a subset of 2^E with the properties ($\mathscr{I}0$), ($\mathscr{I}1$), and ($\mathscr{I}2$).

One can prove that the set $\mathscr{B}(M)$ of bases of a matroid M has the

properties
($\mathscr{B}0$) $A, B \in \mathscr{B}(M) \Rightarrow |A| = |B|$,
($\mathscr{B}1$) $A, B \in \mathscr{B}(M), a \in A \Rightarrow \exists b \in B(M): A \setminus a + b \in \mathscr{B}(M)$,
($\mathscr{B}1^*$) $A, B \in \mathscr{B}, b \in B(M) \Rightarrow \exists a \in A: A \setminus a + b \in \mathscr{B}(M)$,
while the set $\mathscr{C}(M)$ of circuits of a matroid M has the properties
($\mathscr{C}1$) $A, B \in \mathscr{C}(M) \Rightarrow A \not\subseteq B$,
($\mathscr{C}2$) $A, B \in \mathscr{C}(M), x \in A \cap B \Rightarrow \exists C \in \mathscr{C}(M): C \subseteq A \cup B \setminus x$,
($\mathscr{C}2'$) $A, B \in \mathscr{C}, x \in A \cap B, y \in A \setminus B \Rightarrow \exists C \in \mathscr{C}: y \in C \subseteq A \cup B \setminus x$.

One can prove that $\mathscr{B}^*(M) = \{E \setminus B : B \in \mathscr{B}(M)\}$ is the set of bases of a matroid to be denoted by M^*. The matroid M^* is called *dual* to M. A base B^* and a circuit C^* of the matroid M^* are called, respectively a *cobase* and a *cocircuit* of the matroid M. Let $\mathscr{C}^*(M)$ denote the set of all cocircuits of M, so that $\mathscr{B}^*(M) = \mathscr{B}(M^*)$ and $\mathscr{C}^*(M) = \mathscr{C}(M^*)$.

One can show that if $\mathscr{B}(\mathscr{I})$ has properties ($\mathscr{B}0$), ($\mathscr{B}1$), or $\mathscr{C}(\mathscr{I})$ has properties ($\mathscr{C}1$), ($\mathscr{C}2$) or properties ($\mathscr{C}1$), ($\mathscr{C}2'$), then $\mathscr{I}$ has properties ($\mathscr{I}0$), ($\mathscr{I}1$), and ($\mathscr{I}2$), i.e., $(E, \mathscr{I})$ is a matroid.

For $T \subseteq 2^E$ and $X \subseteq E$, the set of sets $\{A : A \in T, A \subseteq X\}$ is said to be the *restriction* $T|_X$ of the set T to the set X.

Obviously, $M|_X$ is a matroid. Denote $M|_X$ by $M \cdot X$ or $M \setminus \overline{X}$, and denote $(M^*|_X)^*$ by $M \times X$ or $M/\overline{X}$. Obviously, $\mathscr{C}(M \cdot X) = \mathscr{C}(M)|_X$ and $\mathscr{C}^*(M \times X) = \mathscr{C}^*(M)|_X$. The matroid $M \cdot X = M \setminus \overline{X}$ is called the *restriction* of M to X and is said to be obtained from M by *deleting* $\overline{X}$. The matroid $M \times X = M/\overline{X}$ is called the *contraction* of M to X and is said to be obtained from M by *contracting* $\overline{X}$.

According to ($\mathscr{B}0$), all bases of a matroid M have the same number of elements, which is called the *rank* of the matroid M and is denoted by $r(M)$. Obvbiously $r^*(M) = |E| - r(M)$ is the rank of M^*; it is called the *corank* of M: $r^*(M) = r(M^*)$.

An integer function $\rho_M = \rho$ on the set 2^E of all subsets of E is called the *rank function of a matroid* M if $\rho(X) = r(M \cdot X)$ for $X \in E$. In other words, $\rho(X)$ is the cardinality of a maximal independent set in X. We write ρ^* instead of ρ^*_M. One can prove that the rank function of a matroid has the following properties:
($\rho 1$) $0 \leq \rho(X) \leq |X|$,
($\rho 2$) $X \subseteq Y \Rightarrow \rho(X) \leq \rho(Y)$ (monotonicity),
($\rho 3$) $\rho(X \cup Y) + \rho(X \cap Y) \leq \rho(X) + \rho(Y)$ (submodularity).

Given $X \subseteq E$, let $\text{cl}(X)$ denote a maximal subset A of E such that $X \subseteq A$ and $\rho(A) = \rho(X)$. It is easy to show that $\text{cl}(X)$ is defined uniquely. The function $\text{cl}: 2^E \to 2^E$ is called the *closure operator* for M. One can prove that the closure operator has the following properties:

(cl 1) $X \subseteq \text{cl}(X)$,
(cl 2) $X \subseteq Y \Rightarrow \text{cl}(X) \in \text{cl}(Y)$,
(cl 3) $\text{cl}(X) = \text{cl}(\text{cl}(X))$, and
(cl 4) if $y \notin \text{cl}(X)$ and $y \in \text{cl}(X + x)$ then $x \in \text{cl}(X + y)$.

The circuit property ($\mathscr{C}2$) implies that if $X \in M$ and $X + x \notin M$ for

$x \in E \setminus X$, then there exists a unique circuit in $X + x$ and it contains x; denote this circuit by $C(M, X + x)$, or just by $C(X + x)$. Obviously, $X + x \setminus y \in M$ if and only if either $X + x \in M$, or $X + x \notin M$ but $y \in C(M, X + x)$.

A circuit (cocircuit) of a matroid M consisting of a single element is called a *loop* of M (a *coloop* or an *isthmus* of M, respectively). The set of all coloops of a matroid M is denoted by $I(M)$. An element e in E is called a *cyclic element of* M if e belongs to some circuit in $\mathscr{C}(M)$. Put $D(M) = E \setminus I(M)$. One can prove that $D(M)$ is the set of all cyclic elements of M: $D(M) = \{e \in E : e \in C \in \mathscr{C}(M)\}$. A matroid M is called *free* if $D(M) = \varnothing$.

2.2. Denote by VG and EG the sets of the vertices and the arcs of a directed graph G [9]. A *path* P in G is a sequence of vertices $(p_0, \ldots, p_k)$ such that $(p_i, p_{i+1}) \in EG$ for any $i \in \{0, \ldots, k-1\}$, and k is the *length of the path* P. If there are no identical vertices in this sequence then the path is called *simple*. For $X \subseteq VG$, denote by $G(X)$ the set of all vertices in G attainable along the paths from X, so that $X \subseteq G(X)$. Put $\mathscr{L}(G) = \{X : X \subseteq VG, G(X) \subseteq X\}$ (we suppose that $\varnothing \in VG$, so that $\varnothing \in \mathscr{L}(G)$). Obviously, a subset of vertices in G belongs to the family $\mathscr{L}(G)$ if and only if no arc of G goes out of X. The set of all arcs of G coming into X (the tail of such an arc is in $VG \setminus X$ and the head is in X) constitutes a so-called *directed cut* of the digraph G. Obviously, $\mathscr{L}(G)$ is a lattice of sets closed with respect to union. We call $\mathscr{L}(G)$ the *lattice of directed cuts of a graph* G. Let $A_1, \ldots, A_n$ be the bicomponents of the graph G. Then $\{G(A_i) : i = 1, \ldots, n\}$ is the set of the so-called *join-indecomposable elements of the lattice* $\mathscr{L}(G)$, and $\{A_1, \ldots, A_n\}$ is a principle partition for the lattice of sets $\mathscr{L}(G)$ [8].

2.3. Let $\mathscr{A} = (A_1, \ldots, A_k)$ be a sequence of systems of sets A_i on E: $A_i \subseteq 2^E$. The system $A = \{X = \bigcup\{X_i \in A_i : i = 1, \ldots, k\}\}$ is called the *union of the sets* $A_1, \ldots, A_k$ and is denoted by $A_1 \vee \cdots \vee A_k$, or $\bigvee_{i=1}^{k} A_i$. A sequence $\lambda = (X_1, \ldots, X_k)$ of sets from E is called an $\mathscr{A}$-*sequence*, or an $\mathscr{A}$-*packing*, if $X_i \in A$ and $X_i \cap X_j = \varnothing$ with $i \neq j$. If $X \in A$, then there exists an $\mathscr{A}$-sequence $\lambda = (X_1, \ldots, X_k)$, in general not unique, such that $X = \bigcup\{X_i : i = 1, \ldots, k\}$. Such an $\mathscr{A}$-sequence is also called an $\mathscr{A}$-*partition*, or an $\mathscr{A}$-*coloring*, of the set $X \in \bigvee_{i=1}^{k} A_i$.

Assume that all A_i in $\mathscr{A}$ are independence sets. Then, obviously, A is also an independence set. For such an $\mathscr{A}$, from each $\mathscr{A}$-sequence $\lambda = (X_1, \ldots, X_k)$, we construct a graph $G_\lambda(\mathscr{A}) = G_\lambda$ in the following way: $VG_\lambda = E$, and for $x, y \in E$, (x, y) is an arc in G_λ if and only if
 (a) $x \notin X_i$ and $y \in X_i$ for some $i \in \{1, \ldots, k\}$, and
 (b) y belongs to the circuit of the independence set A_i lying in $X_i + x$.
A path in the graph $G_\lambda(\mathscr{A})$ is called a λ-*path*.

In considering an $\mathscr{A}$-partition $\lambda = (X_1, \ldots, X_k)$, it is convenient to regard the elements of the set X_i as being colored with color i, and the elements of $E \setminus X$ as being discolored. Then, the process of rearranging the $\mathscr{A}$-partition λ can be interpreted as recoloring of elements of E.

§3. The notion of an augmenting path and a constructive description of circuits of the union of matroids

The notion of the union of matroids plays an important role in matroid theory (see [10]). It turns out to be significant for the problems of maximum packing of matroid bases, of minimum covering of the ground set with bases (or independent sets) of a matroid, and of finding a maximum common independent set of two matroids on the same ground set (see Chapter 8 in [1]). The present section introduces the notion of an augmenting, or active, path, which guarantees the possibility of augmenting any nonmaximum independent set of the union of matroids. A constructive description of circuits of the union of matroids is given, which will be used substantially later on. Using this description of circuits of the union of matroids, we present simple proofs of the some well-known theorems concerning unions of matroids [10].

3.1. Let $\mathscr{M} = (M_1, \ldots, M_k)$ be a sequence of matroids on the set E and $\lambda = (X_1, \ldots, X_k)$ an $\mathscr{M}$-sequence (see 2.3). A λ-path is called *active*, or *augmenting*, if there exists $X_i \in \lambda$ such that $x_m \notin X_i$ and $X_i + x_m \in M_i$. An element x in E is called λ-*active* if it belongs to a λ-active path. In what follows ρ_i is assumed to be the rank function of the matroid M_i.

3.2. LEMMA. *Let* $\mathscr{M} = (M_1, \ldots, M_k)$ *be a sequence of matroids on the set* E, *and* $\lambda = (X_1, \ldots, X_k)$ *an* $\mathscr{M}$-*partition of a set* $X \subseteq M = \bigvee_1^k M_i$. *If an element* $z \in E \setminus X$ *is* λ-*active, then* $X + z \in M$.

PROOF. Let $P = (z = x_0, x_1, \ldots, x_m)$ be an active λ-path, $x_k \in X_{i_k}$, $k = 1, \ldots, m$, and so $x_m + X_p \in M_p$ for some $p \in \{1, \ldots, k\}$, $p \neq i_m$. If $m = 0$, then $X_p + z \in M_p$, and so $X + z = (X_i + z) + \sum \{X_j, j \neq i\} \in M$. Suppose that the assertion holds for active λ-paths of length less than m. Let us show that it holds also for active λ-paths of length m. We construct a new $\mathscr{M}$-partition $\lambda' = (X_1', \ldots, X_k')$ of the set X by putting $X_i' = X_i$ for $i \neq i_m, p$, and $X_{i_m}' = X_{i_m} \setminus x_m$, $X_p' = X_p + x_m$.

Suppose $P^* = (z = x_0, x_1, \ldots, x_{m-1})$ is a λ'-path. Since $x_{m-1} \notin X_{i_m}'$ and $X_{i_m}' + x_{m-1} \in M_{i_m}$, we see that P^* is an active λ'-path of length $m-1$, and therefore, by the inductive hypothesis, $X + z \in M$.

Suppose now that P^* is not a λ'-path. Let c be the minimum index such that the arc (x_c, x_{c+1}) of the path P is not a λ'-arc, and hence $P' = (z = x_0, x_1, \ldots, x_c)$ is a λ'-path. Obviously, (x_c, x_{c+1}) is not a λ'-arc if and only if $x_{c+1}, x_m \in C(M_{i_m}, X_{i_m} + x_c)$. Therefore, $X_{i_m}' + x_c = X_{i_m} + x_c \setminus x_m \in M_{i_m}$, and P' is an active λ-path of length $c \leq m-1$. By the inductive hypothesis, $X + z \in M$. □

3.3. LEMMA. *Let $\mathcal{M} = (M_1, \ldots, M_k)$ be a sequence of matroids on E, and $\beta = (X_1, \ldots, X_k)$ be an $\mathcal{M}$-partition of a set $X \in \bigvee M_i$. Suppose that the elements of a set $Y \subseteq \overline{X}$ are not β-active. Put $A = G_\beta(Y)$. Then $X_i' = X_i \cap A$ is a base of the matroid $M_i' = M_i \cdot A$, $i = 1, \ldots, k$.*

PROOF. Assume that $X_i + x \in M_i$ for some $i \in \{1, \ldots, k\}$ and $x \in A \setminus X$. Since $x \in A$, there exists a β-path P from some $y \in Y$ to x. Since $X_i + x \in M_i$, the β-path P is active, and hence y is a β-active element in Y, a contradiction. Therefore, $x + X_i \notin M_i$ for any $i \in \{1, \ldots, k\}$ and $x \in A \setminus X_i$. Then $C(M_i, X + x) \cap X_i \subseteq X_i'$, and hence $X_i' + x \notin M_i$, so that X_i' is a base in M_i'. □

3.4. COROLLARY. *Under the assumptions of Lemma 3.3, $X \cap A$ is a base of the independence set $\bigvee M_i|_A$.*

3.5. THEOREM. *Let $\mathcal{M} = (M_1, \ldots, M_k)$ be a sequence of matroids on the set E, $M = \bigvee M_i$, let $\beta = (X_1, \ldots, X_k)$ be an $\mathcal{M}$-partition of $X \in M$, and let $a \in \overline{X}$. If a is not β-active, then $X + a$ contains a unique circuit of the independence set M, and this circuit is $A = G_\beta(a)$.*

PROOF. By Lemma 3.3, $A \notin M$. By Lemma 3.2, $A \setminus x \in M$ for any $x \in A$. Thus, A is the unique circuit of the set M in $X + a$. □

3.6. Lemma 3.2 and Theorem 3.5 imply

THEOREM. *Let $\mathcal{M} = (M_1, \ldots, M_k)$ be a sequence of matroids on the set E, let $\beta = (X_1, \ldots, X_k)$ be an $\mathcal{M}$-partition of $X \in \bigvee_1^k M_i$, and let $a \in \overline{X}$. Then $X + a \in \bigvee_1^k M_i$ if and only if the element a is β-active.*

Thus, if $X \in \bigvee_1^k M_i$ is not a maximum independent set in $\bigvee_1^k M_i$, then it can always be augmented with the help of an active path. Actually, the proof of Lemma 3.2 contains an algorithm for augmenting such a set X with the help of an active path. This algorithm rearranges the $\mathcal{M}$-partition β of the set X into a new $\mathcal{M}$-partition β' of the same set X so that a β-active element $a \notin X$ can be added to a set X_i' in β' to obtain the set $X_i' + a$, still independent in M. To obtain an active β-path, one can take the shortest path $P = (z = x_0, x_1, \ldots, x_m)$ in G_β. Then, $P' = (z = x_0, x_1, \ldots, x_{m-1})$ is obviously the shortest active path in the new digraph G_β' (see 3.2). Thus, if the element x_i of the path P is colored with the color of the element x_{i+1}, $i = 1, \ldots, m - 1$, and the element x_m is colored with color p, then each new set of color j for $j = 1, \ldots, k$ appears to be independent in M_j. The above algorithm for augmenting an independent set X in $\bigvee_1^k M_i$ is used in the algorithms solving the covering and packing problems in matroids, which are described below.

3.7. Theorem 3.5 implies several corollaries.

3.7.1. For any $X \in \bigvee_1^k M_i$ and $x \in \overline{X}$, according to 3.5, the set $X + x$ has at most one circuit of the independence family $\bigvee_1^k M_i$, and therefore we get

COROLLARY (see [10]). *The union of matroids is a matroid.*

3.7.2. COROLLARY. *Let B be a base of the matroid $\bigvee_1^k M_i$ and $\gamma = (B_1, \ldots, B_k)$ be an $\mathscr{M}$-partition of B. Then $D(\bigvee_1^k M_i) = G_\gamma(\overline{B})$.*

3.7.3. COROLLARY. *If C is a circuit of the matroid $\bigvee_1^k M_i$, then $|C| = 1 + \sum_1^k \rho_i(C)$.*

A different proof of this statement was given in [11].

3.7.4. COROLLARY. *A subset C of E is a circuit of the matroid $\bigvee_1^k M_i$ if and only if for some $c \in C$ (actually, for every $c \in C$) there exists an $\mathscr{M}$-partition $\gamma = (X_1, \ldots, X_k)$ of the set $X = C \setminus c$ such that X_i is a base of the matroid $M_i \cdot C$, $i = 1, \ldots, k$, and $C = G_\gamma(c)$.*

3.7.5. From Corollary 3.7.3, we derive

COROLLARY. *$C \subseteq E$ is a circuit of the matroid $\bigvee_1^k M_i$ if and only if for every element $c \in C$ there exists an $\mathscr{M}$-partition $(X_1, \ldots, X_k)$ of the set $X = C \setminus c$ such that X_i is a base of $M_i \cdot C$, $i = 1, \ldots, k$.*

3.8. Obviously, $r(\bigvee_1^k M_i) \leq \sum_1^k \rho_i(X) + |\overline{X}|$ for any $X \subseteq E$. Lemma 3.3 and Corollary 3.7.2 indicate that the equality in the above relation is attained at $X = D(\bigvee_1^k M_i)$. Thus, we have a constructive proof of the following well-known theorem which we call *the matroid rank theorem*.

THEOREM (see [10]). *We have*

$$r\left(\bigvee_{i=1}^k M_i\right) = \min\left\{\sum_1^k \rho_i(X) + |\overline{X}| : X \subseteq E\right\}$$
$$= \sum_1^k \rho_i\left(D\left(\bigvee_s^k M_i\right)\right) + \left|E \setminus D\left(\bigvee_s^k M_i\right)\right|.$$

§4. Applying the matroid rank theorem to matroid packing, covering, and intersection problems

4.1. Let M_i, $i = 1, \ldots, k$, be matroids on E and $M^s = \bigvee_1^s M_i$. Obviously, there exist k pairwise disjoint bases of the matroids $M_1, \ldots, M_k$ in E if and only if $r(M^k) = \sum_1^k r(M_i)$. So, the problem of finding k disjoint bases of the matroids $M_1, \ldots, M_k$ in E (called the *problem of packing bases of the matroids $M_1, \ldots, M_k$ in E*) is reduced to the problem of constructing a base of the matroid M^k (an algorithm to solve the latter problem is discussed in §8).

4.1.1. From the matroid rank theorem (see 3.8), we derive the following well-known criterion for the existence of a packing of bases of the matroids $M_1, \ldots, M_k$ in E.

THEOREM (see [1]). *A collection* $B_1, \ldots, B_k$ *of pairwise disjoint bases of the matroids* $M_1, \ldots, M_k$ *in* E *exists if and only if* $\sum_1^k r(M_i \times \overline{A}) \leq |\overline{A}|$ *for any* $\overline{A} \subseteq E$.

PROOF. Assume that there exists a collection $B_1, \ldots, B_k$ of pairwise disjoint bases of the matroids $M_1, \ldots, M_k$ in E, and hence $r(M^k) = \sum_1^k |B_i| = \sum_1^k r(M_i)$. By the matroid rank theorem (see 3.8), $\sum_1^k r(M_i) = r(M^k) = \min\{\sum_1^k \rho_i(A) + |\overline{A}| : A \subseteq E\}$, and hence $\sum_1^k (r(M_i) - \rho_i(A)) \leq |\overline{A}|$ for any $A \subseteq E$. Since $r(M_i) - \rho_i(A) = r(M_i \times \overline{A})$, we have $\sum_1^k r(M_i \times \overline{A}) \leq |\overline{A}|$ for any $\overline{A} \subseteq E$.

Assume $\sum_1^k r(M_i \times \overline{A}) \leq |\overline{A}|$ for any $\overline{A} \subseteq E$. Then $\sum_1^k (r(M_i) - \rho_i(A)) \leq |\overline{A}|$ for any $A \subseteq E$, and therefore

$$\sum_1^k r(M_i) \leq \min\left\{\left(\sum_1^k \rho_i(A) + |\overline{A}|\right) : A \subseteq E\right\}.$$

By the matroid rank theorem (see 3.7.2), the right-hand side of the inequality equals $r(M^k)$, and hence, $\sum_1^k r(M_i) \leq r(M^k)$. But obviously, $\sum_1^k r(M_i) \geq r(M^k)$. Thus, $\sum_1^k r(M_i) = r(M^k)$, and there exists a collection $B_1, \ldots, B_k$ of pairwise disjoint bases of the matroids $M_1, \ldots, M_k$ in E. □

A subset $X \subseteq E$ such that $|X| \leq \sum_1^k r(M_i \times X)$ is said to be *naturally an obstacle to packing bases of the matroids* $M_1, \ldots, M_k$ in E.

4.1.2. From Theorem 4.1.1 we immediately derive

COROLLARY. *A matroid* M *on* E *has* k *pairwise disjoint bases if and only if* $k \cdot r(M \times X) \leq |X|$ *for any* $X \subseteq E$.

4.1.3. Denote by pack M the maximum number of pairwise disjoint bases of a matroid M. Then 4.1.2 implies

COROLLARY.

$$\operatorname{pack} M = \min\left\{\left\lfloor \frac{|X|}{r(M \times X)} \right\rfloor : X \subseteq E\right\}.$$

The problem of finding a collection $B_1, \ldots, B_p$ of $p = \operatorname{pack} M$ pairwise disjoint bases of a matroid M in E is reduced, obviously, to a sequence of problems of packing of a given (and increasing) number of bases of M.

4.2. Let M_i, $i = 1, \ldots, k$, be matroids on E as in 4.1 and let $M^s = \bigvee_1^s M_i$. Clearly, E is the union of k independent sets $X_i \in M_i$, $i = 1, \ldots, k$ ($E = \bigcup_1^k X_i$), if and only if $r(M^k) = |E|$. Therefore, the problem of finding k independent sets $X_1, \ldots, X_k$ of the matroids $M_1, \ldots, M_k$

whose union is E (called the *problem of covering the base set E with independent sets of the matroids $M_1, \ldots, M_k$*) is reduced again to the same problem of constructing a base of the matroid M^k.

4.2.1. The matroid rank Theorem 3.7.2 yields

THEOREM (see [1]). *E can be represented as the union of k independent sets $X_i \subseteq M_i$, $i = 1, \ldots, k$, if and only if $\sum_1^k r(M_i \cdot A) \geq |A|$ for any $A \subseteq E$.*

PROOF. Let $E = \bigcup_1^k X_i$, $X_i \in M_i$, i.e., $r(M^k) = |E|$. Then, by the matroid rank Theorem 3.7.2,

$$|E| = r(M^k) = \min\left\{\sum_1^k r(M_i \cdot A) + |\overline{A}| : A \subseteq E\right\},$$

and hence $|E| - |\overline{A}| = |A| \leq \sum_1^k r(M_i \cdot A)$ for any $A \subseteq E$.

Suppose that $|A| \leq \sum_1^k r(M_i \cdot A)$ for any $A \subseteq E$. Then

$$|E| \leq \min\{r(M_i \cdot A) + |\overline{A}| : A \subseteq E\}.$$

By the matroid rank Theorem 3.7.2, the right-hand side of the above inequality equals $r(M^k)$, so $|E| \leq r(M^k)$. On the other hand, obviously $|E| \geq r(M^k)$. Thus, $|E| = r(M^k)$, which means that $E = \bigcup_1^k X_i$ for some $X_i \in M_i$, $i = 1, \ldots, k$. □

For natural reasons, a subset $X \subseteq E$ such that $|X| > \sum_1^k r(M_i \times X)$ is called an *obstacle to covering the set E with independent sets of the matroids $M_1, \ldots, M_k$*. Clearly, minimal (under inclusion) obstacles are circuits of the matroid $M^k = \bigvee_1^k M_i$.

4.2.2. We derive from Theorem 4.2.1 the following result.

COROLLARY. *Let M be a matroid on E. Then E is the union of k independent sets of the matroid M if and only if $k \cdot r(M \cdot A) \geq |A|$ for any $A \subseteq E$.*

4.2.3. Denote by $\operatorname{cov} M$ the minimum number of independent sets of a matroid M whose union is E.

COROLLARY.
$$\operatorname{cov} M = \max\left\{\left\lceil \frac{|A|}{r(M \cdot A)} \right\rceil : A \subseteq E\right\}.$$

The problem of covering E with the minimum number of independent sets of a matroid M is reduced, obviously, to successive problems of covering E with a given (and increasing) number of independent sets of M.

4.3. Let M_1 and M_2 be matroids on E. It is easy to see that M_1 and M_2 have a common independent set of cardinality r if and only if $r(M_1 \vee M_2^*) \geq r(M_2^*) + r$. So, the problem of finding a set of cardinality r independent both in M_1 and in M_2, or finding out that there are no such sets (called the *intersection problem for the matroids* M_1 *and* M_2), is reduced to the problem of constructing a base of the resultant matroid $M_1 \vee M_2^*$. The problem of finding a maximal set in E independent both in M_1 and M_2 is reduced, obviously, to the matroid intersection problem.

4.3.1. The matroid rank Theorem 3.7.2 yields:

THEOREM (see [1]). *Let* M_1 *and* M_2 *be matroids on* E. *Then* M_1 *and* M_2 *have a common independent set of cardinality* r *if and only if* $r \leqslant r(M_1 \cdot A) + r(M_2^* \cdot A)$ *for any* $A \subseteq E$.

PROOF. Suppose $r(M_1 \vee M_2^*) \geqslant r(M_2^*) + r$. By the matroid rank theorem, $r(M_2^*) + r \leqslant r(M_1 \vee M_2^*) = \min\{r(M_1 \cdot A) + r(M_2^*) + |\overline{A}|: A \subseteq E\}$, so that $r \leqslant r(M_1 \cdot A) + r(M_2 \cdot \overline{A})$ since $\overline{A} - (r(M_2^*) - r(M_2^* \cdot A)) = r(M_2 \cdot \overline{A})$.

Suppose $r \leqslant r(M_1 \cdot A) + r(M_2 \cdot \overline{A})$ for any $A \subseteq E$. Then $r(M_2^*) + r \leqslant \min\{r(M_1 \cdot A) + r(M_2^* \cdot A) + |\overline{A}|: A \subseteq E\}$. By the matroid rank theorem, the right-hand side of the inequality equals $r(M_1 + M_2^*)$. Thus, $r(M_2^*) + r \leqslant r(M_1 \vee M_2^*)$, and hence a set of cardinality r does exist in E. □

§5. Extremal sets of a matroid set

Let M_i be a matroid on E, ρ_i be the rank function of M_i, $i = 1, \ldots, k$, ρ be the rank function of the union $\bigvee_1^k M_i$, and $\mathcal{M} = (M_1, \ldots, M_k)$. A set A is said to be an *extremal set of the matroid set* $M_1, \ldots, M_k$ if the right-hand side of the formula in 3.8 for the rank of the union $\bigvee_1^k M_i$ attains its minimum at A. Denote by $\mathcal{E} = \mathcal{E}(\mathcal{M})$ the class of all extremal sets of the matroid set $\mathcal{M}$. Extremal sets are interpreted naturally as "obstacles" for constructing a maximum possible independent set (base) in the union of the matroids in $\mathcal{M}$. We shall see that $\mathcal{E}(\mathcal{M})$ is a lattice of subsets of E. Moreover, we shall describe, in certain terms, all extremal sets for $\mathcal{M}$.

5.1. As was indicated in 3.8, from 3.4 and 3.7.2 we derive

PROPOSITION. $D(\bigvee_1^k M_i)$ *is an extremal set of the matroid set* $\mathcal{M}$.

5.2. PROPOSITION. $X \subseteq E$ *is an extremal set of the matroid siquence* $M_1, \ldots, M_k$ *if and only if*

$$\rho(X) = \sum_1^k \rho_i(X); \qquad X \subseteq D\left(\bigvee_{i=1}^k M_i\right).$$

PROOF. Obviously, $r(\bigvee M_i) = r((\bigvee_1^k M_i) \cdot X) + r((\bigvee_1^k M_i) \times \overline{X})$. Moreover, $r((\bigvee_1^k M_i) \cdot X) = \rho(X) \leqslant \sum_1^k \rho_i(X)$ and $r((\bigvee_1^k M_i) \times \overline{X}) \leqslant |\overline{X}|$. Therefore, according to 3.8, X is an extremal set if and only if $\rho(X) = \sum_1^k \rho_i(X)$ and $r((\bigvee_1^k M_i) \times \overline{X}) = |\overline{X}|$. The latter means that $\overline{X}$ is a set of coloops of the matroid $\bigvee_1^k M_i$, i.e., $X \supseteq D(\bigvee_1^k M_i)$. □

5.3. From 3.8 and 5.2 we derive:

COROLLARY. $D(\bigvee_1^k M_i)$ is the minimal extremal set of the matroid set $M_1, \ldots, M_k$.

5.4. We need the following simple statement.

PROPOSITION. *Let $X \subseteq E$. Then $\rho(X) = \sum_1^k \rho_i(X)$ if and only if a base of the matroid $\bigvee_1^k M_i \cdot X$ is the union of the disjoint matroids $M_i \cdot X$, $i = 1, \ldots, k$.*

PROOF. Let B_i be a base of $M_i \cdot X$, $i = 1, \ldots, k$, and $B_i \cap B_j = \emptyset$ for $i \neq j$. Then, $\sum_1^k B_i$ is a base of the matroid $(\bigvee_1^k M_i) \cdot X$ and $\rho(X) = |\sum_1^k B_i| = \sum_1^k |B_i| = \sum_1^k \rho_i(X)$. Conversely, let $\rho(X) = \sum_1^k \rho_i(X)$, and let B be a base of $(\bigvee_1^k M_i) \cdot X$. Then there exists an $\mathscr{M}$-partition $(X_1, \ldots, X_k)$ of the base B such that $\sum_1^k \rho_i(X) = \rho(X) = |B| = \sum_1^k |X_i|$. As $X_i \in M_i$, we have $|X_i| \leq \rho_i(X)$ for $i = 1, \ldots, k$. Therefore, $|X_i| = \rho_i(X)$, and hence, X_i is a base of the matroid $M_i \cdot X$. □

5.5. From 5.2 and 5.4 we derive

PROPOSITION. *Let $X \subseteq E$. Then X is an extremal set of the matroid set $M_1, \ldots, M_k$ if and only if*

(a) *a base of the matroid $(\bigvee_1^k M_i) \cdot X$ is the union of disjoint bases of the matroids $M_1 \cdot X, \ldots, M_k \cdot X$, and*

(b) $X \supseteq D(\bigvee_1^k M_i)$.

5.6. Let $\mathscr{Z} = \mathscr{Z}(\mathscr{M})$ be the class of subsets $X \subseteq E$ such that $\rho(X) = \sum_1^k \rho_i(X)$ (then $\mathscr{E}(\mathscr{M}) \subseteq \mathscr{Z}(\mathscr{M})$).

LEMMA. *If $X^1, X^2 \in \mathscr{Z}$, then $X^1 \cup X^2 \in \mathscr{Z}$.*

PROOF. We can assume that $X^1 \cup X^2 = E$, since otherwise we could consider the matroids $M_i(X^1 \cup X^2)$, $i = 1, \ldots, k$, on E instead of the matroids M_i. Since $X^s \in \mathscr{Z}$, 5.4 implies that there exists a packing $(B_1^s, \ldots, B_k^s)$ of bases of the matroids $M_1 \cdot X^s, \ldots, M_k \cdot X^s$, $s = 1, 2$. Clearly, $B_i^1 \setminus X^2$ contains a base R_i^1 of the matroid $M_i \times (X^1 \setminus X^2)$. Then $B_i = R_i^1 + B_i^2$ is a base of $M_i(X^1 \cup X^2)$, so that $(B_1^s, \ldots, B_k^s)$ is a packing of bases of the matroids $M_1, \ldots, M_k$. According to 5.4, $X^1 \cup X^2 \in \mathscr{Z}$. □

5.7. It follows from Lemma 5.6 that there exists a unique maximal (under inclusion) set in the class $\mathscr{Z}(\mathscr{M})$, to be denoted by $H(\mathscr{M})$. By 5.1, $D(\bigvee_1^k M_i) \in \mathscr{Z}(\mathscr{M})$. Therefore, 5.2 implies

THEOREM. *$\mathscr{E}(\mathscr{M})$ is a lattice of subsets of E (under inclusion), $H(\mathscr{M})$ is the maximal set, and $D(\bigvee_1^k M_i)$ is the minimal set of the lattice $\mathscr{E}(\mathscr{M})$.*

5.8. LEMMA. *Let $\lambda = (B_1, \ldots, B_k)$ be an $\mathscr{M}$-partition of a base B of the matroid $\bigvee_1^k M_i$, and X be an extremal set of the set $M_1, \ldots, M_k$. Then $B_i \cap X$ is a base of $M_i \cdot X$, $i = 1, \ldots, k$.*

PROOF. Set $X_i = B_i \cap X$. By 5.2, $X \supseteq D(\bigvee_1^k M_i)$. Therefore, $X \cap B = \sum_1^k X_i$ is a base of $(\bigvee_1^k M_i) \cdot X$, so that $\rho(X) = |X \cap B| = \sum_1^k |X_i|$. By 5.2, $\rho(X) = \sum_1^k \rho_i(X)$. As $X_i \in M_i$ for any $i = 1, \ldots, k$, we have $|X_i| \leq \rho_i(X)$, and $X_i \in M_i$ is a base of $M_i \cdot X$ for any $i = 1, \ldots, k$. □

5.9. Let $\lambda = (B_1, \ldots, B_k)$ be an $\mathcal{M}$-partition of a base B of the matroid $\bigvee_1^k M_i$, and let $K(\lambda)$ denote the set of all non-λ-active elements (vertices in G_λ).

THEOREM. $H(\mathcal{M}) = K(\lambda)$ (hence $K(\lambda)$ depends neither on λ nor on B).

PROOF. Let $H(\mathcal{M}) = H$. By 5.8, $H \cap B_i$ is a base of $M_i \cdot H$. Therefore, each element in H is not λ-active, so that $H \subseteq K(\lambda)$. Now, it remains to prove that $K(\lambda) \in \mathcal{E}(\mathcal{M})$. Since $D(\bigvee_1^k M_i) \subseteq H(\mathcal{M}) \subseteq K(\lambda)$, by 5.5 it suffices to prove that $K_i(\lambda) = B_i \cap K(\lambda)$ is a base of the matroid $M_i' = M_i \cdot K(\lambda)$. Let us consider $x \in K(\lambda) \setminus K_i(\lambda)$ and prove that $K_i(\lambda) + x \notin M_i$. Since x is not λ-active, there exists the circuit $C(M_i, B_i + x)$. If this circuit contains an element y in $B_i \setminus K_i(\lambda)$, then y is λ-active, hence x is λ-active as well, a contradiction. Thus, $C(M_i, B_i + x) \subseteq K_i(\lambda) + x$, i.e., $K_i(\lambda) + x \notin M_i$. □

5.10. Let, as above, $\lambda = B_1, \ldots, B_k$ be an $\mathcal{M}$-partition of a base B of the matroid $\bigvee_1^k M_i$. In 5.7 we concluded that $D(\bigvee_1^k M_i)$ is the minimal extremal set of the matroid set $\mathcal{M}$. In terms of the digraph G_λ, $D(\bigvee_1^k M_i)$ is the set $G_\lambda(\overline{B})$ of all elements-vertices attainable in G_λ from $\overline{B}$ (see (3.7.2)). It turns out that any extremal set of the set $\mathcal{M}$ can be described in similar terms. Put $D(\bigvee_1^k M_i) = D$ and $H(\mathcal{M}) = H$.

THEOREM. $\mathcal{E}(\mathcal{M}) = \{G_\lambda(A) : \overline{B} \subseteq A \subseteq H\}$, where B is an arbitrary base of the matroid $\bigvee_1^k M_i$.

PROOF. By 5.9, H is the set of all non-λ-active elements. Therefore, for $A \subseteq H$, we have $\rho(G_\lambda(A)) = \sum_1^k \rho_i(G_\lambda(A))$. If $A' \subseteq A$, then $G_\lambda(A') \subseteq G_\lambda(A)$. By 3.7.2, $D = G_\lambda(\overline{B})$. Therefore, if $\overline{B} \subseteq A \subseteq H$, then, by 5.2, $G_\lambda(A) \in \mathcal{E}(\mathcal{M})$. And conversely: let $X \in \mathcal{E}(\mathcal{M})$. Then, by 5.2, $\rho(X) = \sum_1^k \rho_i(X)$ and $D \subseteq X$. According to 5.5, $X_i = B_i \cap X$ is a base of the matroid $M_i \cdot X$. Hence, $X = G_\lambda(\overline{B} + (X \setminus D))$. □

REMARK. Proposition 5.7 follows immediately from this theorem.

5.13. In what follows we need a somewhat different notation for the graph G_λ, indicating directly its dependence on the set $\mathcal{M}$. Let $G_\lambda = G(\mathcal{M}, \lambda)$. For $X \subseteq E$, put

$$dX = dX(\mathcal{M}) = X \setminus \left(\bigvee_{i=1}^k M_i\right), \qquad G^X(\mathcal{M}, \lambda) = G(\mathcal{M}, \lambda) - (E \setminus dX),$$

and, as above, $\mathcal{M} \cdot X = (M_1 \cdot X, \ldots, M_k \cdot X)$.

From 5.10, we immediately deduce

THEOREM. (c1) $\mathscr{E}(\mathscr{M}) = \mathscr{E}(\mathscr{M} \cdot H)$.

(c2) *For any* $A \in \mathscr{E}(\mathscr{M})$, *the lattice* $\mathscr{E}(\mathscr{M} \cdot A)$ *of extremal sets of the matroid set* $\mathscr{M} \cdot A$ *is isomorphic to the lattice* $\mathscr{Z}(G^A(\mathscr{M}, \lambda))$ *of directed cuts of the graph* $G^A(\mathscr{M}, \lambda)$ *(obviously,* $\mathscr{E}(\mathscr{M} \cdot A) = \mathscr{E}(\mathscr{M})|_A$).

(c3) *Let* $\{T_i : i = 1, \ldots, n\}$ *be the collection of all bicomponents of the graph* $G^A(\mathscr{M}, \lambda)$. *Then* $\{G_\lambda(\overline{B} + T_i) : i = 1, \ldots, n\}$ *is the collection of all join-indecomposable elements of the lattice* $\mathscr{E}(\mathscr{M} \cdot A)$, *while* $\{D, T_1, \ldots, T_n\}$ *is the principal partition of the set* H *for the lattice* $\mathscr{E}(\mathscr{M} \cdot A)$.

We shall see below that for $A \in \mathscr{E}(\mathscr{M})$, the graph $G^A(\mathscr{M}, \lambda)$ is the graph $G(\mathscr{M}^A, \lambda^A)$ for another matroid set $\mathscr{M}^A$ and some $\mathscr{M}^A$-partition λ^A of some base in the union of matroids from $\mathscr{M}^A$.

5.11. As above, $H(\mathscr{M}) = H$, $D(\bigvee_1^k M_i) = D$, $I(\bigvee_1^k M_i) = I$. For $X \subseteq E$, put $M_i * X = (M_i \cdot X) \times dX$ (so that dX is the ground set of the matroid $M_i * X$), $\mathscr{M} * X = (M_1 * X, \ldots, M_k * X)$, and $\lambda * X = (B_1 \cap dX, \ldots, B_k \cap dX)$.

THEOREM. *Let* A *be an extremal set of the set* $\mathscr{M} = (M_1, \ldots, M_k)$: $A \in \mathscr{E}(\mathscr{M})$, *and let* $\lambda = (B_1, \ldots, B_k)$ *be an* $\mathscr{M}$-*partition of a base* B *of the matroid* $M(\mathscr{M}) = \bigvee_1^k M_i$. *Then:*

(c1) $B_i \cap A$ *is a base of the matroid* $M_i * A$ *so that* $dA = \sum_1^k B_i \cap A$ *is a base of the matroid* $M(\mathscr{M} * A) = \bigvee_1^k (M_i * A)$ *(i.e.,* $D(\bigvee_1^k M(\mathscr{M} * A)) = \varnothing$, *and hence,* $M(\mathscr{M} * A)$ *is a free matroid), and* $\lambda * A$ *is an* $\mathscr{M} * A$-*partition of the base* dA.

(c2) $G^A(\mathscr{M}, \lambda) = G(\mathscr{M} * A, \lambda * A)$.

PROOF. (p0) Obviously, $VG^A(\mathscr{M}, \lambda) = VG(\mathscr{M} * A, \lambda * A) = dA$. Put $M_j * A = M_j'$ and $B_j \cap dA = B_j'$. It suffices to prove that for any $i, j \in \{1, \ldots, k\}$, and any $x \in B_i'$ and $y \in B_j'$, one has $y \in C(M_j, B_j + x) \iff y \in C(M_j', B_j' + x)$.

(p1) Let us first prove the following auxiliary statement. Let $\mathscr{C}_y$ be the set of circuits T of the matroid M_j such that $y \in T$ and $T \cap I \subseteq B_j$ (so that $\varnothing \ne T \setminus B_j \subseteq D$). Then $\mathscr{C}_y = \varnothing$. Assume the converse, i.e., $\mathscr{C}_y \ne \varnothing$, and choose in $\mathscr{C}_y$ a circuit K such that the cardinality of the set $\delta K = K \setminus B_j$ is minimum. Let $z \in \delta K$. By 5.8 (or by 3.3), $B_j^d = B_j \cap D$ is a base of $M_j \cdot D$ so that there exists a circuit $C_z = C(M_j, B_j^d + z)$ such that $z \in C_z \subseteq D$ and, as a result, $y \notin C_z$. By the sharpened circuit axiom ($\mathscr{C}2'$), there exists a circuit K' of the matroid M_j such that $y \in K' \subseteq K \cup C_z \setminus z$. Then $K' \cap I \subseteq B_j$, and hence, $K' \in \mathscr{C}_y$. But $|\delta K'| \le |\delta K \setminus z| < |\delta K|$, which contradicts the choice of the circuit K in $\mathscr{C}_y$.

(p2) Assume that $y \in C(M_j, B_j + x)$. Since $A \in \mathscr{E}(\mathscr{M})$, it follows from 5.8 that $C \subseteq A$, and therefore, C is a circuit of the matroid $M_j \cdot A$. By the property of the contraction operation, there exists a circuit C'_y of the matroid M'_j that contains y and lies in $\hat{C} = C \cap dA$. For this circuit C, there exists a circuit C_y of the matroid $M_j \cdot A$ such that $C_y \cap dA = C'_y$. Then $y \in C_y$. Suppose $x \notin C'_y$. Then $C'_y \cap I \subseteq B_j$, and hence $C_y \in \mathscr{C}_y$, which contradicts (p1). Thus, $x, y \in C'_y$. Since $C'_y \subseteq \hat{C}$, we have $C_y = C(M'_j, B'_j + x)$.

(p3) Conversely, assume that $y \in C(M'_j, B'_j + x) = C'_{xy}$. Consider also the circuit $C = C(M_j, B_j + x)$. Let $y \notin C$. As indicated in (p1), $C \subseteq A$. By the contraction property, there exists a circuit C_{xy} of the matroid M such that $C'_{xy} = C_{xy} \cap dA$. By the sharpened circuit axiom ($\mathscr{C}2'$) (applied to the circuits C and C_{xy}), there exists a circuit C_y such that $y \in C_y \subseteq C \cup C_{xy} \setminus x$. Then $C_y \cap I \subseteq B_j$, and therefore $C_y = \mathscr{C}_y$, which contradicts (p1). Thus, the following statement holds:

5.13. PROPOSITION. *If $\bigvee_1^k (M_i \times A)$, $A \subseteq E$, is a free matroid (i.e., $D(\bigvee_1^k M(\mathscr{M} \times A)) = \varnothing$), then $A \subseteq I(\bigvee_1^k M)$.*

PROOF (by induction on k). For $k = 1$, the assertion follows directly from the properties of the contraction operation. Let the assertion hold for some $k = p \geq 1$; we prove it for $k = p + 1$. Put $\bigvee_1^p M_i = M^1$ and $M_{p+1} = M^2$. Suppose that there exists an element $x \in A \cap D(M^1 \vee M^2)$. By the contraction property, there exist a circuit of the matroid $(M^1 \vee M^2) \times A$ and a circuit C^* of the matroid $M^1 \vee M^2$ such that $x \in C^* = C^* \cap A$. By the construction of circuits in the sum of matroids (see 3.5), $C^* \setminus x = X^1 + X^2$, where X^i is a base of the matroid $M_i \cdot C^*$, $i = 1, 2$. Put $X_1^i = X_i \cap A$. Three cases are possible:

(c1) $X^1 = X^2 = \varnothing$; or

(c2) one of the sets X^1, X^2, say X^1, is empty, and the other, i.e., X^2, is nonempty; or

(c3) both sets X^1 and X^2 are nonempty.

In case (c1), x is a loop of the matroid $\bigvee_{i=1}^2 (M^i \times A) = \bigvee_{i=1}^{p+1} (M_i \times A)$, contradicting the assumptions. In case (c2), x is a loop of the matroid $M^1 \times A$, and each element $y \in X_1^1$ is also a loop of the matroid $M^1 \times A$, since $C(M^1, X^1 + y) \subseteq X^1 + y \subseteq \overline{A} + y$. Thus, the set $X_1^2 + x$ consists of loops of the matroid $M^1 \times A$. Since $M^1 \times A + M^2 \times A$ is a free matroid by the assumption, we see that A is its base, and therefore, $A = A^1 + A^2$, $A^i \in M^i \times A$, $i = 1, 2$. Since $X^2 + x$ consists of loops of the matroid $M^1 \times A$, we obtain $X_1^2 + x \in M^2 \times A$. By the construction of circuits in the union of matroids established in 3.5, there exists an element $z \in X^1 + x$ such that $C(M^2, X^2 + z) \cap X_1^2 \neq \varnothing$, and by the property of the contraction operation, there must exist a circuit of the matroid $M^1 \times A$ that belongs to $X_1^2 + x$; this yields $X_1^2 + x \notin M^2 \times A$, a contradiction.

Thus, only the case (c3) remains. Let $X_2^i \subseteq X_1^i$ be a base of the minor $(M^i \times A) \cdot X_2^i$, $i = 1, 2$. Then $X_2 = \sum_{i=1}^{2} X_2^i \in \sum_{i=1}^{2} (M^i \times A)$. Let us prove that X_2^i is a base of the minor $(M^i \times A) \cdot C$, $i = 1, 2$, i.e., that for any element $y \in C \setminus X_2^i$ there exists a circuit $C(M^i \times A, X_2^i + y)$. Since X^i is a base of the matroid $M^i \cdot C^*$, $X^i + y$ contains the circuit $C(M^i, X^i + y)$. Two cases are possible:

(c3.1) $C(M^i, X^i + y) \cap X_1^i = \varnothing$,

(c3.2) $C(M^i, X^i + y) \cap X_1^i \neq \varnothing$.

In case (c3.1), y is a loop in $M^i \times A$, and therefore it represents the desired circuit $C(M^1 \times A, X_2^i + y)$. In case (c3.2), by the properties of the contraction operation, one can find a circuit C' of the matroid $M^1 \times A$ such that $y \in C'$ and $C' \setminus y \subseteq X_2^i$. Let us show that there exists a circuit $C(M^1 \times A, X_2^i + y)$. Assume the circuit C' is chosen in such a way that the cardinality of the set $C' \setminus X_2^i$ is minimum. If there exists an element $z \in C' \setminus X_2^i$, $z \neq y$, then $z \in X_1^i$, and there exists the circuit $C(M^1 \times A, X_2^i + z)$. Then, by the sharpened circuit axiom ($\mathscr{C}2'$), there exists a circuit C'' of the matroid $M^i \times A$ such that $y \in C'' \subseteq C' \cup C(M^1 \times A, X_2^i + z) \setminus z$. The circuit C'' has the same properties as C', but $|C'' \setminus X_2^i| < |C' \setminus X_2^i|$, which contradicts the choice of the circuit C'. Therefore, $C' \setminus X_2^i = \{y\}$, i.e., $C' = C(M^1 \times A, X_2^i + y)$. Thus, in case (c3), the set $X_2 = \sum_{i=1}^{2} X_2^i$ is a base of the matroid $\bigvee_{i=1}^{2} (M_i \times A) \cdot C$, and therefore, there exists the circuit $C(M^1 \times A + M^2 \times A, X_2 + x)$, which contradicts the assumption. □

5.14. THEOREM. *A subset $X \subseteq E$ is an extremal set of the matroid $\bigvee_{i=1}^{k} M_i$ if and only if $\rho(X) = \sum_{i=1}^{k} \rho_i(X)$ and the matroid $\bigvee_{i=1}^{k} (M_i \times \overline{X})$ is free.*

PROOF. The necessity follows from Propositions 5.2 and 7.3, and the sufficiency from Propositions 5.2 and 5.13. □

5.15. PROPOSITION. *Let $X \subseteq E$ satisfy $\rho(X) = \sum_{i=1}^{k} \rho_i(X)$. Then $\bigvee_{i=1}^{k} (M_i \times \overline{X})$ is a free matroid if and only if $\overline{X} \subseteq I(\bigvee_{i=1}^{k} M_i)$.*

§6. A base sequence of a matroid sequence and its properties

6.1. Let $M_1, \ldots, M_k$ be a matroid sequence on E. We denote by pack$(M_1, \ldots, M_k)$ the problem of finding the longest sequence $(B_1, \ldots, B_p)$ of sets in E such that $B_i \cap B_j = \varnothing$ with $i \neq j$ and B_i is a base of M_i, $i = 1, \ldots, k$. Denote by cov$(M_1, \ldots, M_k)$ the problem of finding the shortest sequence $X_1, \ldots, X_c$ of sets in E such that X_i is an independent set of M_i and $\bigcup_{i=1}^{c} X_i = E$ (if it exists). Clearly, the problems pack$(M_1, \ldots, M_k)$ and cov$(M_1, \ldots, M_k)$ generalize, respectively, the problems pack(M) and cov(M) to the case of matroid sequences. The solution to both problems, similarly to §4, is reduced to constructing bases of the matroids $\bigvee_{i=1}^{s} M_i$, $s = 1, \ldots, k$.

In this section, we define, for a matroid sequence $(M_1, \ldots, M_k)$, a special sequence $(Y_1, \ldots, Y_k)$ of independent sets of the matroids $M_1, \ldots, M_k$ called a *base sequence* of the sequence $(M_1, \ldots, M_k)$. We shall see that base sequences have some interesting properties. In particular, they contain solutions to both problems pack$(M_1, \ldots, M_k)$ and cov$(M_1, \ldots, M_k)$. In the case when all matroids of the sequence $(M_1, \ldots, M_k)$ are identical $(M_i = M)$, and their number is sufficiently large, a base sequence of this sequence describes certain properties of the matroid M and is called a *base sequence of the matroid* M. As mentioned above, the notion of a regular packing of spanning forests of a graph introduced by Polesskiĭ in 1971 [4] serves as a prototype for the notion of a base sequence.

6.2. Consider a sequence $\mathcal{M} = (M_1, \ldots, M_k)$ of matroids on E. As in 3.1, a sequence $\beta = (Y_1, \ldots, Y_k)$ of subsets of E is called an $\mathcal{M}$-sequence if

(c1) $X_i \in M_i$, $i = 1, \ldots, k$, and
(c2) $X_i \cap X_j = \emptyset$ for $i \neq j$, $i, j = 1, \ldots, k$;

If $X \in \bigvee_1^k M_i$, then there exists an $\mathcal{M}$-sequence $\beta = (X_1, \ldots, X_k)$ such that $X = \sum_1^k X_i$. Such an $\mathcal{M}$-sequence β is called an $\mathcal{M}$-*partition* of X, and the $\mathcal{M}$-sequence $\beta = (B_1, \ldots, B_k)$ is called a *base sequence of the matroid sequence* $\mathcal{M} = (M_1, \ldots, M_k)$, provided

(c3) $\sum_{i=1}^{l} B_i$ is a base of the matroid $\bigvee_1^l M_i$ for any $l \in \{1, \ldots, k\}$.

6.3. PROPOSITION. *Each base B of the matroid $\bigvee_1^k M_i$ has an $\mathcal{M}$-partition that is a base sequence of the sequence* $\mathcal{M} = (M_1, \ldots, M_k)$.

PROOF. Let $\mathcal{M}^l = (M_1, \ldots, M_l)$ and $M^l = \bigvee_1^l M_i$, $l = 1, \ldots, k$. Let us prove that for any base B^l of the matroid M^l, there exists an M^l-partition that is a base sequence of M^l. For $i = 1$, the statement is obvious. Since $\bigvee_1^k M_i = (\bigvee_1^{k-1} M_i) \vee M_k$, we have $B = X^{k-1} + X_k$, where $X^{k-1} \in \bigvee_1^{k-1} M_i$, $X_k \in M_k$. Let B^{k-1} be a base of the matroid $(\bigvee_1^{k-1} M_i) \cdot B$ containing X^{k-1}. Then $B = B^{k-1} + B_k$, where $B_k = X_k \setminus B^{k-1}$. Let us prove that B^{k-1} is a base of the matroid $\bigvee_1^{k-1} M_i$. Suppose the converse, i.e., $B^{k-1} + x \in \bigvee_1^{k-1} M_i$ for some $x \in E \setminus B^{k-1}$. If $x \in B_k$, then B^{k-1} is not a base of $(\bigvee_1^{k-1} M_i) \cdot B$, a contradiction. If $x \in \overline{B}$, then $B + x \in \bigvee_1^k M_i$, and therefore B is not a base of $\bigvee_1^k M_i$, a contradiction. By the inductive hypothesis, there exists an $\mathcal{M}^{k-1}$-partition $(B_1, \ldots, B_{k-1})$ of the base B^{k-1} which is a base sequence of $\mathcal{M}^{k-1}$. Then $(B_1, \ldots, B_{k-1}, B_k)$ is an $\mathcal{M}^k$-partition of the base B that is a base sequence of $\mathcal{M}^k$. □

6.4. The arguments of 6.3 immediately imply the following result.

PROPOSITION. *Let $\beta = (B_1, \ldots, B_k)$ be a base sequence of a sequence $\mathcal{M} = (M_1, \ldots, M_k)$. Then $(|B_1|, \ldots, |B_k|)$ is the lexicographical maximum of $\{(|X_1|, \ldots, |X_k|) : (X_1, \ldots, X_k)$ is an $\mathcal{M}$-sequence$\}$. In particular,*

(a) *B_{l+1} is a base of the matroid $M_{l+1} \setminus \sum_1^l B_i$, $l = 0, \ldots, k-1$, and*

(b) *if $\beta' = (B'_1, \ldots, B'_k)$ is a base sequence of $\mathcal{M}$, then $|B_i| = |B'_i|$, $i = 1, \ldots, k$.*

6.5. A base sequence $\beta = (B_1, \ldots, B_k)$ of a matroid sequence $\mathcal{M} = (M_1, \ldots, M_k)$ splits into *blocks* in the following way: $\mathcal{B}_1 = (B_1, \ldots, B_{s_1})$, $\ldots$, $\mathcal{B}_{j+1} = (B_{s_j+1}, \ldots, B_{s_{j+1}})$, $\ldots$, so that $B_i \in \mathcal{B}_1 \iff B_i$ is a base of M_i, and, in general, $B_i \in \mathcal{B}_{j+1} \iff B_i$ is a base of $M_i \setminus \sum \{B_s : s \leq s_j\}$. Denote by $b = b(\beta, \mathcal{M})$ the *number of blocks* of the base sequence β. Certainly, $\mathcal{B}_j$ and s_j depend on β and $\mathcal{M}$: $\mathcal{B}_j = B_j(\beta, \mathcal{M})$ and $s_j = s_j(\beta, \mathcal{M})$. The number of elements $l_j(\beta, \mathcal{M}) = |\mathcal{B}_j(\beta, \mathcal{M})| = s_j(\beta, \mathcal{M}) - s_{j-1}(\beta, \mathcal{M})$ of the block $\mathcal{B}_j$ is called its *length*. Obviously, $\sum \{l_j : j = 1, \ldots, b\} = k$. From 6.4 we conclude that if β and β' are base sequence for $\mathcal{M}$, then $l_1(\beta, \mathcal{M}) = l_1(\beta', \mathcal{M}) = l_1(\mathcal{M})$. Possibly, $l_j(\beta, \mathcal{M}) \neq l_j(\beta', \mathcal{M})$ for $j \geq 2$, i.e., $l_j(\beta, \mathcal{M})$ depends, in general, on β.

6.6. Consider a sequence $\mathcal{M}^k = (M, \ldots, M)$ of k identical matroids M, and let $\beta^k = (B_1, \ldots, B_k)$ be a base sequence of $\mathcal{M}^k$. Then, obviously, $B_i = \emptyset$ with $i = c = \operatorname{cov} M$. A base sequence of the sequence M^c consisting of $c = \operatorname{cov} M$ identical matroids M is said to be a *base sequence of the matroid M*. It is easy to see that each block $\mathcal{B}_i(\beta, \mathcal{M})$ of a base sequence β of the matroid M consists of sets of equal cardinality: $B_p, B_q \in \mathcal{B}_i \Rightarrow |B_p| = |B_q| = r_i(\beta, \mathcal{M})$; the number $r_i(\beta, \mathcal{M})$ is called the *rank of the block $\mathcal{B}_i(\beta, \mathcal{M})$*; the number of blocks, the length and the rank of each block of a base sequence β of the matroid M do not depend on β, and depend merely on M:

$$b(\beta, \mathcal{M}) = b(M), \qquad l_i(\beta, \mathcal{M}) = l_i(M), \qquad r_i(\beta, \mathcal{M}) = r_i(M).$$

Obviously,

$$r_{i+1} < r_i, \qquad \sum_1^b l_i = \operatorname{cov} M; \qquad \sum_1^b l_i r_i = |E|.$$

Thus, a base sequence of a matroid M is a partition of E into the minimum number of independent sets of the matroid M. This partition consists of a sequence of maximum packings of bases of "narrowing" matroids.

§7. Base sequence and extremal sets of matroid sequences

7.1. Let $\mathcal{M} = \mathcal{M}^k = (M_1, \ldots, M_k)$, where each M_i is a matroid on E. Let $A \subseteq E$. For $\lambda = (X_1, \ldots, X_k)$, set $\lambda \cdot A = (X_1 \cap A, \ldots, X_k \cap A)$. The sequence of sets $\lambda \cdot A$ is said to be the *trace of the sequence λ on A*. Set

$\mathscr{M} \cdot A = (M_1 \cdot A, \ldots, M_k \cdot A)$ and $\mathscr{M} \times A = (M_1 \times A, \ldots, M_k \times A)$. If $\beta = (B_1, \ldots, B_k)$ is a base sequence of the matroid sequence $\mathscr{M}$, then the trace $\beta \cdot A$ of the base sequence β, in general, is not a base sequence of the matroid sequence $\mathscr{M} \cdot A$. However, if A is an extremal set of the set $\mathscr{M}^l$, $l \leqslant k$, then the situation differs.

THEOREM. *Let β be a base sequence of a matroid sequence $\mathscr{M}^k$ and A be an extremal set of a matroid sequence $\mathscr{M}^l$, $l \leqslant k$. Then $\beta \cdot A$ is a base sequence of the matroid sequence $\mathscr{M} \cdot A$.*

PROOF. By the definition of base sequence, the set $\sum_1^j B_i$ is a base of the matroid $\bigvee_1^j M_i = M^j$ for each $j = 1, \ldots, k$. According to 5.2, $\overline{A} \subseteq I(M^l) \subseteq I(M^j)$ for $j \geqslant l$. Therefore, $(\sum_1^j B_i) \cap A \sum_1^j (B_i \cap A)$ is a base of the matroid $(\bigvee_1^j M_i) \cdot A = \bigvee_1^j (M^i \cdot A)$. By 5.5, the base $\sum_1^l (B_i \cap A)$ is the union of disjoint bases $B_1 \cap A, \ldots, B_l \cap A$ of the matroids $M_1 \cdot A, \ldots, M_l \cdot A$, respectively. Therefore for $j < l$ the set $\sum_1^j (B_i \cap A)$ is a base of the matroid $\bigvee_1^j (M_i \cdot A)$. □

COROLLARY. *Let $M_i = M$ and A be an extremal set of the matroid sequence $\mathscr{M}^k$, $k = 1, \ldots, \operatorname{cov} M$. If β is a base sequence of the matroid M, then $\beta \cdot A$ is a base sequence of the matroid $M \cdot A$.*

7.2. PROPOSITION. *Let A be an extremal set for $\mathscr{M}^l$, and $(B_1, \ldots, B_k)$ be a base sequence for M^k, $l \leqslant k$. Then $(B_i \cap \overline{A}) \in (M_i \times \overline{A})$, $i = 1, \ldots, k$.*

PROOF. Since $\overline{A} \subseteq I(M^l) \subseteq \sum_1^l B_i$, we have $B_i \cap \overline{A} = \varnothing$ with $i = l+1, \ldots, k$. Assume the converse, i.e., that $B_i \cap \overline{A}$ contains a circuit of the matroid $M_i \times \overline{A}$ for some $i \leqslant l$. By the definition of the contraction, there exists a circuit C of the matroid M_i such that $C' = C \cap \overline{A}$. Among these circuits, we choose the circuit C with the minimal cardinality of the set $C \setminus B_i$. Since $C \nsubseteq B_i$, we have $C \setminus B_i \neq \varnothing$. It follows from 5.5 that $B_i \cap A$ is a base of the matroid $M_i \cdot A$. Therefore, for each element $x \in C \setminus B_i$, there exists the circuit $C(M_i, B_i + x) \subseteq A$. Let us fix $x \in C \setminus B_i$, $y \in C'$. By the sharpened circuit axiom $(\mathscr{C}2)'$, there exists a circuit C'' of the matroid M such that $y \in C'' \subseteq C \cup C(M_i, B_i + x) \setminus x$. Since $y \in C'' \cap \overline{A} \subseteq C'$, we have $C'' \cap \overline{A} = C'$. But $|C'' \setminus B_i| \leqslant |C \setminus B_i \setminus x| < |C \setminus B_i|$, which contradicts the choice of the circuit C. □

7.3. COROLLARY. *Let A be an extremal set for $\mathscr{M}^k$. Then the matroid $\bigvee_1^k (M_i \times \overline{A})$ is free.*

PROOF. Put $l = k$ in Proposition 7.2. Then $\overline{A} = \sum_1^k (B_i \cap \overline{A})$, while $B_i \in \overline{A} \in M_i \times \overline{A}$. □

7.4. COROLLARY. *Let $M_i = M$. Then an extremal set of the matroid set $\mathscr{M}^k$ is closed in the matroid M.*

PROOF. Let $(B_1, \ldots, B_c)$ with $c = \operatorname{cov} M$ be a base sequence of the matroid M. By 7.2, $B_i \cap \overline{A} \in M_i \times \overline{A}$, $i = 1, \ldots, c$. But $\sum_1^c B_i = E$, so that there are no loops of the matroid M in $\overline{A} = \sum_1^c (B_i \cap \overline{A})$, i.e., $\operatorname{cl}(A) = A$. □

REMARK. If a matroid set $\mathscr{M}^k$ consists of nonidentical matroids, then an extremal set A of $\mathscr{M}^k$ need not, in general, be closed in each matroid M_i.

7.5. THEOREM. *Let A be an extremal set of the set $\mathscr{M}^l$, $l \leqslant k$, and $(B_1^0, \ldots, B_k^0)$ and $(B_1^\times, \ldots, B_k^\times)$ be base sequence of the matroid sequences $\mathscr{M} \cdot A = (M_1 \cdot A, \ldots, M_k \cdot A)$ and $\mathscr{M} \times A = (M_1 \times \overline{A}, \ldots, M_k \times \overline{A})$, respectively. Then $(B_1^0 + B_1^\times, \ldots, B_k^0 + B_k^\times)$ is a base sequence of the matroid sequence $\mathscr{M} = (M_1, \ldots, M_k)$.*

PROOF. Since $B_1^0, B_1^\times$ are bases of the matroids $M_1 \cdot A, M_1 \times A$, respectively, we see that $B_1^0 + B_1^\times$ is a base of the matroid M_1. Put $B_i = B_i^0 + B_i^\times$ and $B^j = \sum_1^j B_i$, $i, j = 1, \ldots, k$. Assume the converse, i.e., that $(B_1, \ldots, B_k)$ is not a base sequence for $\mathscr{M}^k$, and let m be the minimal number j such that the set B is not a base of the matroid $M^j = \bigvee_1^j M_i$, so that $j > 1$. Let us prove that $m \leqslant l$. Suppose that $m > l$. Then B^l is a base of the matroid $M^l = \bigvee_1^l M_i$, and therefore, $B^l \supseteq I(M^l) \supseteq \overline{A}$. But $I(M^m) \supseteq I(M^l)$, and hence, $\overline{A} \subseteq I(M^m)$. By the assumption, $\sum_1^m B_i^0$ is a base of the matroid $\bigvee_1^m (M_i \cdot A) = (\bigvee_1^m M_i) \cdot A$. On the other hand, $\sum_1^m B_i^\times = \overline{A}$ is a base of the free matroid $M^m \times \overline{A}$, because all circuits of the matroid M^m lie in A. But then, $\sum_1^m B_i^0 + \sum_1^m B_i^\times = B^m$ is a base of the matroid M^m, a contradiction. So, $m \leqslant l$. Let $(\tilde{B}_1, \ldots, \tilde{B}_m)$ be a base sequence of the sequence $\mathscr{M}^m = (M_1, \ldots, M_m)$, so that $\sum_1^m \tilde{B}_i = \tilde{B}^m$ is a base in M^m.

Put $\hat{B}_i = \tilde{B}_i \cap \overline{A}$. By 7.2, $\hat{B}_i \in M_i \times \overline{A}$, and so $\sum_1^m \hat{M}_i \in \bigvee_1^m (M_i \times \overline{A})$. According to 7.1, the set $\tilde{B}^m \cap A = \sum_1^m (\tilde{B}_i \cap A)$ is a base of the matroid $\bigvee_1^m (M_i \times A)$. Next, $\sum_1^m B_i^\times \subseteq \overline{A}$ by the assumption, and hence $B_m \cap A = \sum_1^m (B_i^0 \cap A)$ represents a base of the same matroid, so we have $|B^m \cap A| = |\tilde{B}^m \cap A|$. Thus, $|\sum_1^m \hat{B}_i| = |\tilde{B}^m| - |\tilde{B}^m \cap A| > |B^m| - |B^m \cap A| = \sum_1^m B_i^\times$, a contradiction. □

7.6. THEOREM. *Let A be an extremal set of the set $\mathscr{M}^l$, $l \leqslant k$, and $(B_1, \ldots, B_k)$ be a base sequence of the sequence $\mathscr{M} = (M_1, \ldots, M_k)$. Then $(B_1 \cap \overline{A}, \ldots, B_k \cap \overline{A})$ is a base sequence of the matroid sequence $\mathscr{M}^k \times \overline{A} = (M_1 \times \overline{A}, \ldots, M_k \times \overline{A})$.*

PROOF. Put $B_i^\times = B_i \cap \overline{A}$ and $\overline{B}_i^0 = B_i \cap A$. By 7.2, $B_i^\times \in M_i \times \overline{A}$. Let m be the least number such that $\sum_1^m B_i^\times$ is not a base of the matroid $\bigvee_1^m (M_i \times \overline{A})$. It is easy to see that $m < l$, since for $m \geqslant l$ the set $\sum_1^m B_i^\times = \overline{A}$ is a base of the free matroid $\bigvee_1^m (M_i \times \overline{A})$. Let $(\hat{B}_1, \ldots, \hat{B}_m)$ be a base sequence of the sequence $\mathcal{M}^m \times \overline{A}$. By 7.1, $(B_1^0, \ldots, B_m^0)$ is a base sequence for $\mathcal{M}^m \circ \overline{A}$. Put $\tilde{B}_i = B_i^0 + \hat{B}_i$, $i = 1, \ldots, m$. By 7.5, $(\tilde{B}_1, \ldots, \tilde{B}_m)$ is a base sequence for $\mathcal{M}^m$. Therefore, $|\sum_1^m \tilde{B}_i| > |\sum_1^m B_i|$, a contradiction. □

§8. An algorithm for constructing a base sequence of a matroid sequence

8.1. Consider a sequence $\mathcal{M}^k = (M_1, \ldots, M_k)$ of matroids on E. We construct a base sequence for $\mathcal{M}^k$ recursively, as follows. Let a base sequence $\beta^l = (X_1, \ldots, X_l)$ be already constructed for $\mathcal{M}^l = (M_1, \ldots, M_l)$, $1 \leqslant l < k$, so that $X^l = \sum_1^l X_i$ is a base of M^l. Let us describe the step of a procedure that finds a base sequence for $\mathcal{M}^{l+1}$. Take a base X_{l+1} of the matroid $M_{l+1} \setminus X^l$. Put $\lambda = (X_1, \ldots, X_l, X_{l+1})$ and $X^{l+1} = \sum_1^{l+1} X_i$. Construct the graph G_λ. Inquire whether an active path from $\overline{X^{l+1}}$ exists in G_λ. If none is found, then, by Theorem 3.5, λ is a base sequence for $\mathcal{M}^{l+1}$. If such a path P is found and it originates from an element $x \in \overline{X^{l+1}}$, then rearrange the $\mathcal{M}^{l+1}$-sequence λ applying the procedure described in 3.6. A new $\mathcal{M}^{l+1}$-sequence $\lambda' = (X_1', \ldots, X_l', X_{l+1}')$ is obtained such that $x \in X_{l+1}'$, $|X_i'| = |X_i|$ for $i = 1, \ldots, l$, and $|X_{l+1}'| = |X_l| + 1$. Repeating this procedure, we finally get a base sequence of the sequence $\mathcal{M}^{l+1}$. After k steps of the procedure, we obtain a base sequence for $\mathcal{M}^k$.

8.2. The amount of computation required by this algorithm is the next question to be discussed. The *independence oracle for a matroid* M is an oracle that detects whether a given subset $X \subseteq E$ is independent in M (i.e., $X \in M$). If X represents a dependent set in M (i.e., $X \notin M$), then the oracle indicates some circuit $C \subseteq X$. Assume that each matroid M_i is given together with its independence oracle O_i and an *operation* is a reference to one of the oracles O_i. Then the number of operations is the total number of calls to the oracles $O_1, \ldots, O_k$. Clearly, the algorithm for finding a base sequence for $\mathcal{M}^k$ described in 8.1 requires no more than $k \cdot |E|$ calls to the oracles $O_1, \ldots, O_k$ to add each new element to the currently constructed $\mathcal{M}$-sequence (or, more precisely, to construct the graph G_λ). Therefore, the above algorithm refers to the oracles $O_1, \ldots, O_k$ no more than $k \cdot |E|^2$ times. To construct a base sequence of a given matroid M (and, at the same time, to solve the packing and covering problems for M) by this algorithm requires $(\operatorname{cov} M) \cdot |E|^2$ calls to the independence oracle $O(M)$ that describes M. Since $\operatorname{cov} M$ does not exceed (and can be of order) $|E|$, we see that the algorithm requires, in general, $O(|E|^3)$ calls to the independence oracle (or independence tests, as we say). All known algorithms that solve the packing and covering problems require the same number $O(|E|^3)$ of independence tests [1]. Now, we describe a modification of the above algorithm that requires the smallest known number $O(|E|^2)$ of independence tests to construct a base sequence of a given matroid (see also [2]). The algorithm of finding regular packings in graphs [4] serves as a prototype for this modified algorithm.

8.3. To reduce the amount of computations required by the above algorithm, the following "fork" lemma is important.

LEMMA. *Let* $M^k = \bigvee_1^k M_i$, $M_i = M$, *and let* B^k *be a base in* M^k, *namely* $B^k = \sum_1^k B_i$, $B_i \in M_i$, $i = 1, \ldots, k$, *while* X *is a base of the matroid* $M \setminus B^k$, *and* $\lambda = (B_1, \ldots, B_k, X)$. *Then for any arc* (y, z) *in* G_λ, $y, z \notin X$, *there exists a "fork" of two arcs* (y, x) *and* (x, z) *with* $x \in X$.

PROOF. Let $z \in B_i$. Since (y, z) is an arc in G_λ, there exists a unique circuit $C_{yz} = C(M, B+y)$ such that $z \in C_{yz}$. As X is a base in $M \setminus B^k$, we see that $X + y$ contains a unique circuit $C = C(M, X+y)$. Since B^k is a base of M^k, we see that for any $a \in E \setminus B^k$, there exists a unique circuit $C_a = C(M, B+a)$. We have to prove that for some $x \in C \cap X = C \setminus y$, there exists an arc (x, z) in G_λ, i.e., there exists a circuit C_x such that $z \in C_x$. Suppose the converse. Consider the circuit C_x, $x \in C \setminus y$. By the sharpened circuit axiom ($\mathscr{C}2'$), there exists a circuit A in M such that $y \in A \subseteq C \cup C_x \setminus x$, and therefore $y \in A \subseteq C \cup B_i \setminus z$. Let $\mathscr{A} = \{A: y \in A \subseteq C \cup B_i \setminus z\}$, and let $C^* \in \mathscr{A}$ and $|C^* \cap X| = \min\{|A \cap X|: A \in \mathscr{A}\}$. If $C^* \cap X = \varnothing$, then $C^* \subseteq B_i + y$, and hence, $C^* = C_{yz}$. But $z \in C_{yz} \setminus C^*$, a contradiction. Thus, $\varnothing \neq C^* \cap X \ni a$. By the sharpened circuit axiom $(\mathscr{C}2)'$, there exists a circuit C' such that $y \in C' \subseteq C^* \cup C_a \setminus a$, and hence, $y \in C' \subseteq C \cup B_i$. Since $C_a \not\ni z$ by the assumption, we obtain $C' \not\ni z$, and $C' \in \mathscr{A}$. But $C' \cap X \subseteq C^* \cap X \setminus a$, whence $|C' \cap X| < |C^* \cap X|$, which contradicts the choice of C^*. □

8.4. Now we describe a modification of the algorithm from 8.1 that requires fewer independence tests to construct a base sequence of a matroid M on E. Let $\mathscr{M} = (M_1, \ldots, M_c)$, $\mathscr{M}^k = (M_1, \ldots, M_k)$, $k \leq c$, $M^k = \bigvee_1^k M_i$, as in 8.3, but now $M_i = M$, $t = 1, \ldots, c$, $c = \operatorname{cov} M$. As in 8.1, the algorithm of finding a base sequence is recurrent. It consists of c steps $T_1, \ldots, T_c$. At step T_k, a base sequence β^k is constructed for $\mathscr{M}^k$. In particular, a base B of the matroid M is constructed at step T_1, and $\beta^1 = \{B\}$. Let us describe step T_{k+1}, $k \geq 1$, that follows after step T_k, which yields an $\mathscr{M}^k$-sequence $\beta^k = (B_1, \ldots, B_k)$. Step T_{k+1} consists of consecutive iterations $T_{k+1, r}$. At iteration $T_{k+1, r}$, an $\mathscr{M}^{k+1}$-sequence $\lambda^r = (B_1^r, \ldots, B_{k+1}^r)$ is constructed such that $(B_1^r, \ldots, B_{k+1}^r)$ is an $\mathscr{M}^k$-sequence. Iteration $T_{k+1, 0}$ is trivial and results in an $\mathscr{M}^{k+1}$-sequence $\lambda^0 = (B_1^0, \ldots, B_{k+1}^0) = (B_1, \ldots, B_k, \varnothing)$.

Let us describe the iteration $T_{k+1,r+1}$, $r \geqslant 0$, that follows the iteration $T_{k+1,r}$, which yields an $\mathcal{M}^k$-sequence $\lambda^r = (B_1^r, \ldots, B_{k+1}^r)$. Put $B^r(k) = \sum_1^k B_i^r$. First, construct a base X of the matroid $M \setminus B^r(k)$ containing B_{k+1}^r, and a list of circuits $\mathscr{C}_{k+1,r+1} = \{C(M, X+z): z \in E \setminus B^r(k) \setminus X\}$. Put $\lambda = (B_1^r, \ldots, B_k^r, X)$. Obviously, $|X| \leqslant |B_k^r|$. If $X = E \setminus B^r(k)$ or $|X| = |B_k^r|$, then $\beta^{k+1} = \lambda$. Suppose that $X \subset E \setminus B^r(k)$ and $|X| < |B_k^r|$. Put $X_1 = X \cap (\bigcup \{C: C \in \mathscr{C}_{k+1,r+1}\})$, so that X_1 is the set of all cyclic elements of the matroid $M \setminus B^r(k)$ lying in X. We determine the sets $Y_{i+1} = G_\lambda(X_i)$ and $X_{i+1} = (X_i \cup G_\lambda(Y_{i+1})) \cap X$ recurrently. Denote by G^i and G_{i+1} the sets of arcs in G_λ leading from X_i to Y_{i+1} and from Y_{i+1} to X_{i+1}, respectively. Let $l = l(k+1, r+1)$ be a minimum s such that either

(c1) $X_{s+1} = X_s$, and therefore, $Y_{s+1} = Y_s$, or
(c2) Y_{s+1} contains an element y such that $X + y \in M$.

In case (c1), we have $\beta^{k+1} = \lambda$, and $D(M^{k+1}) = E \setminus B^r(k) \setminus X + X_l + Y_{l+1}$. In case (c2), using the sets G^l, G_{l+1}, and $\mathscr{C}_{k+1,r+1}$, we can find a λ-active path, or more precisely, an alternating λ-path $P = (z, x_1, y_1, x_2, \ldots, x_r, y_r = y)$, where $z \in E \setminus B^r \setminus X$, $x \in X_l$, $y_i \in Y_{l+1}$, $i = 1, \ldots, r$, and $X + y \in M$. With the help of this λ-path, we rearrange the $\mathcal{M}^{k+1}$-sequence λ into a new $\mathcal{M}^{k+1}$-sequence $\lambda^{r+1} = (B_1^{r+1}, \ldots, B_k^{r+1}, B_{k+1}^{r+1})$ such that $(B_1^{r+1}, \ldots, B_k^{r+1})$ is a new $\mathcal{M}^k$-sequence, $z \in B_k^{r+1}$ and $|B_k^{r+1}| = |X| + 1$. Obviously, β^{k+1} results from some iteration $T_{k+1,p}$ as λ^p.

8.5. Let us estimate the number of calls to the independence oracle in the algorithm for finding a base sequence of a matroid M described in 8.4. Let the algorithm produce a base sequence $(K_1, \ldots, K_c)$ of the matroid M, where $c = \operatorname{cov} M$. Let us estimate the number $p_{k+1,r+1}$ of calls at iteration $T_{k+1,r+1}$. To construct a base X of the matroid $M \setminus B^r(k)$ containing B_{k+1}^r, as well as to create X_1 and the list of circuits $\mathscr{C}_{k+1,r+1}$, we need $p'_{k+1,r+1}$ calls to the oracle. To construct X_l, Y_{l+1}, G^l, and G_{l+1}, where $l = l(k+1, r+1)$, we need

$$p''_{k+1,r+1} = k \cdot |X_1| + |Y_2| + \sum_{i=2}^{l}[k \cdot (|X_i| - |X_{i-1}|) + |Y_{i+1}| - |Y_i|]$$
$$= k|X_l| + |Y_{l+1}| \leqslant k|X| + |B^r(k)|$$

calls. Since $|X| \leqslant |K_{k+1}|$, step T_{k+1} involves at most $|K_{k+1}| + 1$ iterations. Therefore, the number p_{k+1} of calls at step T_{k+1} is estimated as

$$p_{k+1} \leqslant (|K_{k+1}| + 1)(k|K_{k+1}| + 2|E|) = k|K_{k+1}|^2 + (2|E| + k)|K_{k+1}| + 2|E|.$$

Now, the number p of calls to the oracle required by the entire algorithm is estimated as

$$p = \sum_{k=1}^{c} p_k \leq \sum_{k=1}^{c}(k-1)|K_k|^2 + 2|E|\sum_{k=1}^{c}|K_k| + \sum_{k=1}^{c}(k-1)|K_k| + 2c|E|.$$

Obviously,

$$\sum_{k=1}^{c}|K_k| = |E|, \qquad |K_k| \geq |K_{k+1}|, \qquad c \leq |E|.$$

It is easy to see that if $x_1 \geq x_2 \geq \ldots \geq x_n \geq 0$ and $\sum_{k=1}^{n} x_k \leq a$, then

$$\sum_{k=1}^{n} k x_k^2 \leq a^2.$$

Therefore,

$$p \leq 3 \cdot |E|^2 + 3c|E| \leq 6|E|^2.$$

§9. Comparing different algorithms for packing, covering, and intersection in matroids

As indicated in §4, any algorithm for packing, covering, and intersection in matroids actually constructs a base of the union of matroids. However, any algorithm constructing a base of the union of matroids terminates upon reaching an "obstacle", which is an extremal set of the set of matroids. As shown in §5, the set $\mathscr{E}(\mathscr{M})$ of the extremal sets for a matroid set $\mathscr{M} = (M_1, \ldots, M_k)$ constitutes a lattice (under inclusion). It turns out that every known algorithm for constructing a base of the union $\bigvee_1^k M_i$ runs across an obstacle which is either the maximal or the minimal extremal set in the lattice $\mathscr{E}(\mathscr{M})$, and hence all known algorithms can be classified in two groups by this qualitative sign. In this section, we examine some known algorithms from this point of view. The matroid algorithms of §§3 and 8 stop on reaching the minimal set in $\mathscr{E}(\mathscr{M})$.

9.1. Consider the Edmonds algorithm for decomposing a set E into the minimum number of independent sets of a matroid [12]. If M has loops, then the problem, obviously, has no solution. Therefore, assume that M has no loops. Let $M_i = M$ be a matroid on E, $i = 1, 2, \ldots$, $\mathscr{M}^k = (M_1, \ldots, M_k)$, and $M^k = \bigvee_1^k M_i$. Suppose that a subset $X \subseteq E$ can be decomposed into at least k independent sets of the matroid M, and such a partition $(X_1, \ldots, X_k)$ has already been constructed, so that $\sum_1^k X_i = X$. The goal of the next step of the algorithm from [12] is to find whether there exists $x \in E \setminus X$ such that $X' = X + x \in M^k$. If it does exist, then the set X' can be decomposed into at least k independent sets from M as well, and the algorithm constructs such a partition $(X'_1, \ldots, X'_k)$, so that $\sum_1^k X'_i = X'$.

In the opposite case, k is replaced by $k + 1$. Namely, for $X' = X + x$, where $x \in E \setminus X$, $(X_1, \ldots, X_k, X_{k+1} = \{x\}) = (X'_1, \ldots, X'_k, X'_{k+1})$ is a minimum $\mathscr{M}^{k+1}$-partition. Then, if $X' \neq E$, this step is repeated, based now on the minimum partition of the new set X' obtained.

Now we describe a step of the algorithm on the basis of the minimum partition $(X_1, \ldots, X_k)$ of the set $X = \sum_1^k X_i$. Construct a sequence $T_0, T_1, \ldots, T_p$ of sets closed in M, $T_0 = E$, $T_i = \mathrm{cl}(T_{i-1} \cap X_{s_i})$ for some s_i such that $|T_{i-1} \cap X_{s_i}| < \rho(T_{i-1})$. The last number p is specified by the moment when either

(a1) $|T_p \cap X_i| = \rho(T_p)$ for any $i \in \{0, \ldots, p\}$, or
(a2) $E \setminus (T_p \cup X) \neq \varnothing$.

It is easy to see that $T_0 \supset T_1 \supset \cdots \supset T_p$ and such a moment p does exist. In case (a1), $|X| = k\rho(T_p) + |E \setminus T_p|$, and, by Theorem 5.9, X is a base of the matroid M^k, while T_p is an extremal set of the matroid set $\mathscr{M}^k$. In case (a2), $X + x \in M^k$ for any $x \in E \setminus (T_p \cup X)$, and an $\mathscr{M}^k$-partition of the set $X + x = (X_1, \ldots, X_k)$ is constructed in some way with the help of the sequence $T_0, T_1, \ldots, T_k$.

PROPOSITION. *If a step of the Edmonds algorithm ends by case* (a1), *then T_p is the maximal set $H(M^k)$ of the lattice of extremal sets of the set $\mathscr{M}^k$.*

PROOF. Observe first that $X \cap H(M^k)$ is a base of the matroid $M^k \cdot H(M^k)$, because X is a base of the matroid M^k and $H(M^k) \supseteq D(M^k)$. Therefore, according to 5.8, $X_i' = X_i \cap H(M^k)$ is a base of the matroid $M_i \cdot H(M^k)$, $i = 1, \ldots, k$. We shall show that $H(M^k) \subseteq T_i$, $i = 0, \ldots, p$, by induction on i. Clearly, $H(M^k) = E = T_0$. Suppose that $H(M^k) \subseteq T_t$, $0 \leqslant t < p$. Then $X_i' \subseteq X_i \cap T_t$ for any $i = 1, \ldots, k$. Moreover, $H(M^k) = \mathrm{d}(X_i')$. Since $\mathrm{cl}(X_i') \subseteq \mathrm{cl}(X_{s_{t+1}} \cap T_t) = T_{t+1}$, we have $H(M^k) \subseteq T_{t+1}$. □

9.2. Consider the Lawler-Iri-Tomizawa algorithm used to find a maximum common independent set of two matroids [13], [14]. Let M_1 and in M_2 be matroids on E. The problem is to find a maximum set A^* independent both in M_1 and in M_2, i.e., $A^* \in M_1 \cap M_2$ and $|A^*| = \max\{|A|: A \in M_1 \cap M_2\}$. In 4.3 we have shown how to reduce this problem to the problem of constructing a base of the sum of two matroids. Therefore, it can be solved by the algorithms described in §8. Another algorithm is suggested in [13] and [14] (see also [7]). It looks as follows. If a common independent set A_k is already constructed at the kth step, then at the $(k + 1)$st step, either a common independent set A_{k+1} of cardinality greater than $|A_k|$ is constructed from A_k, or $A_k = A^*$ is shown to be a maximum common independent set of the matroids M_1 and M_2. Let us describe the $(k + 1)$st step of the algorithm. On the basis of the set $A = A_k$, one constructs an auxiliary directed graph Γ_A with the set of vertices E and arcs of two types for $x \in A$ and $y \in E \setminus A$:

(a1) (y, x) is an arc of the graph Γ_A if $x \in C(M_1, A + y)$;

(a2) (x, y) is an arc in Γ_A if $x \in C(M_2, A+y)$.

In the digraph Γ_A, one looks for a path from a vertex of the set $E \setminus \text{cl}_2 A$ (the sources) to vertices of the set $E \setminus \text{cl}_1 A$ (the sinks). Here, cl_i is the closure operator in the matroid M_i. If there is no such a path, then the algorithm terminates, and one can prove that A_k is a maximum common independent set of the matroids M_1 and M_2. Otherwise, one finds the shortest path P, deletes from A all the elements that belong to P, and adds elements of P that do not belong to A, i.e., $A_{k+1} = A_k + P \setminus (A_k \cap P)$. Then, obviously, $|A_{k+1}| = |A_k| + 1$.

Let us compare this algorithm with the algorithm from 3.6. Let B_2^* be a base of the matroid M_2^*, and B a base of the matroid $M_1 \setminus B_2^*$, and let $\lambda = (B_2^*, B)$. The digraphs Γ_B and G_λ have the same set of vertices E. All arcs of type (a1) of the digraph Γ_B are present in the digraph G_λ as well. However, an arc of type (a2) belongs to the digraph G_λ only if $y \in B_2^*$, because $E \setminus B_2^*$ is a base of the matroid M_2. There are no other arcs in the digraph G_λ, and therefore, G_λ is just a subgraph of Γ_B.

The algorithm described in 3.6 suggests that we should look for active paths in the digraph G_λ starting from elements of the set $E \setminus B_2^* \setminus B$. An element $x \in E \setminus B_2^* \setminus B$ is active if there is a path in the digraph G_λ starting at x and ending at $y \notin \text{cl}_1 B$ (i.e., y is a sink in the digraph Γ_B). Obviously, $x \in \text{cl}_2 B$, since $E \setminus B_2^*$ is a base of the matroid M_2, and $E \setminus B_2 \supseteq (B+x)$. Thus, x is a source in the digraph Γ_B.

Lemma 5.8 implies that the Lawler-Iri-Tomizawa algorithm constructs the set $H(M_2^* \vee M_1)$.

References

1. D. J. A. Welsh, *Matroid theory*, Academic Press, New York, 1976.
2. A. K. Kelmans, M. V. Lomonosov, and V. P. Polesskii, *On minimum coverings in matroids*, Problemy Peredachi Informatsii **12** (1976), no. 3, 94–107; English transl. in Problems Inform. Transmission **12** (1976).
3. V.P Polesskii, *On one way of constructing a structure-reliable communication network*, Discrete Automata and Communication Networks (V. G. Lazarev, editor), "Nauka", Moscow, 1970, pp. 13–19. (Russian)
4. _____, *On one lower bound on the reliability of information networks*, Problemy Peredachi Informatsii **7** (1971), no. 2, 88–96; English transl. in Problems Inform. Transmission **7** (1971).
5. _____, *Covering a finite graph with minimal forests*, Problemy Peredachi Informatsii **12** (1976), no. 2, 76–82; English transl. in Problems Inform. Transmission **12** (1976).
6. _____, *The structure of isthmuses of the sum of matroids*, Problemy Peredachi Informatsii **12** (1976), no. 2, 95–104; English transl. in Problems Inform. Transmission **12** (1976).
7. N. Nakamura and M. Iri, *A structural theory for submodular functions, polymatroids and polymatroid intersections*, Research Memorandum RMI 81-66, August, 1981.
8. M. Iri, *A review of recent work in Japan on principal partitions of matroids and their applications*, Ann. New York Acad. Sci. **319** (1979), 306–319.
9. F. Harari, *Graph theory*, Addison–Wesley, Reading, MA, 1967.
10. C. St. J. A. Nash-Williams, *An application of matroids to graph theory*, Théorie des Graphes, Congr. Internat. d'Études (Rome, 1966), Dunod, Paris, 1967, pp. 263–265.

11. L. Lovasz and A. Recski, *On the sum of matroids*, Acta Math. Acad. Sci. Hungar **24** (1973), 329–333.
12. J. Edmonds, *Minimum partition of a matroid into independent subsets*, J. Res. Nat. Bur. Standards Sect. B **69B** (1965), 67–72.
13. E. L. Lawler, *Matroid intersection algorithms*, Math. Programming **9** (1975), 31–56.
14. M. Iri and N. Tomizawa, *An algorithm for finding an optimal independent assignment*, J. Oper. Res. Soc. Japan **19** (1976), 32–57.

Optimal Distribution Sorting

E. V. KENDYS, V. M. MAKAROV,
A. R. RUBINOV, AND E. M. TISHKIN

1. Introduction

Put $N_k = \{1, 2, \ldots, k\}$ and consider natural numbers M and P and a sequence $I = (i_1, i_2, \ldots, i_n)$, where $i_j \in N_M$ with $j \in N_n$. The numbers i_j are said to be *keys*, the cardinality of the set $\Omega(I) \subset N_M$ of all different keys of the sequence I (denoted by $\|I\|$) is said to be the *cardinality* of the sequence I, and the number $|I| = n$ is said to be the *length* of the sequence I. A *sorting* of the sequence I is a rearrangement that yields a nondecreasing sequence. The *distribution sorting* is a sorting that consists of a sequence of similar operations called *distributions*. To describe the distribution operation, consider P given sequences $I_1, I_2, \ldots, I_P$. Initially, let us assume that $I_1 = I$, while for $k \geqslant 2$ the I_k are empty. The distribution operation looks as follows. One chooses an arbitrary number $p_0 \in N_P$, decomposes the sequence I_{p_0} into P disjoint subsequences $J_1, J_2, \ldots, J_P$, and then regards the sequence I_{p_0} as empty. Next, for all $p \in N_P$, one replaces the sequence I_p by the sequence $I_p \cdot J_p$. Here, the operation $S_1 \cdot S_2$ means the concatenation of two sequences S_1 and S_2, when a sequence of length $|S_1| + |S_2|$ results on adding the sequence S_2 to the right of S_1. After the last distribution, one of the sequences I_p must represent the sorted sequence I, while the other sequences must be empty.

Let, for example, $M = 6$, $P = 3$, $n = 14$,

$$I = (2, 5, 2, 1, 3, 6, 1, 4, 2, 6, 5, 5, 2, 3). \tag{1}$$

Then we choose $p_0 = 1$ for the first distribution, include the 4th and 7th elements of the sequence I_1 (i.e. the elements with key 1) into the sequence J_1, include keys 2, 4, 5 in J_2, and keys 3, 6 in J_3 (Table 1).

1991 *Mathematics Subject Classification*. Primary 68P10.

Translation of Vestnik Vsesoyuz. Nauchno-Issled. Inst. Zheleznodorozh. Transporta **1989**, no. 8, 1–8.

Table 1

Distr. number	I_1	I_2	I_3
1	(2, 5, 2, 1, 3, 6, 1, 4, 2, 6, 5, 5, 2, 3) ↓	—	—
2	(1, 1)	(2, 5, 2, 4, 2, 5, 5, 2) ↓	(3,6,6,3)
3	(1, 1) · (2, 2, 2, 2)	(5, 5, 5)	(3, 6, 6, 3) · (4) ↓
4	(1, 1) · (2, 2, 2) · (3, 3) · (4)	(5, 5, 5) · (6, 6) ↓	—
	(1, 1) · (2, 2, 2) · (3, 3) · (4) · (5, 5, 5) · (6, 6)	—	—

We choose $p_0 = 2$ for the second distribution, include key 2 into the sequence J_1, key 5 into J_2, and key 4 into J_3. For the third distribution, we choose $p_0 = 3$, include keys 3 and 4 into J_1, key 6 into J_2, while J_3 remains empty. In the last, 4th distribution, we concatenate just I_2 with I_1, i.e. we choose $p_0 = 2$ and let $J_1 = I_2$, $J_2 = J_3 = \varnothing$. In Table 1, the sequence I_{p_0} subject to distribution is marked by a down arrow (↓), and each sequence I_p is separated by parentheses into groups of elements whose previous "distribution history" is similar.

Certainly, this method of sorting sequence (1) is not unique. Another method, given below, requires three distributions. The general problem is to construct distribution sorting algorithms requiring the minimal number of distributions.

Observe that each partition of the sequence I_{p_0} into subsequences J_1, J_2, ..., J_P by the above algorithm of sorting sequence (1) is induced by a partition of the key set $\Omega(I_{p_0})$ into P subsets, i.e. this algorithm can be specified by choosing a number p_0 for the next distribution of the sequence I_{p_0} and defining a partition of the key set $\Omega(I_{p_0})$ into P subsets $\Omega(J_p)$, $p \in N_P$. Algorithms that admit the above description and allow to sort any subsequence I with given M and P are called *universal*. The number of distributions required by universal algorithms to sort a sequence I is independent of the structure of the sequence I and is determined by the numbers M and P.

A universal algorithm requiring the minimal number of distributions with given M and P is suggested in [1], where distribution sorting is employed to model an external sorting on $P+1$ magnetic tapes. In this case the choice of the class of universal algorithms is natural, because the length n of a sequence I under sorting is assumed by far greater than the RAM space, so that a preliminary analysis of the sequence I may require much time. The present paper suggests distribution sorting algorithms that minimize the number of distributions for sorting each particular sequence, i.e. optimal *individual* algorithms.

For example, with $M = 8$ and $P = 2$, one can sort the sequence $I = (5, 1, 6, 2, 7, 3, 8, 4)$ in two distributions. The first distribution separates keys 1, 2, 3, 4 and 5, 6, 7, 8 into different sequences. The second distribution concatenates the sequence having the greater keys with the sequence having the lesser keys. The best universal algorithm requires five distribution operations with $M = 8$ and $P = 2$.

The problem of constructing optimal individual algorithms arises naturally in modelling the process of sorting trains. Consider a railroad yard that has P switching tracks. Let one track be occupied by a train whose structure is described by a sequence I, where each key i_j represents the number of the recipient of the jth car (the numeration starts from the end of the train). After sorting, the train visits the recipients (empty tracks) successively and the corresponding cars are uncoupled each time from the end of the train. It is assumed impossible to extract a car from the middle of the train during a visit to a recipient. Then the distribution operation corresponds to one switching operation, at which the train standing at a switching track is pulled to the hump and then distributed into p tracks. After the last switching operation, a correctly sorted train must occupy one of the tracks. Train lengths are relatively small, while the switching operation time is considerable. Therefore we need to find the best individual algorithm.

Prior to constructing individual algorithms, we describe the optimal universal algorithm in a form convenient for further argument.

2. The optimal universal algorithm

Let us define a sequence of numbers $F_P(t)$, where $P \geq 2$ and $t = 1, 2 \ldots$, with the help of the recurrence relation

$$F_P(t) = \sum_{k=t-P}^{t-1} F_P(k), \qquad t = 1, 2, \ldots, \tag{2}$$

and the initial conditions $F_P(0) = 1$, $F_P(-k) = 0$, $k = 1, 2, \ldots, P-1$. The numbers $F_P(t)$ are called the *generalized Fibonacci numbers of order* P. For $P = 2$, the generalized Fibonacci numbers coincide with the ordinary Fibonacci numbers.

Let us describe now the optimal universal distribution sorting algorithm, assuming first that $|\Omega(I)| = M = F_P(T)$. The algorithm is constructed in two stages. Tables 2 and 3 on the next page show a sample sorting with $P = 2$, $T = 5$, $M = F_2(5) = 8$. At the first, preliminary stage, we construct an algorithm permuting the elements of the sequence I in such a way that identical keys are placed in succession. At the second stage, we transform this algorithm into the required sorting algorithm. We assume the cyclic numeration on the set N_P of the indices of sequences, i.e. the sequence I_P is followed by I_1 and the sequence I_{p+kP} coincides with I_p (J_{p+kP} coincides with J_p).

Let the number p_0 of the sequence subject to the tth distribution operation be equal to t, i.e. let sequences be distributed in the cyclic order. Before the tth distribution operation, the sequence I_t is the concatenation of subsequences: $I_t = S_1 \cdot S_2 \cdots S_{k(t)}$ such that $\|S_k\| = F_P(T-t+1)$, $k \in N_{k(t)}$ (the algorithm is well defined since this property is retained while executing the algorithm). At the tth distribution operation of the preliminary stage, we include $F_P(T-t)$ minimal keys of each sequence $S_1, S_2, \ldots, S_{k(t)}$ into the

Table 2

Distribution number	I_1	I_2
1	(1, 2, 3, 4, 5, 6, 7, 8) ↓	—
2	(6, 7, 8)	(1, 2, 3, 4, 5) ↓
3	(6, 7, 8) · (1, 2, 3) ↓	(4, 5)
4	(8) · (3)	(4, 5) · (6, 7) · (1, 2) ↓
5	(8) · (3) · (4) · (6) · (1) ↓	(5) · (7) · (2)
	—	(5) · (7) · (2) · (8) · (3) · (4) · (6) · (1)

Table 3

Distribution number	I_1	I_2
1	(1, 2, 3, 4, 5, 6, 7, 8) ↓	—
2	(2, 4, 7)	(1, 3, 5, 6, 8) ↓
3	(2, 4, 7) · (3, 5, 8) ↓	(1, 6)
4	(4) · (5)	(1, 6) · (2, 7) · (3, 8) ↓
5	(4) · (5) · (6) · (7) · (8) ↓	(1) · (2) · (3)
	—	(1) · (2) · (3) · (4) · (5) · (6) · (7) · (8)

sequence J_{t+1}, then we include the next $F_P(T-t-1)$ keys of each sequence $S_1, S_2, \ldots, S_{k(t)}$ into the sequence J_{t+2}, and so on. Finally, we include $F_P(T-t+1-P)$ maximal keys of the sequences $S_1, S_2, \ldots, S_{k(t)}$ into the sequence J_{t+P}. After the last, Tth distribution operation, the sequence I_{T+1} is the concatenation of $M = F_P(T)$ sequences of cardinality 1, while the other sequences I_p are empty (see Table 2). Unlike Table 1, Table 2 indicates the sets $\Omega(I_p)$ instead of the sequences I_p, i.e. a record like (4, 5) may refer, for example, to the sequence (5, 5, 4, 5, 4, 4).

In the sequence I_{T+1} obtained at the preliminary stage, identical keys follow in succession, but their sequence fails to be increasing. Denote this sequence by $L = (L(1), L(2), \ldots, L(F_P(T)))$. Observe that if we replace the $L(i)$th key by the ith key (i.e. distribute the ith keys, $i \in N_{F_P(T)}$, always into the sequence I_p where the $L(i)$th keys are distributed), then the above algorithm sorts the original sequence I correctly in T distribution operations (see Table 3).

If strict inclusions occur in $\Omega(I) \subset N_M \subset N_{F_P(T)}$, then the above algorithm is still correct. Here some keys are just missing. Therefore, the above algorithm requires $F_P^{-1}(\|I\|)$ distributions to sort an arbitrary sequence I, where the number $F_P^{-1}(m)$ is specified by the relations $F_P(F_P^{-1}(m) - 1) < m \leq F_P(F_P^{-1}(m))$.

The following statement is true.

THEOREM 1. *The above algorithm is unimprovable in the class of universal algorithms.*

To prove the theorem, it suffices to indicate for any M a sequence I such that $\|I\| = M$ and no algorithm can sort I in less then $F_P^{-1}(M)$ distributions. The sequence $I^M = (M, M-1, \ldots, 1)$ can serve this purpose.

LEMMA 1. *If a distribution sorting algorithm can sort the sequence I^M in T distributions, then $F_P(T) \geqslant M$.*

PROOF. Let us prove the lemma by induction on T. For $T = 2$, the assertion is obvious. Let the assertion of the lemma hold for $T < T_0$, and suppose that for $M_0 > F_P(T_0)$ there exists a distribution sorting algorithm capable of sorting I^{M_0} in T_0 distributions. Let the first distribution performed by this algorithm result in sequences $I_1, I_2, \ldots, I_P$, ordered in nonincreasing lengths. Then $|I_1| = \|I_1\| \leqslant F_P(T_0-1)$, because otherwise the sequence I_1 ordered reversely cannot be sorted in the remaining $T_0 - 1$ distributions by the inductive hypothesis. We have also the inequality $|I_2| \leqslant F_P(T_0 - 2)$, because otherwise one of the sequences I_1 and I_2 is left undistributed after the second distribution and cannot be sorted in the remaining (T_0-2) distributions. Similarly, $|I_p| \leqslant F_P(p)$ with $p \in N_P$. Therefore, it follows from relation (2) that $M_0 \leqslant F_P(T_0)$, which contradicts the assumption $M_0 > M_P(T_0)$. The lemma is proved.

3. The optimal individual algorithm

The main concept of constructing the optimal individual distribution sorting algorithm for a sequence I is as follows. Suppose that the sequence I is split into m disjoint subsequences $S_1, S_2, \ldots, S_m$ such that the sequence $S_1 \cdot S_2 \cdots S_m$ is nondecreasing and $m \leqslant M$. Such a partition is said to be *reducing*. If one assigns to each element of the sequence I a conventional key equal to the number of the subsequence S_k to which it belongs, and sorts the sequence I in conventional keys with the help of the universal algorithm described above, then the sequence I appears to be sorted in original keys as well, because the universal algorithm does not change the relative position of elements with identical keys. If at least one generalized Fibonacci number occurs between the numbers m and M, i.e. $m \leqslant F_P(T) < M$, then the number of distributions of such an algorithm is less than that of the universal algorithm. Below, we describe an algorithm for constructing a reducing partition with the minimal number m of conventional keys and prove that the individual algorithm constructed in this way cannot be improved in the class of all distribution sorting algorithms.

We say that an element i_{j_1} of a sequence I is placed *irregularly* with respect to an element i_{j_2} (and write $i_{j_1} \prec i_{j_2}$) if $i_{j_1} > i_{j_2}$ and one of the following conditions holds:

(S1) $j_1 < j_2$, or

(S2) $j_1 > j_2$ and there exists an element i_{j_3} such that $i_{j_1} > i_{j_3} > i_{j_2}$ and $j_3 < j_2$.

Table 4

Element number	1	2	3	4	5	6	7	8	9	10	11	12	13	14
Original key	2	5	2	1	3	6	1	4	2	6	5	5	2	3
Conventional key	2	4	2	1	2	4	1	3	1	4	3	3	1	2

LEMMA 2. *If $i_{j_1} \prec i_{j_2}$, then the following statements hold*:

(a) *In the process of distribution sorting of the sequence I performed by an arbitrary algorithm, the elements i_{j_1} and i_{j_2} occur, at least once, in different subsequences I_p.*

(b) *The elements i_{j_1} and i_{j_2} cannot have identical conventional keys with any reducing partition of the sequence I.*

PROOF. Statement (a) follows from the fact that if a distribution operation does not take the elements i_{j_1} and i_{j_2} to different sequences I_p, then their relative order does not change (case S1), while the set of elements lying between them can be only reduced (case S2). Statement (b) is an obvious corollary of (a). The lemma is proved.

Observe that the relations $i_{j_1} \prec i_{j_2}$, $i_{j_2} \prec i_{j_3}$ lead to $i_{j_1} \prec i_{j_3}$. Therefore, if $i_{j_1} \prec i_{j_2} \prec \cdots \prec i_{j_m}$ is a chain of elements placed mutually irregularly, then assertion (b) of Lemma 2 implies that any reducing partition contains at least m subsequences. Let $m^0(I)$ be the maximal length of a chain as above, and let $m_0(I)$ be the minimal number of subsequences S_k in reducing partitions of the sequence I. Then

$$m^0(I) \leqslant m_0(I). \tag{3}$$

We are going to construct a reducing partition and a chain of elements placed mutually irregularly such that each subsequence of the partition contains an element of the chain. By inequality (3), this reducing partition is minimal. Table 4 illustrates the construction of the minimal reducing partition for sequence (1).

The algorithm for constructing the minimal reducing partition consists of $m_0(I)$ left to right examinations of the sequence I. At each examination, we form one subsequence S_k. At the first examination, we include all keys equal to 1 into S_1. Then we add to S_1 all keys equal to 2 and lying to the right of the last key 1 (see Table 4). If there are some keys 2 left, i.e. at least one of them was found to the left of the last key 1, then the construction of S_1 terminates. Otherwise, we proceed adding keys 3 that lie to the right of the last key 2, and so on, until we complete constructing the subsequence S_1. Denote by $q(1)$ the maximal key of those included in the subsequence S_1 completely. Denote by j_1 the index of the last element of I included in S_1 among the elements with key $q(1)$. Performing the second examination, we include in S_2 all keys $q(1)+1$ not included in S_1, then add all keys $q(1)+2$ that lie to the right of the last key $q(1)+1$ included in S_2, and so on. Denote by $q(2)$ the maximal key of those included in the subsequences S_1 and S_2 completely. Denote by j_2 the index of the last element with key $q(2)$ included in S_2. The process continues until the entire sequence I is exhausted at a certain examination (in Table 4, we have $q(1) = 1$, $j_1 = 7$, $q(2) = 3$, $j_2 = 14$, $q(3) = 4$, $j_3 = 8$, $q(4) = 6$, $j_4 = 10$, and sequence (1) can be sorted in 3 distributions since $F_3(3) = 4$). Let the number of the subsequences S_k in the above reducing partition be equal to m. Observe that $i_{j_k} \prec i_{j_{k+1}}$ for $k = 1, 2, \ldots, m - 1$, and therefore $m = m_0(I) = m^0(I)$. The minimal reducing partition constructed above is called *canonical*, while the element i_{j_k} is called the *leading* element of the subsequence S_k, $k \in N_{m_0(I)}$.

To prove the optimality of the individual sorting algorithm suggested here and based on sorting in conventional keys, consider an arbitrary sequence J^M that satisfies the conditions $\Omega(J^M) = N_M$ and $|J^M| = \|J^M\| = M$, along with an arbitrary distribution sorting algorithm A for the sequence J^M. We apply to the sequence $I^M = \{M, M-1, \ldots, 1\}$ all the operations performed by algorithm A while sorting the sequence J^M, i.e. we distribute the same keys into the same subsequences I_p as algorithm A does. The algorithm that transposes the elements of the sequence I^M in this way (we call it B), generally, is not bound to be an algorithm of sorting the sequence I^M.

LEMMA 3. *If algorithm A for any $k = 1, 2, \ldots, M - 1$ takes keys k and $k+1$ into different sequences at least once, then algorithm B sorts the sequence I^M correctly.*

PROOF. When algorithm A takes keys k and $k + 1$ from different sequences I_p into the same one for the last time, key $k + 1$ occurs to the right of key k, and their mutual arrangement does not change with further operations. Therefore, algorithm B takes key $k + 1$ to the right of key k as well. The lemma is proved.

THEOREM 2. *The minimal number of distributions required by an arbitrary distribution sorting algorithm to sort a sequence I equals $F_P^{-1}(m_0(I))$.*

PROOF. Consider an arbitrary distribution sorting algorithm C that sorts the sequence I. Algorithm C restricted to the leading elements of the canonical partition is called *algorithm A*. It follows from assertion (a) of Lemma 2 and Lemma 3 that one can construct, on the basis of algorithm A, an algorithm B sorting the sequence $I^{m_0(I)}$. From this and Lemma 1 we infer that the number of distributions required by algorithm B, and thus by algorithms A and C, is not less than $F_P^{-1}(m_0(I))$. The theorem is proved.

4. Related problems of distribution sorting.

Treating distribution sorting as sorting trains, we arrive naturally at interesting problems close to the above problem on the minimal number of distributions. They differ in criteria, restrictions, and distribution operation definitions.

A more general definition of the distribution operation differs from the above one in the following: the sequence I_{p_0} is represented as the concatenation $I_{p_0} = S_1 \cdot S_2$ of two subsequences S_1 and S_2, after which I_{p_0} is set equal to S_1 and S_2 is split to disjoint subsequences $J_1, J_2, \ldots, J_p$. This definition admits that one can operate at the hump with a part of the train from track p_0 as well as with the entire train. If the length of the sequence S_2 under distribution is limited, then the problem on the minimal number of distributions is modified considerably. In practice, such a limitation means a restriction on the switching lead. Restrictions on lengths of switching tracks, i.e. on the lengths of the sequences I_p, are also possible.

The number of cars rearranged in sorting can serve as an optimization criterion instead of the number of distributions. This implies that each distribution operation enters the criterion with a weight equal to the length of the sequence under distribution. Such a problem makes sense with both the original and generalized definition of the distribution operation.

The number of uncouplings can serve as another criterion. This implies that each distribution operation enters the criterion with a weight equal to the number of continuous segments that constitute the sequence under distribution and are distributed entirely to one sequence J_p. This criterion refers to the situation when there is no hump and cars are directed to switching tracks by "backing".

Combinations of diverse criteria and restrictions are also possible. Let us describe, for example, a practical algorithm for the two-criterion problem of minimizing the sum of lengths of sequences under distribution among all the algorithms that use the minimal number of distributions. The possibility of minimizing the second criterion is attributed to the following facts. Observe, first of all, that in the universal algorithm described above, different conventional keys participate in different numbers of distributions. Let $r(k)$ be the number of distributions involving conditional keys k. Then the value of the second criterion is determined by the formula $V = \sum_{k \in N_{m_0(I)}} r(k)v(k)$, where $v(k)$ is the number of conventional keys equal to k. In the case when $m_0(I)$ lies between two successive Fibonacci numbers, the above proposal to treat the keys with numbers greater than $m_0(I)$ missing is not the best one. If all the numbers $v(k)$ are identical, then we must renumber conventional keys to make the indices with maximal $r(k)$ missing (the sequence $r(k)$ is not monotone). In the general case, one can choose the optimal numeration of conventional keys by applying the dynamical programming method. Another possibility for minimizing the value v is associated with the ambiguity in choosing a reducing partition of a sequence I. If we examine, for instance, sequence (1) from right to left and form the partition elements beginning from the last one, then sequence (1) appears to split into four subsequences in a way different from the canonical partition. If $m_0(I) \neq F_p(T)$, then we must solve the problems of numeration and the construction of a reducing partition simultaneously. This can be accomplished with the help of the dynamical programming method as well. Finally, one more possibility for minimizing v is associated with the ambiguity in choosing a universal algorithm that sorts conventional keys. At the last, Tth distribution of the preliminary stage of the universal algorithm described in §2, we must concatenate the Tth subsequence with the $(T+1)$st subsequence. If we concatenate the $(T+1)$st subsequence with the Tth subsequence instead, then another universal algorithm results, providing a lesser value of v with $m_0(I) = F_p(T)$ and identical $v(k)$. We suggest the following algorithm to solve the above two-criteria problem. At the first stage, we determine $m_0(I)$ and construct the two universal algorithms described above. At the second stage, we solve the joint problem of choosing a conventional numeration and constructing a reducing partition for each of the two algorithms. Then we choose the better of the two variants. The shortcoming of this algorithm is that it does not allow for the possibility of the generalized definition of the distribution operation, while with $m_0(I) \neq F_p(T)$ there can exist, in general, universal algorithms distinct from the above two.

The authors know only heuristic alternative solutions to all the problems of this section related to the main problem of minimizing the number of distributions.

Reference

1. Donald E. Knuth, *The art of computer programming.* Vol. 3, Addison-Wesley, Reading, MA, 1973.

Branching Packing in Weighted Graphs

P. A. PEVZNER

1. Introduction

We consider the problem of branching packing, which generalizes the problem of a maximum flow in a network. Maximum flow problem corresponds to packing of paths from the source s to the sink t (each such path corresponds to transmission of a signal from s to t), while the problem under consideration corresponds to packing of branchings (each branching corresponds to transmission of a signal from the source s to all vertices of a network). Thus, while the maximum flow problem can be considered as a problem of information transmission from the source to the sink, the problem of branching packing corresponds to information transmission from the root to all vertices of a network. The "max=min" property for the branching packing was proven by Edmonds [1]. His proof is constructive and is based on a rather complicated algorithm. Tarjan [2], [3] and Lovász [4] suggested polynomial branching packing algorithms for the case of the unit arc capacity. However, these algorithms require polynomial time for increasing the packing weight by 1; thus, their complexity substantially depends on arc capacities. In the present paper we describe a polynomial algorithm for branchings packing and estimate the minimum number of branchings in the maximum packing.

2. Statement of the Problem and Basic Definitions

2.1. Let $G(V, E)$ be a directed graph with the finite set of vertices $V = VG$ and the set of arcs $E = EG$. Each arc e has *beginning* $t(e) \in V$ and *end* $h(e) \in V$. In the graph G there is a marked vertex r (*root*), while on the set of arcs E a nonnegative *capacity* function c is defined. The graph G together with the root r and the capacity function c is called a *network* and denoted $N(G, c)$.

1991 *Mathematics Subject Classification.* Primary 05B40, 68R10.
Translation of Combinatorial Methods in Flow Problems (A. V. Karzanov, editor), Vsesoyuz. Nauchno-Issled. Inst. Sistem. Issled., Moscow, 1979, pp. 113–127; Zbl. **494**, 90024.

A *path* in the graph G from $v \in V$ to $w \in V$ is a sequence $v = t(e_1)$, e_1, $h(e_1) = t(e_2)$, e_2, ..., $h(e_{n-1}) = t(e_n)$, e_n, $h(e_n) = w$ consisting of vertices and arcs of G.

A *branching* is a subgraph of the graph G in which there is exactly one path from r to any other vertex $v \in V$. To each branching B we assign a positive number β, called the *weight* of the branching B, and a function $\overline{B}$ on E defined by the relation

$$\overline{B} = \begin{cases} 1, & \text{if } e \in EB \\ 0, & \text{if } e \notin EB. \end{cases}$$

A set $\mathfrak{B} = \{(\beta_1, B_1), \ldots, (\beta_k, B_k)\}$ of branchings of the graph G is called a *packing* in the network $N(G, c)$ if $\sum_{i=1}^{k} \beta_i \overline{B}_i \leq c$. The number $\sum_{i=1}^{k} \beta_i$ is called the *weight* of the packing $\mathfrak{B}$, and k is its *power*. A maximum weight packing is called simply *maximal*, while a maximal packing of the minimum power is called *perfect*.

Let $r \in R \subset V$. A *cut* $[R, \overline{R}]$ is a set of arcs of the type $\{e \mid t(e) \in R, h(e) \notin R\}$, while the *cut capacity* is the sum of capacities of all its vertices. The "max=min" property for branching packing is established by the following theorem.

THEOREM (Edmonds [1]). *The weight of the maximal branching packing equals the minimum cut capacity.*

2.2. Network partitioning.

2.2.1. In what follows we admit the existence of multiple arcs in the graph $G(V, E)$. Denote by V^2 the family of ordered pairs of the set V.

Let $f_\alpha: V \to P$ be a mapping Denote by $f_\alpha(X)$ the image of a set $X \subseteq V$ and by $f_\alpha^{-1}(X)$ the inverse image of a set $X \subseteq P$. The mapping f_α induces the mapping $F_\alpha : V^2 \to P^2$, thus defining a graph $G_\alpha(f_\alpha(V), F_\alpha(E))$ with the root $f_\alpha(r)$ (loops and arcs coming to the root can arise in this graph; however, they are insubstantial and will be ignored). Defining arc capacities in the graph G_α by the rule $c_\alpha(e) = c(F_\alpha^{-1}(e))$, $e \in F_\alpha(E)$, we obtain a network $N_\alpha(G_\alpha, c_\alpha)$. The arc $F_\alpha(e)$ in the graph G_α is called the *image* of an arc $e \in E$.

2.2.2. Let $r \in R \subset V$ and $S = V \setminus R$. Denote by z a vertex not belonging to V and let $R' = R \cup z$, $S' = S \cup z$.

Consider mappings $f_R : V \to R'$ and $f_S : V \to S'$ defined by

$$f_R(v) = \begin{cases} v, & \text{if } v \in R, \\ z, & \text{if } v \in S, \end{cases} \qquad f_S(v) = \begin{cases} z, & \text{if } v \in R, \\ v, & \text{if } v \in S. \end{cases}$$

A *partition* of the network $N(G, c)$ (of the graph G) by a cut $[R, S]$ is an operation transforming the network $N(G, c)$ into two networks $N_R(G_R, c_R)$

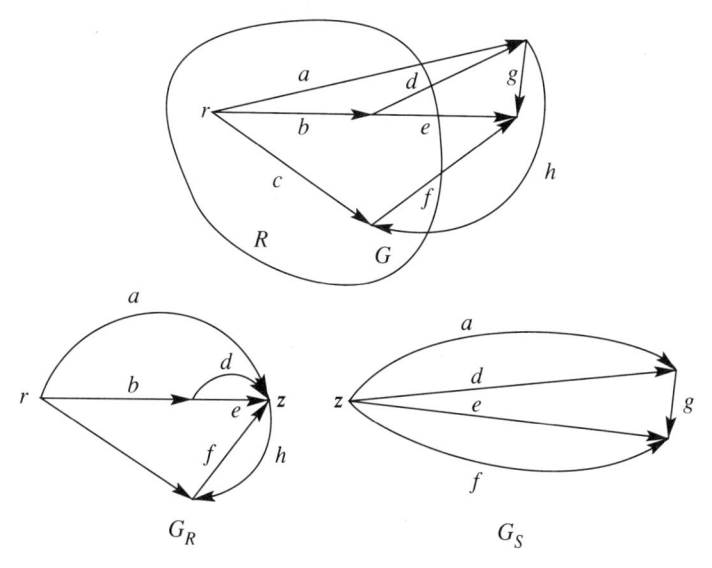

FIGURE 1

and $N_S(G_S, c_S)$ (of the graphs G_R and G_S) with the roots r and z respectively. In other words, G_R and G_S are obtained from G by identifying the sets of vertices S and R respectively (Figure 1).

2.2.3. We say that a set of vertices $U \subseteq EG$ in a graph G with root r is *connected* if for any $v \in VG$ there exists a path from r to v consisting of arcs from U.

The capacity of a cut $[S, \overline{S}]$ in the network $N(G, c)$ is denoted by $c[S, \overline{S}]$ or $R_S(c)$, while the sets S and $\overline{S}$ are called *segments* of the cut. The capacity of a minimal cut in the network $N(G, c)$ is denoted by $R(c)$, while the set of minimal cuts is denoted by $\mathscr{R}(c)$.

To each arc $e \in EG$ we associate a function $\overline{e}$ on the set EG assuming the value 1 on the arc e and 0 on the remaining arcs. The arcs of the network $N(G, c)$ are the arcs of the graph G whose capacity is positive. The set of arcs of the network N is denoted by EN.

The part of a path γ between $v \in VG$ and $w \in VG$ is denoted by v, γ, w. Below, the number of vertices in the graph G is denoted by n and the number of arcs is denoted by p; $[x]$ is defined as the largest integer not exceeding $x \in \mathbb{R}$.

3. Main theorems

3.1. THEOREM. *Let $B_1(R', U_1)$ and $B_2(S', U_2)$ be branchings in graphs G_R and G_S, respectively. If in the branching B_2 there is a single arc e from the root z and $R_R(F_S^{-1}(e)) \in U_1$, then the set of vertices $U = F_R^{-1}(U_1) \cup F_S^{-1}(U_2)$ forms a branching $B(V, U)$ in the graph G.*

PROOF. 1. Let us prove that the arc set of U is connected.

(a) For any vertex $v \in R$ there exists a path γ from r to v in the graph G_R, all arcs of which belong to B_1. If this path does not pass through z, then its inverse image, the path $F_R^{-1}(\gamma)$, is the desired path from r to v in the graph G. Let the path γ come through z, and let e' and e'' be arcs of it such that $h(e') = t(e'') = z$. Since $f_R(t(F_R^{-1}(e''))) = t(F_R(F_R^{-1}(e''))) = z$, it follows that $t(F_R^{-1}(e'')) \in S$ and thus there exists a path η in the graph G_S from z to $t(F_R^{-1}(e''))$, all arcs of which belong to B_2. This path starts with the arc e, since in B_2 there is only one arc from the root z. Since there also is only one arc of the branching B_1 coming into z, we have $F_R(F_S^{-1}(e)) = e'$. From the above consideration it follows that

$$r, F_R^{-1}(\gamma), t(F_R^{-1}(e')), F_S^{-1}(e), h(F_S^{-1}(e)), \eta,$$
$$t(F_R^{-1}(e'')), F_R^{-1}(e''), h(F_R^{-1}(e'')), F_R^{-1}(\gamma), v$$

is the desired path from r to v in the graph G.

(b) For any vertex $v \in S$ there exists a path γ from z to v in the graph G_S. Since this path starts with the arc e and $t(F_R(F_S^{-1}(e))) \in R$, B_1 contains a path η from r to $t(F_R(F_S^{-1}(e)))$. Clearly, η does not pass through z, and thus

$$r, F_R^{-1}(\eta), t(F_R(F_S^{-1}(e))), F_S^{-1}(e), h(F_S^{-1}(e)), \gamma, v$$

is a path from r to v in the graph G.

2. Since B_1 and B_2 are branchings, into each vertex from VB_1 and VB_2 (except the roots r and z) come single arcs from the sets U_1 and U_2, respectively. Since $F_R(F_S^{-1}(e)) \in U_1$, into each vertex of the set VG (except the root r) also comes a single arc of the set U.

From 1 and 2 it follows that $B(U, V)$ is a branching. □

3.2. THEOREM. *Let $[R, S]$ be a minimal cut in the network $N(G, c)$ and let $\mathfrak{B}_R = \{(\beta_R^1, B_R^1), \ldots, (\beta_R^d, B_R^d)\}$ and $\mathfrak{B}_S = \{(\beta_S^1, B_S^1), \ldots, (\beta_S^m, B_S^m)\}$ be maximal branching packings in the networks N_R and N_S, respectively. Then there exists a maximal branching packing $\mathfrak{B} = \{(\beta^1, B^1), \ldots, (\beta^l, B^l)\}$, with cardinality not exceeding $d + m$, such that for any $k = 1, \ldots, l$ there exist $j = 1, \ldots, d$ and $i = 1, \ldots, m$ satisfying*

$$EB^k = F_R^{-1}(EB_R^j) \cup F_S^{-1}(EB_S^i).$$

PROOF. 1. The proof is by induction on the total number $d + m$ of branchings in the packings $\mathfrak{B}_R$ and $\mathfrak{B}_S$. For $d + m = 0$ the assertion is obvious. We assume that for all maximal $\mathfrak{B}_R$ and $\mathfrak{B}_S$ with $d + m < A$ the theorem holds, and prove it for $d + m = A$.

2. It is clear that for any two sets $r \in X \subset R'$ and $z \in Y \subset S'$ the following equalities hold:

$$c_R[X, R'\backslash X] = c\left[f_R^{-1}(X), V\backslash f_R^{-1}(X)\right],$$

$$c_S[Y, S'\backslash Y] = c\left[f_S^{-1}(Y), V\backslash f_S^{-1}(Y)\right],$$

and thus the cuts $[R'\backslash z, z]$ and $[z, S'\backslash z]$ are minimal in the networks N_R and N_S, respectively.

3. Since $\mathfrak{B}_S$ is a minimal packing and $[z, S'\backslash z]$ is a minimal cut in N_S, the Edmonds theorem implies that each branching B_S^i in the packing $\mathfrak{B}_S$ passes the cut $[z, S\backslash z]$ only once, and thus in B_S^i there is exactly one arc e from the root z. Since $[R'\backslash z, z]$ is a minimal cut in N_R, by the Edmonds theorem all its arcs are filled by the packing $\mathfrak{B}_R$ (i.e. for any $u \in R'\backslash[z, z]$ we have $\sum_{k=1}^{d} \beta_R^k \overline{B}_R^k(u) = c_R(u)$) and $F_R(F_S^{-1}(e))$ belongs to some branching B_R^j. By Theorem 3.1, the arcs of the set $F_R^{-1}(EB_R^j) \cup F_S^{-1}(EB_S^i)$ form a branching B in the graph G.

4. Without loss of generality, it can be assumed that $\beta_R^j \leq \beta_S^i$. It is simple to demonstrate that the cut $[R, S]$ is minimal in the network $N(G, c - \beta_R^j \overline{B})$ while the packings

$$\overline{\mathfrak{B}}_R' = \left\{ (\beta_R^1, B_R^1), \ldots, (\beta_R^{j-1}, B_R^{j-1}), (\beta_R^{j+1}, B_R^{j+1}), \ldots, (\beta_R^d, B_R^d) \right\}$$

and

$$\overline{\mathfrak{B}}_S' = \{ (\beta_S^1, B_S R^1), \ldots, (\beta_S^{i-1}, B_S^{i-1}), (\beta_S^i - \beta_R^j, \beta_S^i)$$
$$(\beta_S^{i+1}, B_S^{i+1}), \ldots, (\beta_S^m, B_S^m) \}$$

are maximal in the networks $N_R(G_R, (c - \beta_R^j \overline{B})_R)$ and $N_S(G_S, (c - \beta_S^i \overline{B})_S)$, respectively.

Since the total number of branchings in the packings $\overline{\mathfrak{B}}_R'$ and $\overline{\mathfrak{B}}_S'$ have decreased at least by 1 as compared to the packings $\overline{\mathfrak{B}}_R$ and $\overline{\mathfrak{B}}_S$, by the induction assumption there exists a maximal packing of branchings $\overline{\mathfrak{B}}'$ in the network $N(G, c - \beta_R^j B)$ of complexity lower than $d + m$, possessing the required property. Adding the branching (β_R^j, B) to the packing $\overline{\mathfrak{B}}'$, we obtain the desired branching packing. □

If a cut $[R, S]$ is minimal in the network $N(G, c)$, then according to Theorem 3.2 the problem of branching packing in N can be reduced to the problems of branching packings in the networks N_R and N_S. If the number of vertices in each segment of the cut $[R, S]$ exceeds 1 (such cuts are called *nontrivial*), then the number of vertices in each graph G_R and G_S is smaller than the number of vertices in G.

4. Algorithm for branching packing

4.1. Preprocessing of the network.

4.4.1. An arc e in a network $N(G, c)$ is called *thick* if it does not belong to any minimal cut. A network $N(G, c)$ is called *critical* if it contains no thick arcs.

Without loss of generality we can assume that the initial network $N(G, c)$ is critical. Indeed, if some arc is thick, its capacity can be decreased by some amount Δ without changing the minimal cut value. After that either $c(e)$ would be equal to 0 (and thus the number of arcs in the network N would decrease by 1), or the arc e would belong to some minimal cut (and thus the cardinality of the set of thick arcs would decrease by at least 1) in $N(G, c - \Delta \cdot e)$. Δ is determined from the relation

$$\Delta = R(c - c(e) \cdot \overline{e}) + c(e) - R(c).$$

Thus, in order to find Δ it is sufficient to determine $R(c - c(e) \cdot \overline{e})$, which requires estimating the maximum flow from r to $h(e)$ in the network $N(G, c - c(e) \cdot \overline{e})$. The number of operations for transforming the initial network into a critical one is $O(pF_{n,p})$ (here $F_{n,p}$ denotes the number of operations for construction a maximal flow in a network with n vertices and p arcs).

4.1.2. A vertex v in a network $N(G, c)$ ($v \neq r$) is said to be *next to the root* if all arcs of the network N entering v come from the root r. Define a mapping $f_\alpha : V \to V \backslash v$ by the rule

$$f_\alpha = \begin{cases} w, & \text{if } w \neq v, \\ r, & \text{if } w = v, \end{cases}$$

and consider the network $N_\alpha(G_\alpha, c_\alpha)$ with the root r. Each branching B in the graph G_α corresponds to a branching $F_\alpha^{-1}(EB) \cup (r, v)$ in the graph G. Since $R(c_\alpha) \geq R(c)$, we can, given an arbitrary maximal branching packing in N_α, construct a branching packing of the weight $R(c)$ in the network N. Having this in mind, we shall assume that there are no next to the root vertices in the initial network.

4.1.3. In a network N there exist exactly n trivial cuts, namely, $n-1$ vertex cuts of the type $[V \backslash v, v]$ and one root cut $[r, V \backslash r]$.

LEMMA. *If a network N contains no nontrivial minimal cuts, then all vertex cuts in this network are minimal.*

PROOF. Let $v \in V \backslash r$. Since the network N does not contain next to the root vertices, there exists a vertex $w \in V$, $w \neq r$, such that the arc (w, v) belongs to N. Since the network N is critical, there exists a minimal cut $[R, \overline{R}]$ passing through (w, v), which is trivial by the assumption of the lemma. Since $r, w \in R$, and $v \notin R$, we have $\overline{R} = \{v\}$. $\square$

4.2. Problem A.

4.2.1. Below we consider only branchings satisfying restrictions on the the capacity, i.e. branchings (β, B) such that $\beta \overline{B} \leq c$.

A branching (β, B) is called *permissible* in a network $N(G, c)$ if $R(c - \beta \overline{B}) = R(c) - \beta$ (if the branching weight B is not given, we consider B permissible if $(1, B)$ is a permissible branching). A branching packing $\mathfrak{B} = \{(\beta_1, B_1), \ldots, (\beta_m, B_m)\}$ is *permissible* in a network $N(G, c)$ if for all $i = 1 \ldots, m$ the branching (β_i, B_i) is permissible in the network $N(G, c - \sum_{j=1}^{i-1} \beta_j \overline{B}_j)$ (it is clear that the order of branchings in a packing does not affect its permissibility).

Denote $c_{\mathfrak{B}} = c - \sum_{i=1}^{m} \beta_i \overline{B}_i$. In §4.3 an algorithm for solving the following problem will be presented.

PROBLEM A. *Construct a permissible branching packing $\mathfrak{B}$ in a network $N(G, c)$ such that in the network $N(G, c_{\mathfrak{B}})$ there exists a nontrivial minimal cut $[S, \overline{S}]$, and find this cut.*

REMARK. It is assumed that the network N contains more than three vertices (in a network with three or fewer vertices all cuts are trivial).

4.2.2. A *subbranching* in a graph $G(V, E)$ is a subgraph of G on a set of vertices W in which there exists exactly one path from r to any other vertex $w \in W$. A subbranching B is *permissible* in a network $N(G, c)$ if $R(c - \overline{B}) = R(c) - 1$.

THEOREM. *Let B be a permissible subbranching in a network $N(G, c)$ and let $VG \setminus VB \neq \varnothing$. Then there exists an arc $e \in EG \setminus EB$ such that $B'(VB \cup h(e), EB \cup e)$ is a permissible subbranching.*

This theorem first was proved by Tarjan [3] using the Edmonds theorem, and then by Fulkerson and Harding [5] and by Lovácz [4] without using the Edmonds theorem.

4.2.3. Each subbranching B defines a *partial ordering* on the set of its arcs by the rule: $u \succ e$ if and only if in the subbranching B there exists an directed path from $h(u)$ to $t(e)$. The *closure* of a subset W in a partially ordered set U is the set $W^* = \{u \in U \mid \text{either } u \in W, \text{ or there exists } w \in W \text{ such that } w \succ u\}$. Let B be a subbranching and let $W \subset EB$. By $B_W(VB_W, EB_W)$ we denote the maximal (by inclusion) subbranching contained in the set $EB \setminus W$. Obviously, $EB_W = EB \setminus W^*$.

4.2.4. If not all vertex cuts in a network $N(G, c)$ are minimal, then by Lemma 4.1.3 the network N contains a nontrivial minimal cut (and from the proof of the lemma it is clear how to construct it), which we need to find for the solution of Problem A. Therefore, below we assume that all vertex cuts in the initial network $N(G, c)$ are minimal. For convenience we assume that the root cut also is minimal (cf. the remark at the end of §4.3).

4.3. An algorithm for construction of a nontrivial minimal cut.

4.3.1. Let B_1 be a nonpermissible subbranching in a network $N(G, c)$ that passes through the root cut only once. In this section we use B_1 to find a permissible branching B such that in the network $N(G, c - \overline{B})$ there exists a nontrivial minimal cut.

We say that a subbranching B and a cut $[S, \overline{S}]$ in a network $N(G, c)$ satisfy *Condition* A if $R_S(c - \overline{B}) = R(c) - 1$.

Let $[S_1, \overline{S}_1]$ be a minimal cut in the network $N(G, c - \overline{B}_1)$. Since B_1 is a nonpermissible subbranching, we have $R_{S_1}(c - \overline{B}_1) < R(c) - 1$. Since B_1 passes through each trivial cut not more than once, the capacity of each trivial cut in the network $N(G, c - \overline{B}_1)$ is at least $R(c) - 1$; hence $[S_1, \overline{S}_1]$ is a nontrivial cut. Consider the partially ordered set of arcs $[S_1, \overline{S}_1] \cap EB_1$ (the partial ordering on the set EB_1 induces the partial ordering on all of its subsets) and select in it a closed set W_1 of cardinality $R(c) - 1 - R(c - \overline{B}_1)$. Consider the closure W_1^* of the set W_1 in the entire set EB_1. Obviously, the set of arcs $EB_1 \setminus W_1^*$ constitutes a subbranching

$$B_2 = (VB_2, EB_1 \setminus W_1^*).$$

Since $W_1^* \subset [S_1, \overline{S}_1] = W_1$, it follows that

$$R_{S_1}(c - \overline{B}_2) = R_{S_1}(c - \overline{B}_1) + |W_1^* \cap [S_1, \overline{S}_1]| = R(c) - 1,$$

and thus Condition A holds for the subbranching B_2 and the nontrivial cut $[S_1, \overline{S}_1]$. If B_2 is a nonpermissible subbranching, then we repeat the procedure described above and obtain a subbranching B_3 and a cut $[S_2, \overline{S}_2]$ for which Condition A holds, and so on, until we get a permissible subbranching $B_k(VB_k, EB_k)$ and a nontrivial cut $[S_{k-1}, \overline{S}_{k-1}]$ for which Condition A holds in the network $N(G, c)$ (this process terminates after at most h steps, since $|VB_i| - |VB_{i+1}| \geq 1$). Theorem 4.2.2 implies that in the network $N(G, c)$ there exists a permissible branching B such that $EB_k \subset EB$ (in order to find this branching one can use Tarjan algorithm [3]). Obviously, the packing $\{(1, B)\}$ and the cut $[S_{k-1}, \overline{S}_{k-1}]$ provide a solution of Problem A.

4.3.2. The set of weights β of a branching B, for which the weighted branching (β, B) is permissible, fills a segment $[0, \beta^{\max}]$ ($\beta^{\max}$ can be 0). Below we describe an effective algorithm for determining $\beta^{\max}$.

The slope of a line $ax + b = d$ is the number a. Denote by $L(x, i)$ the line with the slope $-i$ passing through the point $(x, R(c - x\overline{B}))$, and denote by $g(x, i)$ ($i \neq 1$) the x-coordinate of the intersection point of the lines $L(x, i)$ and $L(0, 1)$. The flow-chart of the algorithm for determining $\beta^{\max}$ is presented in Figure 2. The proof of convergence of this algorithm and the estimate of its effectivity is given in §4.4.

4.3.3. Here we present an algorithm that solves Problem A. In a network $N(G, c)$, construct a branching B passing through the root cut exactly once

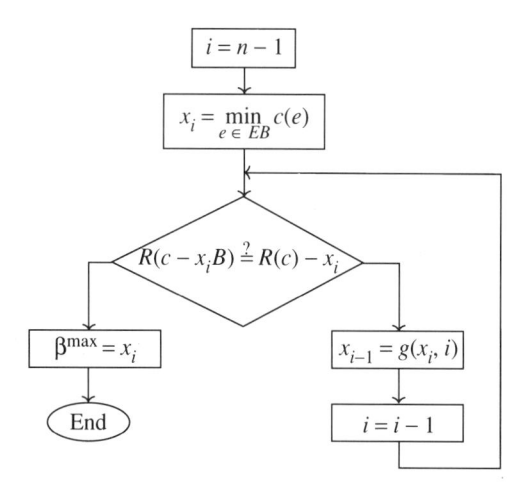

FIGURE 2

(if the root cut is minimal, such a branching exists). Depending on the type of the branching B, the following three cases may occur:

A1. The branching B is nonpermissible. For this case an algorithm for solving Problem A was described in §4.3.1.

A2. The branching B is permissible and $\beta^{\max} < \min_{e \in EB} c(e)$. Obviously, in this case the branching B in the network $N(G, c - [\beta^{\max}]\overline{B})$ (recall that we consider integer branching packings) is nonpermissible and we have case A1.

A3. The branching B is permissible and $\beta^{\max} = \min_{e \in EB} c(e)$. In this case replace the network $N(G, c)$ by $N(G, c - \beta^{\max}\overline{B})$ and repeat the process until one of the cases A1 or A2 occurs.

Note that in case A3 the number of arcs in the network $N(G, c - \beta^{\max}\overline{B})$ is less than in $N(G, c)$, and thus the number of occurrences of case A3 does not exceed p.

The flow-chart of the described algorithm is presented in Figure 3, and the estimate of its complexity is presented in §4.4.

REMARK. We have assumed for convenience that the root cut in the initial network $N(G, c)$ is minimal. This assumption guaranteed that the network contains a branching B passing through the root cut at only one arc, and was used only in 4.3.1. It is clear how to modify the algorithm from 4.3.1 in order to obtain a branching B for which one of the following conditions holds:

(1) in the network $N(G, c - \overline{B})$ there exists a nontrivial minimal cut;
(2) in the network $N(G, c - \overline{B})$ the root cut is minimal.

4.4. Convergence and complexity of the algorithm for solving Problem A.

4.4.1. Let B be a branching and let $[S, \overline{S}]$ be a cut in a network $N(G, c)$.

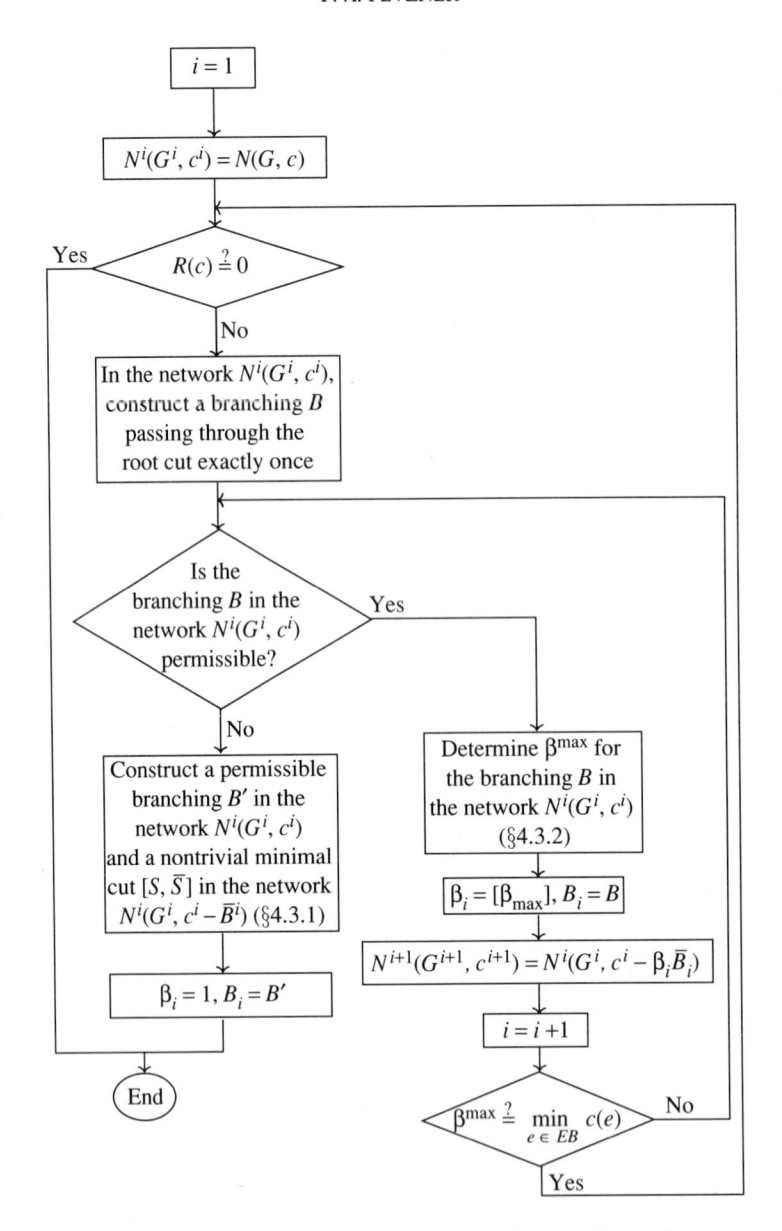

FIGURE 3. The algorithm for solving Problem A

Denote by $\Delta(S, B)$ the cardinality of the set $[S, \overline{S}] \cap EB$. Obviously, $1 \leq \Delta(S, B) \leq n - 1$. Consider the function $R(c - x\overline{B})$ in the variable x on the segment $[0, \min_{e \in EB} c(e)]$. Denote by $\dot{R}(c - x\overline{B})$ the right derivative of this function at the point x. Since $\dot{R}(c - x\overline{B}) = -\max_{[S, \overline{S}] \in \mathscr{R}(c - x\overline{B})} \Delta(S, B)$, it follows that $\dot{R}(c - x\overline{B})$ takes values in the set $\{-1, -2, \ldots, -(n-1)\}$.

LEMMA. $\dot{R}(c - x\overline{B})$ *is nonincreasing on* $[0, \min_{e \in EB} c(e)]$.

PROOF. Let $x < y$. In $N(G, c - x\overline{B})$ and $N(G, c - y\overline{B})$ there exist minimal cuts $[S, \overline{S}]$ and $[T, \overline{T}]$, respectively, such that $\dot{R}(c - x\overline{B}) = -\Delta(S, B)$ and $\dot{R}(c - y\overline{B}) = -\Delta(T, B)$. Since

$$R(c - y\overline{B}) \leq R_S(c - x\overline{B} - (y - x)\overline{B}) = R(c - x\overline{B}) - (y - x) \cdot \Delta(S, B)$$

and

$$R(c - x\overline{B}) \leq R_T(c - y\overline{B} - (y - x)\overline{B}) = R(c - y\overline{B}) + (y - x) \cdot \Delta(T, B),$$

we have

$$\Delta(S, B) \leq \frac{R(c - x\overline{B}) - R(c - y\overline{B})}{y - x} \leq \Delta(T, B),$$

so that $\dot{R}(c - x\overline{B}) \geq \dot{R}(c - y\overline{B})$. □

4.4.2. LEMMA. $\dot{R}(c - x_i\overline{B}) \geq -i$, where x_i is the number obtained at the ith step of the algorithm for determining $\beta^{\max}$ (see Figure 2).

PROOF. The lemma will be proved by induction. For $i = n - 1$ the lemma holds. We assume that it holds for all $i \in \{k + 1, \ldots, n - 1\}$, and prove it for $i = k$.

Lemma 4.4.1 implies that

$$\dot{R}(c - x_k\overline{B}) \geq \frac{R(c - x_{k+1}\overline{B}) - R(c - x_k\overline{B})}{x_{k+1} - x_k}.$$

Since $R(c - x_k\overline{B}) < R(c) - x_k$, and since the points $(x_k, R(c) - x_k)$ and $(x_{k+1}, R(c - x_{k+1}\overline{B}))$ belong to the line $L(x_{k+1}, k+1)$, by assumption we have

$$\dot{R}(c - x_k\overline{B}) > \frac{R(c - x_{k+1}\overline{B}) - (R(c) - x_k)}{x_{k+1} - x_k} = -(k + 1).$$

Thus $\dot{R}(c - x_k\overline{B}) \geq -k$. □

4.4.3. Let us prove that the algorithm described in §4.3.2, terminating at some step $i \geq 2$, indeed determines $\beta^{\max} = x_i$. For this is suffices to show that any $\beta \in [x_i, x_{i+1}]$ leads to a nonpermissible branching.

By Lemmas 4.4.1 and 4.4.2,

$$R(c - \beta\overline{B}) = R(c - x_{i+1}\overline{B} + (x_{i+1} - \beta)\overline{B})$$
$$\leq R(c - x_{i+1}\overline{B}) - \dot{R}(c - x_{i+1}\overline{B})(x_{i+1} - \beta)$$
$$\leq R(c - x_{i+1}\overline{B}) + (i + 1)(x_{i+1} - \beta).$$

Since the point $(\beta, R(c - x_{i+1}\overline{B}) + (i + 1)(x_{i+1} - \beta))$ lies on the line $L(x_{i+1}, i+1)$ between the lines $(x_i, R(c - x_i\overline{B}))$ and $(x_{i+1}, R(c - x_{i+1}\overline{B}))$, then its y-coordinate is smaller than $R(c) - \beta$, and so the branching (β, B) is nonpermissible.

4.4.4. Now we estimate the number of operations performed in one step of solving Problem A in cases A1, A2, and A3.

In case A1 (see §4.3.3) during the ith iteration it is necessary to determine the cut $[S_i, \overline{S}_i]$ and to find in $[S_i, \overline{S}_i] \cap EB_i$ a closed set of cardinality $R(c) - 1 - R(c - \overline{B}_i)$, which requires $O(n \cdot F_{n,p})$ operations. Since the number of iterations does not exceed n, the total estimate in case A1 is $O(n^2 \cdot F_{n,p})$.

Lemma 4.4.2 implies that in cases A2 and A3 the number of iterations of the algorithm for determining $\beta^{\max}$ does not exceed $n - 2$. Since in the ith iteration $R(c - x_i \overline{B})$ should be determined (see 4.3.2), the estimate of the number of operations per step in cases A2 and A3 is $O(n^2 \cdot F_{n,p})$.

Thus the complexity of the algorithm for solving Problem A is
$$O(n^2 F_{n,p} + kn F_{n,p}),$$
where k is the number of occurrences of case A3 during the working of the algorithm (recall that $k \leq p$).

5. b-tree of a branching packing

5.1. A vertex v of a branching B is called a *descendant* of a vertex w if in B there exists a path from w to v, and it is called a *direct descendant* if this path consists of one arc. A branching B is said to be a *b-tree* if each vertex of B either has no descendants or has exactly two direct descendants. To avoid confusion of b-trees with branchings, vertices of b-trees will be called *nodes*. A node of a b-tree is called a *branching node* if it has descendants, and a *leaf* otherwise. Obviously the number of leaves in a b-tree exceeds the number of branching nodes by 1.

5.2. While solving Problem A for a network $N(G, c)$ and decomposing the network by the constructed cut, we obtain a permissible branching packing $\mathfrak{B}$ in $N(G, c)$, a cut $[R, S]$, and two networks $N_R(G_R, c_R)$ and $N_S(G_S, c_S)$. Since the packing $\mathfrak{B}$ is permissible, for a given maximal packing $\mathfrak{B}'$ in the network $N(G, c_\mathfrak{B})$ it is possible to construct a maximal packing $\mathfrak{B}_{\max}$ in the network $N(G, c)$, taking for $\mathfrak{B}_{\max}$ the union of the packings $\mathfrak{B}$ and $\mathfrak{B}'$. By Theorem 3.2, in order to construct a maximal packing in the network $N(G, c_\mathfrak{B})$ it is sufficient to construct maximal branching packings in the networks N_R and N_S. Therefore, after a solution of Problem A is found, the construction of the maximal packing in N reduces to the construction of optimal branching packings in N_R and N_S.

Denote by T_0 the b-tree consisting of the root only (see Figure 4a). Consider the b-tree T_1 (Figure 4b) whose root corresponds to the network N and where leaves correspond to N_R and N_S. If some leaf of this b-tree (e.g. N_R) corresponds to a network containing more than three vertices (below we use the term "leaf with more than three vertices"), then for this network it is possible to solve Problem A, obtain a packing $\mathfrak{B}_1$ and a nontrivial minimal cut $[R_1, S_1]$, and decompose the network by this cut. After this we obtain the b-tree T_2 in Figure 4c (N_{R_1} and N_{S_1} denote the networks obtained after decomposition of the network $N_R(G_R, (c_R)_{\mathfrak{B}_1})$ by the cut $[R_1, S_1]$).

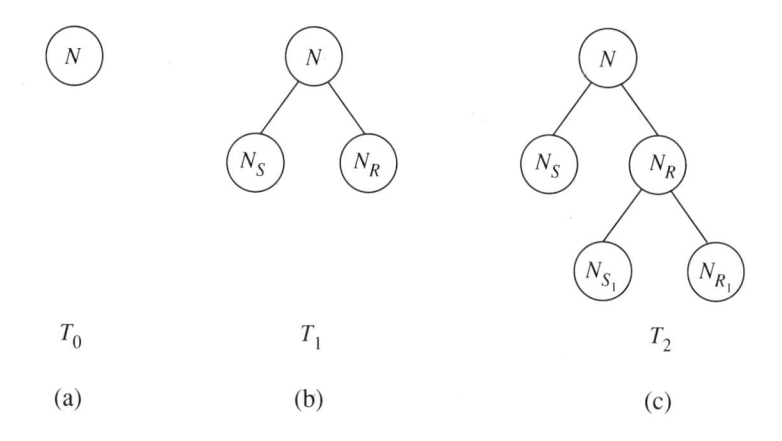

FIGURE 4

Repeating this process, we obtain a sequence of b-trees $T_0, T_1, \ldots, T_m$, where T_{i+1} is obtained from T_i ($0 \le i \le m-1$) by adding two direct descendants to some leaf of the b-tree T_i with more than three vertices (the network corresponding to this network is denoted by $N_i(G_i, (c_i)_{\mathfrak{B}_i})$), while each leaf of the b-tree T_m consists of three vertices. The number of branching nodes in T_m equals exactly m, and each branching node N_i of T_m corresponds to a permissible branching packing $\mathfrak{B}_i$ and a nontrivial minimal cut $[R_i, S_i]$ in $N_i(G_i, (c_i)_{\mathfrak{B}_i})$ constructed in the course of solving Problem A for the network $N_i(G_i, c_i)$.

5.3. The algorithm for constructing an optimal packing consists of m steps. Prior to the ith ($i = 0, \ldots, m-1$) step, maximal branching packings are constructed in all leaves of the b-tree T_{m-i}. The step consists of constructing maximal branching packings in all leaves of the b-tree $T_{m-(i+1)}$. Note that the b-tree T_{m-i} is obtained from $T_{m-(i+1)}$ by adding two descendants $N'_{m-(i+1)}$ and $N''_{m-(i+1)}$ to the leaf $N_{m-(i+1)}$ of $T_{m-(i+1)}$. Since $N'_{m-(i+1)}$ and $N''_{m-(i+1)}$ are leaves of T_{m-i}, prior to the ith step packings in the networks $N'_{m-(i+1)}$ and $N''_{m-(i+1)}$ are already constructed. Given these maximal packings and a permissible branching packing $\mathfrak{B}_{m-(i+1)}$ in $N_{m-(i+1)}$, it is possible, using Theorem 3.2, to construct a maximal branching packing $\mathfrak{B}^{\max}_{m-(i+1)}$ in the network $N_{m-(i+1)}$. This being done, the ith step is completed (previously constructed maximal branching packings in leaves of T_{m-i} and $\mathfrak{B}^{\max}_{m-(i+1)}$ provide maximal branching packings in all leaves of $T_{m-(i+1)}$).

Obtaining consecutively maximal branching packings in all leaves of the b-trees $T_m, T_{m-1}, \ldots, T_1, T_0$, at the $(m-1)$st step we get a maximal branching packing in the initial network $N(G, c)$.

6. Complexity of the algorithm for branching packing

It is easy to prove by induction that the number of branching nodes of the b-tree T_m equals $n-3$. In each branching node N_i Problem A is solved, that is, a packing $\mathfrak{B}$ and a minimal nontrivial cut $[R_i, S_i]$ are determined. The number of operations required for solving Problem A at the ith step is $O(n^2 F_{n,p} + k_i n F_{n,p})$, where k_i is the number of occurrences of case A3 in the course of solving Problem A for the network N_i (see 4.4.4). Therefore, the total number of operations for constructing T_m is

$$O\left(n^3 F_{n,p} + \left(\sum_{i=0}^{n-4} k_i\right) n F_{n,p}\right).$$

Let us prove that

$$\sum_{i=0}^{n-4} k_i = O(n^2).$$

Assume that after decomposition of a graph G by a cut $[R, S]$ graphs G_R and G_S are obtained with roots r and z respectively. Let us estimate the total number of arcs in the graphs G_R and G_S (without multiplicities). Set $U_R = EG_R \setminus [VG_S \setminus z, z]$ and $U_S = EG_S \setminus [VG_S \setminus z, z]$. Obviously $F_R^{-1}(U_R) \cap F_S^{-1}(U_S) = \varnothing$ (recall that in G_R and G_S loops and arcs coming to the root are deleted). Therefore $|U_R| + |U_S| \le |EG|$. Hence

$$\begin{aligned}|EG_R| + |EG_S| &= |U_R| + |U_S| + |[VG_R \setminus z, z]| + |[VG_S \setminus z, z]| \\ &\le |EG| + |R| + |S| = |EG| + |VG|.\end{aligned} \quad (*)$$

Denote by p_i the total number of arcs in leaves of the b-tree T_i. The b-tree T_{i+1} is formed by adding to the leaf $N_i(G_i, c_i)$ of T_i two descendants $N_i'(G_i', c_i')$ and $N''(G_i'', c_i'')$, which are obtained in the decomposition of the network $N_i^*(G_i, (c_i)_{\mathfrak{B}_i})$ by a minimal cut. Since in the construction of the packing $\mathfrak{B}_i$ case A3 occurs k_i times and in each occurrence of case A3 the number of arcs in the network decreases by at least 1, we have $|EN_i| - |EN_i^*| \ge k_i$. This fact together with $(*)$ implies that

$$\begin{aligned}p_{i+1} &= p_i + |EN_i'| + |EN_i''| - |EN_i| \\ &= p_i + (|EN_i'| + |EN_i''| - |EN_i^*|) - (|EN_i| - |EN_i^*|) \le p_i + n - k_i.\end{aligned}$$

Since $p_0 = p$, we have

$$p_m = \sum_{i=0}^{m}(p_{i+1} - p_i) + p_0 \le \sum_{i=0}^{n-4}(n - k_i) + p = p + n(n-3) - \sum_{i=0}^{n-4} k_i.$$

Therefore, $\sum_{i=0}^{n-4} k_i = O(n^2)$, and the total number of operations in the construction of T_m is $O(n^3 F_{n,p})$.

After constructing T_m, it is necessary to construct the maximal branching packings in all of its leaves, and then, as described in §5.3, to construct consecutively maximal packings in leaves of the b-trees T_i. Since T_m has $n-2$ leaves, and each leaf contains 3 vertices, the number of operations in the construction of maximal packings at all leaves of T_m is small (compared to $O(n^3 F_{n,p})$). It is easy to prove that the construction of maximal packings at leaves of T_i from maximal packings at leaves of T_{i+1} also requires a small number of operations, so that the total estimate of effectivity for our algorithm is $O(n^3 F_{n,p})$. Since $F_{n,p}$ can be estimated by $O(n^3)$ [6], we obtain the estimate $O(n^6)$.

7. Cardinality of a branching packing

An arc e'' in a node N'' of the b-tree T_m is called a *descendant* of an arc e' in a node N' if there exist a sequence of nodes $N' = N^1, \ldots, N^k = N''$ of T_m and a sequence of arcs $e' = e^1, \ldots, e^k = e''$ such that:

(1) $e^i \in EN^i$ ($i = 1, \ldots, k$),
(2) N^{i+1} is a direct descendant of N^i ($i = 1, \ldots, k-1$), and
(3) e^{i+1} is the image of e^i ($i = 1, \ldots, k-1$).

Denote the set of descendants of an arc e by $D(e)$.

Networks corresponding to branching nodes of the b-tree T_m are already enumerated as $N_0, \ldots, N_{n-4}$. We now enumerate the networks corresponding to leaves of T_m as $N_{n-3}, \ldots, N_{2n-6}$. In each network N_i ($i = 0, \ldots, 2n-6$) a permissible branching packing $\mathfrak{B}_i = \{(\beta_i^1, B_i^1), \ldots, (\beta_i^{r_i}, B_i^{r_i})\}$ of cardinality r_i is constructed (contrary to the previous assumptions, now it is necessary to count arcs with multiplicities). Theorems 3.2 and 5.2 imply that the cardinality of the branching packing constructed by the algorithm does not exceed $\sum_{i=0}^{2n-6} r_i$.

We say that a branching (β_i^j, B_i^j) *drives out* an arc e from a network $N_i(G_i, c_i)$ if $e \in EB_i^j$ and $c_i(e) = \sum_{k=1}^{j} \beta_i^k B_i^k$. In constructing a packing $\mathfrak{B}_i$ in a network N_i each of cases A1 and A2 occurs at most once, while each occurrence of case A3 is accompanied by driving at least one arc out of the network N_i (see 4.3.3). Since the number of occurrences of case A1 (A2) does not exceed the number of branching nodes in T_m, we have $\sum_{i=0}^{2n-6} r_i \leq 2(n-3) + k$, where k is the number of arcs driven out during the working of the algorithm. Denote by $k(e)$ the number of arcs driven out of the set $D(e)$. If at some step i an arc $e' \in D(e)$ is driven out of the network N_i, then descendants of the arc e are not present in descendants of the node N_i in the b-tree T_m. Therefore the number of arcs driven out of the set $D(e)$ does not exceed the number of leaves in T_m, so that $k(e) \leq n-2$. Since $k = \sum_{e \in EG} k(e)$, the total estimate on the number of branching in the packing we have constructed is $O(np)$. However, the author does not know examples when this bound is achieved, and the problem of finding a tighter upper bound for the cardinality of a perfect packing remains open.

References

1. J. Edmonds, *Edge-disjoint branchings*, Combinatorial Algorithms (R. Rustin, editor), Academic Press, New York, 1972, pp. 91–96.
2. R. E. Tarjan, *Finding edge-disjoint spanning trees*, Proc. Eighth Hawaii Internat. Conf. System Sci. (Honolulu, HA, 1975), Western Periodicals, North Hollywood, CA, 1975, pp. 251–252.
3. _____, *A good algorithm for edge-disjoint branchings*, Inform. Process. Lett. **3** (1974), 51–53.
4. L. Lovácz, *On two minimax theorems in graphs*, J. Combin. Theory Ser. B **21** (1976), 96–103.
5. D. R. Fulkerson and G. C. Harding, *On edge-disjoint branchings*, Networks **6** (1976), 97–104.
6. A. V. Karzanov, *Finding an optimal flow in a network by the method of pre-flows*, Dokl. Akad. Nauk SSSR **215** (1974), 49–52; English transl. in Soviet Math. Dokl. **15** (1974).

Non-3-crossing Families and Multicommodity Flows

P. A. PEVZNER

We consider families of subsets $\mathscr{M} \subset 2^V$ of a finite set V on n elements. Two sets $A, B \subset V$ are called *crossing* if each of the following four sets is not empty (here $\bar{A} = V \setminus A$):

$$A \cap B, \quad \bar{A} \cap B, \quad A \cap \bar{B}, \quad \bar{A} \cap \bar{B};$$

otherwise A and B are called *laminar*. A family $\mathscr{M}$ is called *k-crossing* if there are k members of $\mathscr{M}$ each two of which are crossing. A family $\mathscr{M}$ is called *non-k-crossing* if every k members of $\mathscr{M}$ contain a laminar pair. Non-k-crossing families were introduced by A. V. Karzanov [1] in connection with the *generalized MFMC property* and *cut multicommodity flow problems* (see [2]–[4]). Non-3-crossing anticlique families provide a complete characterization of cut multiflow problems. Such characterization implies a powerful generalization of the Ford–Fulkerson theorem for multiflow problems, including, in particular, the theorems of Hu [5] and Lovász and Cherkasskiĭ [6], [7]. In [1] the following theorem about *multiflow locking* is proved: a family $\mathscr{M}$ is lockable if and only if it is non-3-crossing. On the basis of the locking problem, a full description of cut multiflow problems was obtained: a multiflow problem with *commodity graph* S not containing isolated vertices has a generalized MFMC property if and only if the anticliques of S are non-3-crossing [1], [8], [9]. For details and recent results on the links between non-3-crossing families see Karzanov [10] and Frank et al.[11].

In [1], in the course of constructing an effective algorithm for solution of the locking problem there arose the problem of the maximum number $f(n)$ of sets in a non-3-crossing family. Karzanov [1] noticed that $f(n) = O(n^2)$ and conjectured that $f(n) = O(n)$. M. V. Lomonosov noted that

---

1991 *Mathematics Subject Classification*. Primary 05C38, 90B10.

Translation of Problems of Discrete Optimization and Methods for Their Solution, Tsentral. Èkonom.-Mat. Inst. Akad. Nauk SSSR, Moscow, 1987, pp. 136–142.

The author is grateful to A. R. Rubinov and D. S. Fleĭshman for useful discussions.

$f(n) = O(n \log n)$. In the present paper we prove Karzanov's conjecture: $f(n) = O(n)$.

1

The relation of inclusion defines a *partial order* on the family $\mathscr{M}$. A subfamily $\mathscr{N}$ of $\mathscr{M}$ is called an *antichain* if no two elements of it are comparable. The *trace* $\mathscr{S}(\mathscr{N})$ of a family $\mathscr{N}$ is the family

$$\mathscr{S}(\mathscr{N}) = \{ A \in \mathscr{M} : \exists B \in \mathscr{N} : A \leq B \}.$$

Let $p(\mathscr{N}) = \sum_{A \in \mathscr{N}} |A|$.

In what follows we consider families $\mathscr{M}$ without complements, i.e. not containing simultaneously pairs A, $\bar{A}$. Assume also that all sets in $\mathscr{M}$ contain not more than $n/2$ elements, i.e., for any $A, B \in \mathscr{M}$,

$$A \cup B \neq V. \tag{1}$$

LEMMA 1. *Let $\mathscr{M}$ be a non-k-crossing family and let $\mathscr{N}$ be an antichain in $\mathscr{M}$. Then $p(\mathscr{N}) \leq (k-1)n$.*

PROOF. If $p(\mathscr{N}) > (k-1)n$, then there exists $v \in V$ belonging to at least k sets from $\mathscr{N}$. From the incomparability of elements of $\mathscr{N}$ and condition (1) it follows that these sets are pairwise crossing, which contradicts the condition that $\mathscr{N}$ is non-k-crossing.

LEMMA 2. *Let $\mathscr{N}$ be the maximal (in number of elements) antichain in a non-k-crossing family $\mathscr{M}$. Then $|\mathscr{S}(\mathscr{N})| \leq (k-1)n$.*

PROOF. By Dilworth's theorem there exists a decomposition of $\mathscr{M}$ into $|\mathscr{N}|$ nonintersecting chains $C_1, C_2, \ldots, C_m$. Obviously each chain C_i contains exactly one element A_i from $\mathscr{N}$. Consider families $C_i^- = \{ B \in C_i : B \leq A_i \}$. The condition $\mathscr{S}(\mathscr{N}) = \bigcup_{i=1}^m C_i^-$ and Lemma 1 imply

$$|\mathscr{S}(\mathscr{N})| = \sum_{i=1}^m |C_i^-| \leq \sum_{i=1}^m |A_i| = p(\mathscr{N}) \leq (k-1)n.$$

Let $\mathscr{N}$ and $\mathscr{R}$ be families of sets on V. $\mathscr{N}$ is said to be *imbedded* in $\mathscr{R}$ if each set from $\mathscr{N}$ is contained in some set from $\mathscr{R}$. Consider the family $\mathscr{K}$ of antichains in $\mathscr{M}$. The relation of imbedding defines a partial order in $\mathscr{K}$, namely: if $\mathscr{N}, \mathscr{R} \in \mathscr{K}$, then $\mathscr{N} < \mathscr{R}$ if $\mathscr{N}$ is imbedded in $\mathscr{R}$. A sequence of (distinct) antichains $P = (\mathscr{N}_0, \mathscr{N}_1, \ldots, \mathscr{N}_l)$ is called *primal* if it defines in $\mathscr{K}$ a chain of maximum length between $\mathscr{N}_0$ and $\mathscr{N}_l$.

LEMMA 3. *Let P be a primal chain in $\mathscr{K}$. Then, for any $\mathscr{N}_i \in P$,*

$$\mathscr{S}(\mathscr{N}_{i+1}) \setminus \mathscr{S}(\mathscr{N}_i) = \mathscr{N}_{i+1} \setminus \mathscr{N}_i.$$

PROOF. Let $A \in \mathscr{S}(\mathscr{N}_{i+1}) \setminus \mathscr{S}(\mathscr{N}_i)$. Consider a family of sets $\mathscr{T} = \{ B \in \mathscr{N}_i : B < A \}$ in $\mathscr{N}_i$. Obviously $\mathscr{N} = (\mathscr{N}_i \setminus \mathscr{T}) \cup \{A\}$ is an antichain set of $\mathscr{M}$. If in addition A is not contained in $\mathscr{N}_{i+1}$, then $\mathscr{N}_{i+1} > \mathscr{N} > \mathscr{N}_i$, and $(\mathcal{N}_i, \mathcal{N}, \mathcal{N}_{i+1})$ defines a path of length 2 between $\mathcal{N}_i$ and $\mathcal{N}_{i+1}$ in $\mathcal{H}$, which contradicts the assumption that P is primal. Thus $A \in \mathcal{N}_{i+1}$ and $\mathscr{S}(\mathcal{N}_{i+1}) \setminus \mathscr{S}(\mathcal{N}_i) \subseteq \mathcal{N}_{i+1} \setminus \mathcal{N}_i$. The inclusion $\mathscr{S}(\mathcal{N}_{i+1}) \setminus \mathscr{S}(\mathcal{N}_i) \supseteq \mathcal{N}_{i+1} \setminus \mathcal{N}_i$ is obvious.

2

Let $A \subseteq V$. Denote by $\mathcal{M}|A$ a family of subsets of $V \setminus A$:

$$\mathcal{M}|A = \{ B : \exists C \in \mathcal{M} : B = C \setminus A \}.$$

A *contraction* of $\mathcal{M}$ by a set A is a family $\mathcal{M}|(A \setminus a)$ for any $a \in A$. A set A is said to be *contractible in $\mathcal{M}$* if each set in $\mathcal{M}$ either contains A or does not intersect A. A set $L \subset V$ is called an *ith level lens* (or *i-lens*) for a sequence of antichains $P = (\mathcal{N}_0, \mathcal{N}_1, \ldots, \mathcal{N}_l)$ if in $\mathcal{N}_i$ there exist different sets A and B such that (i) $L = A \cap B$, and (ii) A and B are contained in a set from $\mathcal{N}_{i+1}$. The set of i-lenses is denoted by $LN(i)$.

LEMMA 4. *Each i-lens is contractible in a system of sets $\mathcal{M} \setminus \mathscr{S}(\mathcal{N}_i)$ where $\mathcal{M}$ is a non-3-crossing family.*

PROOF. Let $L = A \cap B$ be an i-lens and let $A, B \in \mathcal{N}_i$. Let $C \in \mathcal{M} \setminus \mathscr{S}(\mathcal{N}_i)$ intersect with L. We prove that C contains L. The sets A, B, C are not all pairwise crossing, they intersect, and C is not contained in $\mathscr{S}(\mathcal{N}_i)$. Thus C contains A or B, and so $L \subset C$.

LEMMA 5. *Lenses of the same level do not intersect.*

PROOF. Let L and Q be intersecting, but not coinciding, i-lenses. Then in $\mathcal{N}_i$ there are sets A, B, C, D such that $L = A \cap B$ and $Q = C \cap D$. Among the sets A, B, C, D there are three that do not coincide. These three sets belong to an antichain and have a common element, and so they are pairwise crossing, which contradicts $\mathcal{M}$ being non-3-crossing.

LEMMA 6. *Lenses of different levels do not coincide.*

PROOF. Let L be simultaneously an i-lens and a j-lens, $i < j$; let $A_1, B_1 \in \mathcal{N}_i$, $A_2, B_2 \in \mathcal{N}_j$, and $A_1 \cap B_1 = A_2 \cap B_2 = L$. Since $\mathcal{N}_i$ is imbedded into $\mathcal{N}_j$, by the definition of an i-lens there exists $C \in \mathcal{N}_j$ such that A_1 and B_1 are contained in C. Since A_2, B_2 and C are non-3-crossing, C concides with either A_2 or B_2, say with A_2. Consider the triple A_1, B_1, B_2. Since

$$L = C \cap B_2 \supset (A_1 \cup B_1) \cap B_2 = (A_1 \cap B_2) \cup (B_1 \cap B_2),$$

B_2 contains neither A_1 nor B_1. Thus A_1, B_1, B_2 are 3-crossing.

An immediate corollary of Lemmas 4, 5, and 6 is

LEMMA 7. *Lenses form a laminar system without repeated elements.*

From Lemma 4 follows

LEMMA 8. *Let $\mathscr{M}$ contain no single-element sets, and let L be an i-lens in $\mathscr{M}$. Then contraction of $\mathscr{M} \setminus \mathscr{S}(\mathscr{N}_i)$ by L generates a non-3-crossing family without single-element sets, whose number of elements is equal to that of $\mathscr{M} \setminus \mathscr{S}(\mathscr{N}_i)$.*

Let $\mathscr{N}$ be an antichain in $\mathscr{M}$, and for each $A \in \mathscr{N}$ let a family $\mathscr{T}(A) \subset \mathscr{M}$ be defined. The system of families $\mathscr{T}(A)$ is called a *decomposition* of $\mathscr{N}$ if the following conditions are satisfied:
 (i) $|\mathscr{T}(A)| \leq |A| - 1$;
 (ii) all sets from $\mathscr{T}(A)$ are contained in A.

The set $\mathscr{L}(\mathscr{N}) = \mathscr{S}(\mathscr{N}) \setminus \bigcup_{A \in \mathscr{N}} \mathscr{T}(A)$ is called the *remainder* of a decomposable set $\mathscr{N}$. From Lemma 1 follows

LEMMA 9. *$|\mathscr{S}(\mathscr{N})| \leq 2n + |\mathscr{L}(\mathscr{N})|$ for any antichain $\mathscr{N}$.*

3

THEOREM. *A non-3-crossing family on n elements contains not more than $6n$ sets.*

The proof is performed by constructing the primal decomposable sequence of sets $P = (\mathscr{N}_0, \mathscr{N}_1, \ldots, \mathscr{N}_l)$, where $\mathscr{N}_0$ is the maximal (in number of elements) antichain in $\mathscr{M}$ and $\mathscr{N}_l$ is a family of maximal sets in $\mathscr{M}$, i.e. $\mathscr{S}(\mathscr{N}_l) = \mathscr{M}$. We will consider the sequence of remainders $\mathscr{L}_0, \mathscr{L}_1, \ldots, \mathscr{L}_l$ of the families $\mathscr{N}_i$ and will prove that $|\mathscr{L}_i| = O(n)$. The proof of the latter statement is based on setting up a correspondence between elements of sets $\mathscr{L}_i$ and j-lenses ($j < i$), the number of which is estimated as $O(n)$ by Lemma 7. Consequently, by Lemma 9, $|\mathscr{M}| = O(n)$. The transition from $\mathscr{N}_i$ to $\mathscr{N}_{i+1}$ during the proof is accompanied by contraction of i-lenses. Sets of elements remaining after contractions of i-lenses ($i = 1, \ldots, l$) form the sequence $V = V_0, V_1, \ldots, V_l$.

PROOF. Delete all single-element sets from $\mathscr{M}$ and take as $\mathscr{N}_0$ the maximal antichain in $\mathscr{M}$. Consider the minimal chain decomposition $C_1, \ldots, C_m$ of the above family (Lemma 2). As a decomposition of a set A_i from the family $\mathscr{N}_0$ take the sets $\mathscr{T}_0(A_i) = C_i^-$. Since $\mathscr{M}$ does not contain single-element sets, $|\mathscr{T}_0(A_i)| \leq |A| - 1$.

Denote by l_i the number of lenses of level $i - 1$ and lower contained in V_i, by u_i the number of deleted vertices after contraction of lenses of level $i - 1$ and lower, and by w_i the number of contractions of lenses of level $i - 1$ and lower. Let $l_0 = u_0 = w_0 = 0$.

Assume now that decompositions $\mathscr{T}_i(A)$ of elements of the family $\mathscr{N}_i$ such that

$$|\mathscr{L}_i| \leq l_i + 2u_i + w_i \tag{2}$$

are already constructed. Construct a decomposition of the family $\mathscr{N}_{i+1}$ satisfying (2) (note that $|\mathscr{L}_0| = 0$, and thus (2) is satisfied for $\mathscr{N}_0$).

Let $\mathcal{N}_{i+1} = (A_1, \ldots, A_p)$. Since $\mathcal{N}_i$ is imbedded in $\mathcal{N}_{i+1}$, then $\mathcal{N}_i$ can be decomposed into nonintersecting families $\mathcal{P}(A_1), \ldots, \mathcal{P}(A_p)$ such that all sets from $\mathcal{P}(A_i)$ are contained in A_i. Now we describe how decompositions of elements of $\mathcal{N}_{i+1}$ are constructed, given decompositions of $\mathcal{N}_i$. Depending on the number of elements in $\mathcal{P}(A)$, three cases are possible.

1. $|\mathcal{P}(A)| = 0$. Then set $\mathcal{T}_{i+1}(A) = \{A\}$. By Lemma 8, A is not a single-element set, and thus $|\mathcal{T}_{i+1}(A)| \leq |A| - 1$.

2. $|\mathcal{P}(A)| = 1$. If $\mathcal{P}(A) = \{A\}$, then set $\mathcal{T}_{i+1}(A) = \mathcal{T}_i(A)$. If $\mathcal{P}(A) = \{B\}$, $B \neq A$, then set $\mathcal{T}_{i+1} = \mathcal{T}_i(B) \cup \{A\}$. Since A contains B, we have $|\mathcal{T}_{i+1}(A)| = |\mathcal{T}_i(B)| + 1 \leq |B| - 1 + 1 \leq |A| - 1$.

3. $|\mathcal{P}(A)| \geq 2$. Let $\mathcal{P}(A) = (B_1, \ldots, B_t)$. Since $B_1, \ldots, B_t$ comprise an antichain in A, by Lemma 1 we get $p(\mathcal{P}(A)) = \sum_{j=1}^{t} |B_j| \leq 2|A|$. Denote $p(\mathcal{P}(A)) - |A| = d$. Obviously, d does not exceed the total number of elements in the lenses formed by the sets $B_1, \ldots, B_t$. Form a family $\mathcal{T}_{i+1}(A)$ from the family $\bigcup_{j=1}^{t} \mathcal{T}_i(B_j)$ by deleting from it some family $\mathcal{D}(A)$ consisting of d elements and adding the set A itself:

$$\mathcal{T}_{i+1}(A) = \left(\bigcup_{j=1}^{t} \mathcal{T}_i(B_j) \cup \{A\}\right) \setminus \mathcal{D}(A).$$

Then

$$|\mathcal{T}_{i+1}(A)| = \sum_{j=1}^{t} |\mathcal{T}_i(B_j)| + 1 - d \leq \sum_{j=1}^{t}(|B_j| - 1) + 1 - \left(\sum_{j=1}^{t} |B_j| - |A|\right) \leq |A| - 1.$$

Consideration of these three cases completes the description of the decomposition for the family $\mathcal{N}_{i+1}$. Now we prove (2) for $|\mathcal{L}_{i+1}|$.

From Lemma 3 and the decomposition construction algorithm for $\mathcal{N}_{i+1}$ it follows that

$$|\mathcal{L}_{i+1}| \leq |\mathcal{L}_i| + \sum_{A \in \mathcal{N}_i} |\mathcal{D}(A)| \leq |\mathcal{L}_i| + \sum_{L \in LN(i)} |L|. \tag{3}$$

Contract all multi-element sets in $LN(i)$. Denote by $LN1(i)$ the set of single-element i-lenses and by $LN2(i)$ the set of multi-element i-lenses. From Lemma 7 it follows that

$$l_{i+1} \geq l_i + |LN(i)| - \sum_{L \in LN2(i)} (|L| - 1) \tag{4}$$

(in contracting a lens L we delete $|L| - 1$ elements). Obviously

$$u_{i+1} = u_i + \sum_{L \in LN2(i)} (|L| - 1), \tag{5}$$

$$w_{i+1} = w_i + |LN2(i)|. \tag{6}$$

By substituting (4), (5) and (6) into (3), we obtain

$$|\mathscr{L}_{i+1}| \leq l_{i+1} - |LN1(i)| + \sum_{L \in LN2(i)} (|L| - 1) + 2 \left(u_{i+1} - \sum_{L \in LN2(i)} (|L| - 1) \right)$$
$$+ w_{i+1} - |LN2(i)| + \sum_{L \in LN(i)} |L| = l_{i+1} + 2u_{i+1} + w_i.$$

Obviously, $l_l \leq n$, and $u_l + w_l \leq n$. Thus $|\mathscr{L}_l|$ does not exceed $3n$, and so, by Lemma 9, $|\mathscr{M}| \leq 5n$. Taking single-element sets into account, we obtain $|\mathscr{M}| \leq 6n$. The theorem is proved.

4

We now formulate two open problems related to non-k-crossing families.

1. One can show that the maximum number $f_k(n)$ of sets in a non-k-crossing family is bounded by $O(n \log n)$. Is it true that $f_k(n) = O(n)$?

2. Is it true that any non-k-crossing family on n elements can be decomposed into r non-$(k-1)$-crossing families (r is independent of n, $k > 3$)?

Families that can be represented as a union of two laminar ones were considered in [8] and [11]. It is possible to show that for $k = 3$ the answer to the second problem is negative (an example of an r-indecomposable non-3-crossing family is a family of stars in a graph without triangles with a chromatic number exceeding r).

References

1. A. V. Karzanov, *Combinatorial methods to solve cut-determined multiflow problems*, Combinatorial Methods for Flow Problems, no. 3 (A. V. Karzanov, editor), Vsesoyuz. Nauchno-Issled. Inst. Sistem. Issled., Moscow, 1979, pp. 6–69. (Russian)
2. _____, *A generalized MFMC-property and multicommodity cut problems*, Finite and Infinite Sets. Vol. II (Proc. Sixth Hungarian Combinatorial Colloq., Eger, 1981; A. Hajnal et al., editors), Colloq. Math. Soc. János Bolyai, vol. 37, North-Holland, Amsterdam, 1984, pp. 443–486.
3. M. V. Lomonosov, *Combinatorial approaches to multiflow problems*, Discrete Appl. Math. **11** (1985), 1–94.
4. A. V. Karzanov, *Metrics and undirected cuts*, Math. Programming **32** (1985), 183–198.
5. T. G. Hu, *Multi-commodity network flows*, Oper. Res. **11** (1963), 344–360.
6. L. Lovász, *On some connectivity properties of Eulerian graphs*, Acta Math. Acad. Sci. Hungar. **28** (1976), 129–138.
7. B. N. Cherkasskiĭ, *Solution of a problem on multi-commodity flows in networks*, Èkonom. i Mat. Metody **13** (1977), 143–151. (Russian)
8. A. V. Karzanov and M. V. Lomonosov, *Flow systems in undirected networks*, Mathematical Programming Etc., Sb. Trudov Vsesoyuz. Nauchno-Issled. Inst. Sistem Issled. **1978**, no. 1, 59–66. (Russian)
9. A. V. Karzanov and P. A. Pevzner, *A characterization of the class of cut-non-determined maximum multiflow problems*, Combinatorial Methods for Flow Problems, no. 3 (A. V. Karzanov, editor), Vsesoyuz. Nauchno-Issled. Inst. Sistem. Issled., Moscow, 1979, pp. 70–81. (Russian)
10. A. V. Karzanov, *Maximization over the intersection of two compatible greedy polyhedra*, Report 91732-OR, Inst. Diskrete Math., Bonn, 1991, pp. 1–25.
11. A. Frank, A. V. Karzanov, and A. Sebo (in press).

The Vector Shortest Path Problem in the l_∞-norm

A. D. VAĬNSHTEĬN

Consider the following generalization of the well-known shortest path problem. Let G be a graph with the vertex set V and the edge set E. To each edge $e \in E$, assign a k-tuple of nonnegative numbers $a(e) = (a_1(e), \ldots, a_k(e))$. Let $S = v_1 \cdots v_m$, $v_i \in V$, be an arbitrary path in G; we assign to S the following vector:

$$a(S) = \sum_{i=1}^{m-1} a(v_i, v_{i+1}) = \left(\sum_{i=1}^{m-1} a_1(v_i, v_{i+1}), \ldots, \sum_{i=1}^{m-1} a_k(v_i, v_{i+1}) \right).$$

Given a norm $\nu(x)$ in $\mathbb{R}^k$, the *length* of the path S is calculated as

$$l(S) = \nu(a(S)). \tag{1}$$

The vector shortest path problem in the ν-norm consists in finding a path joining vertices $u, v \in V$ and such that functional (1) attains its minimum on this path; such a path is said to be *optimal*.

The complexity of the problem depends heavily on the norm $\nu(x)$. For example, in the case of the l_1-norm, that is, for $\nu(x_1, \ldots, x_k) = \sum_{j=1}^{k} |x_j|$, the vector problem is equivalent to the ordinary shortest path problem with edge weights $w(e) = \nu(a(e))$. Below we consider the case of the l_∞-norm, i.e.,

$$\nu(x_1, \ldots, x_k) = \max_{1 \leq j \leq k} |x_j|.$$

For $k = 1$ we have the classical shortest path problem, for which several polynomial algorithms are known (see, for example, [3]). For $k > 1$ the problem turns out to be more complicated. Indeed, even for $k = 2$, the Bellman principle does not apply to the problem. This means that if S is an optimal path joining v_1 and v_2 and a vertex v belongs to S, then the parts of S between v_1 and v and between v and v_2 are not necessarily optimal

---

1991 *Mathematics Subject Classification.* Primary 05C38.
Translation of Èkonom. i Mat. Metody **21** (1985), 1132–1137; MR **87f:**05101.

©1994 American Mathematical Society
0065-9290/94 $1.00 + $.25 per page

paths between the corresponding vertices. Therefore, the set of optimal paths joining v with all the other vertices of G is not necessarily a tree. Besides, this means that the powerful algebraic approaches to path problems given in [1] and [2] also do not apply.

A reason for these complications lies in the following result.

THEOREM 1. *The vector shortest path problem in the l_∞-norm is NP-complete for $k > 1$.*

PROOF. We shall find a polynomial reduction of the well-known partition problem to our problem. Let $a_1, \ldots, a_n$ be an input of the partition problem. We put $k = 2$ and define the graph G containing $2n + 2$ vertices $v_1, \ldots, v_{n+1}, u_1, \ldots, u_{n+1}$ and the following edges: (v_i, v_{i+1}), $1 \leq i \leq n$; (u_i, u_{i+1}), $1 \leq i \leq n$; (v_i, u_i), $1 \leq i \leq n+1$. Next, we set

$$a(v_i, v_{i+1}) = (a_i, 0), \quad 1 \leq i \leq n,$$
$$a(u_i, u_{i+1}) = (0, a_i), \quad 1 \leq i \leq n,$$
$$a(v_i, u_i) = (0, 0), \quad 1 \leq i \leq n+1.$$

The initial partition problem possesses a solution if and only if the length of an optimal path joining v_1 and v_{n+1} in G is $\frac{1}{2}\sum_{i=1}^n a_i$. At this, a_i belongs to the first subset of the required partition if $(v_i, v_{i+1}) \in S$, and to the second if $(u_i, u_{i+1}) \in S$. It is easy to see that the reduction is polynomial. Finally, the problem belongs to NP for any k. The proof is completed.

In view of the above result, we arrive at the problem of finding approximate algorithms to construct vector shortest paths. A natural performance bound for approximate algorithms is given by

$$U_A(k) = \sup_B \frac{l_A(B)}{l_0(B)}, \qquad (2)$$

where $l_0(B)$ is the length of an optimal path for an input B, $l_A(B)$ is the length of the path produced by an algorithm A applied to an input B, and the supremum is taken over all inputs B of the vector shortest path problem for fixed k.

Consider the following class of approximate algorithms.

ALGORITHM A_f. 1. To each edge e of the initial graph, assign a weight $f(e) = f(a_1(e), \ldots, a_k(e))$, where $f: \mathbb{R}_+^k \to \mathbb{R}_+$ is a function invariant under rearrangement of its arguments.

2. Find in G a path joining u and v that minimizes the ordinary length functional (with respect to weights $f(e)$).

The path thus obtained is said to be f-optimal.

The complexity of the algorithm is defined by that of step 2; it is equal to $O(np \log n)$, where $p = |E|$, and $n = |V|$.

The performance bound (2) for the algorithm A_f is denoted by $U_f(k)$.

We are also interested in the bound

$$L(k) = \inf_f U_f(k), \qquad (3)$$

where the infimum is taken over all the functions f satisfying the above conditions.

First, we consider the case when the weights $a_i(e)$ are arbitrary nonnegative numbers.

THEOREM 2. *Let* $0 \leq a_i(e) < \infty$, $e \in E$, $1 \leq i \leq k$. *Then* $L(k) \geq k$.

PROOF. For an arbitrary positive integer n, consider the cycle on $2kn - 1$ vertices $v_1, \ldots, v_{2kn-1}$; let us denote it by G_n. The edges of G_n have the following vector weights: $a(v_i, v_{i+1}) = (1, 0, \ldots, 0) \in \mathbb{R}_+^k$, $1 \leq i \leq kn - 1$; $a(v_{km+i}, v_{km+i+1}) = (0, \ldots, 0, 1, 0, \ldots, 0) \in \mathbb{R}_+^k$, $0 \leq i \leq k-1$, $n \leq m \leq 2n - 1$, where the only nonzero entry is the $(i + 1)$st (we assume $v_{2kn} \equiv v_1$). There are exactly two simple paths joining v_1 and v_{kn} in G_n, namely, $S_1 = v_1 \cdots v_{kn}$ and $S_2 = v_1 v_{2kn-1} v_{2kn-2} \cdots v_{kn}$. Obviously, we have $l(S_1) = kn - 1$ and $l(S_2) = n$; hence, S_2 is the optimal path. However, for an arbitrary function f invariant under rearrangement of its arguments, we have $\sum_{e \in S_1} f(e) \leq \sum_{e \in S_2} f(e)$. Therefore, any algorithm A_f produces an f-optimal path S_1. Now from (2) and (3) we obtain

$$L(k) \geq \sup_{n>0} \frac{kn-1}{n} = k,$$

and the assertion follows.

It turns out that the lower bound in this theorem is attainable. Below we provide a complete description of the class of functions f such that $U_f(k) = k$.

THEOREM 3. *Let there exist a positive constant c such that, for any vector* $(x_1, \ldots, x_k) \in \mathbb{R}_+^k$,

$$\max_{1 \leq j \leq k} x_j \leq cf(x_1, \ldots, x_k) \leq \sum_{j=1}^k x_j. \qquad (4)$$

Then $U_f(k) = k$.

PROOF. Consider an arbitrary pair of vertices u, v in an arbitrary graph G. Let an optimal path S_0 joining u and v contain m edges, and an f-optimal path S_f contain n edges. Then

$$\sum_{i=1}^m f(x_1^i, \ldots, x_k^i) \geq \sum_{i=1}^n f(y_1^i, \ldots, y_k^i), \qquad (5)$$

where $(x_1^i, \ldots, x_k^i)$ is the weight vector assigned to the ith edge of the path S_0, while $(y_1^i, \ldots, y_k^i)$ is the weight vector assigned to the ith edge of S_f.

Without loss of generality, we can assume that

$$l(S_0) = \max_{1 \leq j \leq k} \sum_{i=1}^{m} x_j^i = \sum_{i=1}^{m} x_1^i, \qquad l(S_f) = \max_{1 \leq j \leq k} \sum_{i=1}^{m} y_j^i = \sum_{i=1}^{m} y_1^i. \qquad (6)$$

According to (4),

$$\sum_{j=1}^{k} \sum_{i=1}^{m} x_j^i \geq c \sum_{i=1}^{m} f(x_1^i, \ldots, x_k^i).$$

Hence, taking (6) into account, we obtain

$$\sum_{i=1}^{m} x_1^i \geq \frac{c}{k} \sum_{i=1}^{m} f(x_1^i, \ldots, x_k^i). \qquad (7)$$

Using (4) once more, we arrive at

$$\sum_{i=1}^{n} y_1^i \leq \sum_{i=1}^{n} \max_{1 \leq j \leq k} y_j^i \leq c \sum_{i=1}^{n} f(y_1^i, \ldots, y_k^i). \qquad (8)$$

Relations (5)–(8) yield the following inequalities:

$$\frac{l(S_f)}{l(S_0)} = \frac{\sum_{i=1}^{n} y_1^i}{\sum_{i=1}^{m} x_1^i} \leq \frac{c \sum_{i=1}^{n} f(y_1^i, \ldots, y_k^i)}{\frac{c}{k} \sum_{i=1}^{m} f(x_1^i, \ldots, x_k^i)} \leq k.$$

This holds for any pair of vertices in any graph; hence $U_f(k) \leq k$, provided (4) holds. The inequality $U_f(k) \geq k$ follows from Theorem 2.

To complete the description of the class of functions under consideration, we prove the following statement.

THEOREM 4. *Let there be no c such that f satisfies* (4) *in* $\mathbb{R}_+^k$. *Then* $U_f(k) > k$.

PROOF. Put

$$\gamma = \sup_{(x_1, \ldots, x_k) \in \mathbb{R}_+^k} \frac{\max_{1 \leq j \leq k} x_j}{f(x_1, \ldots, x_k)}. \qquad (9)$$

If $\gamma = \infty$, one can easily find graphs for which the ratio $l(S_f)/l(S_0)$ becomes arbitrarily large.

Let $\gamma < \infty$. Since (4) is false with any c, there exists a point $(\bar{x}_1, \ldots, \bar{x}_k) \in \mathbb{R}_+^k$ such that

$$\sum_{j=1}^{k} \bar{x}_j = (\gamma - \varepsilon) f(\bar{x}_1, \ldots, \bar{x}_k), \qquad \varepsilon > 0. \qquad (10)$$

Suppose that the supremum in (9) is attained at a point $(\tilde{x}_1, \ldots, \tilde{x}_k) \in \mathbb{R}_+^k$. Since $\gamma < \infty$, the values $f(\bar{x}_1, \ldots, \bar{x}_k)$ and $f(\tilde{x}_1, \ldots, \tilde{x}_k)$ are positive, and so there exist positive integers t_1 and t_2 such that

$$t_1 k (\gamma - \varepsilon) f(\bar{x}_1, \ldots, \bar{x}_k) < t_2 \gamma f(\tilde{x}_1, \ldots, \tilde{x}_k) < t_1 k \gamma f(\bar{x}_1, \ldots, \bar{x}_k). \qquad (11)$$

Consider the graph consisting of two vertex-disjoint chains between two given vertices u and v. The first chain consists of t_2 edges with vector weights $(\tilde{x}_1, \ldots, \tilde{x}_k)$, the second consists of t_1 identical groups of k edges, and the weights of the edges in such a group are

$$(\bar{x}_1, \ldots, \bar{x}_2), (\bar{x}_2, \ldots, \bar{x}_k, \bar{x}_1), \ldots, (\bar{x}_k, \bar{x}_1, \ldots, \bar{x}_{k-1}).$$

By the second inequality in (11) and the invariance of f under rearrangement of its arguments, the algorithm A_f produces the first of the above chains as an f-optimal path. Then from (9), (10), and the first inequality in (11) we obtain

$$\frac{l(S_f)}{l(S_0)} \geq \frac{t_2 \max_{1 \leq j \leq k} \tilde{x}_j}{t_1 \sum_{j=1}^k \bar{x}_j} = \frac{t_2 f(\tilde{x}_1, \ldots, \tilde{x}_k)}{t_1 f(\bar{x}_1, \ldots, \bar{x}_k)} \cdot \frac{\max_{1 \leq j \leq k} \tilde{x}_j}{f(\tilde{x}_1, \ldots, \tilde{x}_k)} \cdot \frac{f(\bar{x}_1, \ldots, \bar{x}_k)}{\sum_{j=1}^k \bar{x}_j}$$

$$= \frac{\gamma t_2 f(\tilde{x}_1, \ldots, \tilde{x}_k)}{(\gamma - \varepsilon) t_1 f(\bar{x}_1, \ldots, \bar{x}_k)} > k,$$

and hence $U_f(k) > k$.

The case when the supremum in (9) is not attained is studied similarly. In this case, we choose a point $(\tilde{x}_1, \ldots, \tilde{x}_k) \in \mathbb{R}_+^k$ such that

$$\max_{1 \leq j \leq k} \tilde{x}_j = (\gamma - \xi) f(\tilde{x}_1, \ldots, \tilde{x}_k),$$

where $0 < \xi < \varepsilon$, ε is defined by (10), and the numbers t_1 and t_2 satisfy the inequalities

$$t_1 k(\gamma - \varepsilon) f(\bar{x}_1, \ldots, \bar{x}_k) < t_2(\gamma - \xi) f(\tilde{x}_1, \ldots, \tilde{x}_k) < t_1 k(\gamma - \xi) f(\bar{x}_1, \ldots, \bar{x}_k).$$

The proof is completed.

Next, let us consider the case when the coordinates of the weight vectors $a(e)$ satisfy certain restrictions. First of all, let us study the behavior of the function $L(k)$.

THEOREM 5. *Let $a \leq a_i(e) \leq A$, $e \in E$, $1 \leq i \leq k$. Then*

$$L(k) \geq \frac{k}{1 + (k-1)\delta}$$

with $\delta = a/A$.

PROOF. The proof is similar to that of Theorem 2, save only that one must replace 0 by a and 1 by A in the vector weights.

As before, the above lower bound is attainable, yet the analog of Theorem 3 is no longer true. Indeed, put $k = 2$ and consider the function $f(x_1, x_2) = a + A/2$. Let $\delta = a/A > 1/2$; then

$$\max\{x_1, x_2\} \leq A < a + A/2 = f(x_1, x_2) < 2a \leq x_1 + x_2,$$

and so (4) holds for any point $(x_1, x_2) \in [a, A] \times [a, A]$. Since f is a constant, an f-optimal path must contain the minimal number of edges. In particular, a path containing $n - 1$ edges of weight (A, A) is "better" than a path containing n edges of weight (a, a). Hence

$$U_f(2) \geqslant \sup_{n>0} \frac{(n-1)A}{na} = \frac{1}{\delta},$$

while Theorem 5 for $k = 2$ yields the lower bound $2/(1+\delta)$, which is less than $1/\delta$ for $\delta > 1/2$.

The fact that the lower bound of Theorem 5 is attainable follows from the next statement.

THEOREM 6. *Let* $a \leqslant a_i(e) \leqslant A$, $e \in E$, $1 \leqslant i \leqslant k$, $f(x_1, \ldots, x_k) = \sum_{j=1}^{k} x_j$. *Then*

$$U_f(k) = \frac{k}{1 + (k-1)\delta}$$

with $\delta = a/A$.

PROOF. Consider an arbitrary pair of vertices u, v in an arbitrary graph G. Let an optimal path S_0 joining u and v contain m edges, and let an f-optimal path S_f contain n edges. Then

$$\sum_{i=1}^{m} \sum_{j=1}^{k} x_j^i \geqslant \sum_{i=1}^{n} \sum_{j=1}^{k} y_j^i, \tag{12}$$

where $(x_1^i, \ldots, x_k^i)$ is the vector weight of the ith edge of the path S_0 and $(y_1^i, \ldots, y_k^i)$ is the vector weight of the ith edge of S_f. The inequalities $a \leqslant y_j^i \leqslant A$, $1 \leqslant j \leqslant k$, $1 \leqslant i \leqslant n$, yield

$$l(S_f) = \max_{1 \leqslant j \leqslant k} \sum_{i=1}^{n} y_j^i \leqslant \min\left\{nA, \sum_{i=1}^{n} \sum_{j=1}^{k} y_j^i - (k-1)na\right\}. \tag{13}$$

Next, taking (12) into account, we obtain

$$l(S_0) = \max_{1 \leqslant j \leqslant k} \sum_{i=1}^{m} x_j^i \geqslant \frac{1}{k} \sum_{j=1}^{k} \sum_{i=1}^{m} x_j^i \geqslant \frac{1}{k} \sum_{j=1}^{k} \sum_{i=1}^{n} y_j^i. \tag{14}$$

Let us study the behavior of the ratio $l(S_f)/l(S_0)$. We distinguish two cases.

CASE 1. Let

$$\sum_{i=1}^{n} \sum_{j=1}^{k} y_j^i \geqslant nA + (k-1)na.$$

Then (13) and (14) yield

$$\frac{l(S_f)}{l(S_0)} \leqslant \frac{nA}{\frac{1}{k}\sum_{i=1}^{n}\sum_{j=1}^{k} y_j^i} \leqslant \frac{knA}{nA + (k-1)na} = \frac{k}{1 + (k-1)\delta}.$$

CASE 2. Let
$$\sum_{i=1}^{n}\sum_{j=1}^{k} y_j^i \leqslant nA + (k-1)na.$$

Then, by (13) and (14),

$$\frac{l(S_f)}{l(S_0)} \leqslant \frac{\sum_{i=1}^{n}\sum_{j=1}^{k} y_j^i - (k-1)na}{\frac{1}{k}\sum_{i=1}^{n}\sum_{j=1}^{k} y_j^i} = k - \frac{k(k-1)na}{\sum_{i=1}^{n}\sum_{j=1}^{k} y_j^i}$$

$$\leqslant k - \frac{k(k-1)na}{nA + (k-1)na} = k - \frac{k(k-1)\delta}{1+(k-1)\delta} = \frac{k}{1+(k-1)\delta}.$$

Thus, in either case we have

$$\frac{l(S_f)}{l(S_0)} \leqslant \frac{k}{1+(k-1)\delta},$$

and hence

$$U_f(k) \leqslant \frac{k}{1+(k-1)\delta}.$$

The opposite inequality follows from Theorem 5.

Therefore, the bound of Theorem 5 is attained when the second inequality in (4) turns into an equality. It appears that for $k = 2$ the bound is attained also when the first inequality in (4) turns into an equality.

THEOREM 7. *Let* $a \leqslant a_i(e) \leqslant A$, $e \in E$, $i = 1, 2$, *and* $f(x_1, x_2) = \max\{x_1, x_2\}$. *Then* $U_f(2) = 2/(1+\delta)$ *with* $\delta = a/A$.

PROOF. Consider an arbitrary pair of vertices u, v in an arbitrary graph G. Let an optimal path S_0 joining u and v contain $m+n$ edges of weights $(x_1^1, x_2^1), \ldots, (x_1^{m+n}, x_2^{m+n})$, and let an f-optimal path S_f contain $r+s$ edges of weights $(y_1^1, y_2^1), \ldots, (y_1^{r+s}, y_2^{r+s})$. Without loss of generality, we can assume that the edges of S_0 and S_f are renumbered to satisfy the following conditions:

$$\begin{aligned} x_1^i \geqslant x_2^i, & \quad 1 \leqslant i \leqslant m, & x_1^{m+i} \leqslant x_2^{m+i}, & \quad 1 \leqslant i \leqslant n, \\ y_1^i \geqslant y_2^i, & \quad 1 \leqslant i \leqslant r, & y_1^{r+i} \leqslant y_2^{r+i}, & \quad 1 \leqslant i \leqslant s, \\ \sum_{i=1}^{m+n} x_1^i \geqslant \sum_{i=1}^{m+n} x_2^i, & & \sum_{i=1}^{r+s} y_1^i \geqslant \sum_{i=1}^{r+s} y_1^i. & \end{aligned} \quad (15)$$

Now the f-optimality of the path S_f yields

$$\sum_{i=1}^{m} x_1^i + \sum_{i=1}^{n} x_2^{m+i} \geqslant \sum_{i=1}^{r} y_1^i + \sum_{i=1}^{s} y_2^{r+i}. \quad (16)$$

Let us introduce the following notation:

$$B_j^x = \sum_{i=1}^{m} x_j^i, \quad C_j^x = \sum_{i=1}^{n} x_j^{m+i}, \quad B_j^y = \sum_{i=1}^{r} y_j^i, \quad C_j^y = \sum_{i=1}^{s} y_2^{r+i}, \quad j = 1, 2.$$

By (15) and (16) we have

$$C_1^x \leqslant C_2^x, \quad C_1^y \leqslant C_2^y, \quad B_1^x + C_1^x \geqslant B_2^x + C_2^x,$$
$$B_1^y + C_1^y \geqslant B_2^y + C_2^y, \quad B_1^x + C_2^x \geqslant B_1^y + C_2^y. \tag{17}$$

Let us study the ratio $l(S_f)/l(S_0)$. According to (17),

$$\frac{l(S_f)}{l(S_0)} = \frac{B_1^y + C_1^y}{B_1^x + C_1^x} \leqslant \frac{B_1^y + C_1^y}{B_1^x + C_1^x} \cdot \frac{B_1^x + C_2^x}{B_1^y + C_2^y} \leqslant \frac{B_1^x + C_2^x}{B_1^x + C_1^x}.$$

Evidently, the latter ratio does not exceed the optimal value of the problem

$$\frac{\alpha + \beta}{\alpha + \gamma} \to \max$$
$$\alpha \geqslant \theta, \quad \gamma \leqslant \beta, \quad \alpha + \gamma \geqslant \beta + \theta,$$
$$ma \leqslant \alpha \leqslant mA, \quad ma \leqslant \theta \leqslant mA, \tag{18}$$
$$na \leqslant \beta \leqslant nA, \quad na \leqslant \gamma \leqslant nA.$$

Let us introduce uniform variables $\xi = \gamma/\alpha$, $\zeta = \theta/\alpha$, $\eta = \beta/\alpha$. Then (18) implies

$$\xi = \frac{\gamma}{\alpha} = \frac{\gamma \zeta}{\theta} \leqslant \frac{nA\zeta}{ma} = \frac{t\zeta}{\delta}, \quad \xi = \frac{\gamma}{\alpha} \geqslant \frac{na}{mA} = t\delta,$$

where $t = n/m$. Similar inequalities hold for η. Therefore, we arrive at the problem

$$(1 + \eta)/(1 + \xi) \to \max$$
$$\delta \leqslant \zeta \leqslant 1, \quad \xi \leqslant \eta, \quad 1 + \xi \geqslant \zeta + \eta, \tag{19}$$
$$t\delta \leqslant \xi \leqslant t\zeta/\delta, \quad t\delta \leqslant \eta \leqslant t\zeta/\delta,$$

whose optimal value is not less than that for (18). Denote by Ω the domain defined by the inequalities in (19). Then

$$\max_{\Omega} \frac{1 + \eta}{1 + \xi} \leqslant \max_{\delta \leqslant \zeta \leqslant 1} \max_{t \geqslant 0} \max_{t\delta \leqslant \xi \leqslant t\zeta/\delta} \frac{1 + \min\{1 - \zeta + \xi, t\zeta/\delta\}}{1 + \xi}$$
$$= \max_{\delta \leqslant \zeta \leqslant 1} \max_{t \geqslant 0} \frac{1 + \min\{1 - \zeta + t\delta, t\zeta/\delta\}}{1 + t\delta}$$
$$= \max_{\delta \leqslant \zeta \leqslant 1} \frac{1 + \zeta(1-\zeta)/(\zeta - \delta^2)}{1 + \delta^2(1-\zeta)(\zeta - \delta^2)} = \max_{\delta \leqslant \zeta \leqslant 1} \frac{2\zeta - \zeta^2 - \delta^2}{\zeta(1 - \delta^2)} = \frac{2}{1 + \delta}.$$

Therefore,

$$\frac{l(S_f)}{l(S_0)} \leqslant \frac{B_1^x + C_2^x}{B_1^x + C_1^x} \leqslant \frac{2}{1 + \delta}.$$

Hence, $U_f(2) \leqslant 2/(1 + \delta)$. The opposite inequality follows from Theorem 5.

REMARK. Recently, the author has obtained several new results in the same direction. First, Theorems 1 and 2 hold for some other vector optimization problems, such as the vector minimum spanning tree problem and the vector minimum perfect matching problem [4]. Second, the natural analogs of Theorems 1–5 hold for the vector shortest path problem in the l_p-norm [5]. Precise formulations can be also found in [6].

References

1. Bernard A. Carré, *Graphs and networks*, Clarendon Press, Oxford, 1979.
2. M. Gondran, *Path algebra and algorithms*, Combinatorial Programming: Methods and Applications (Versailles, 1974; B. Roy, editor) NATO Adv. Study Inst. Ser. Ser. C: Math. and Phys. Sci., Reidel, Dordrecht, 1975, pp. 137–148.
3. Bernard Mahr, *A bird's-eye view to path problems*, Group-theoretic Concepts in Computer Science (Proc. Sixth Internat. Workshop, Bad Honnef, 1980), Lecture Notes in Computer Sci., vol. 100, Springer-Verlag, Berlin, 1981, pp. 335–353.
4. A. D. Vaĭnshteĭn, *Vector optimization problems on graphs*, Software for Solving Optimal Planning Problems, Moscow, 1990, pp. 97–98. (Russian)
5. _____, *The vector shortest path problem in the l_p-norm*, Modelling and Optimization of Complex Systems, Omsk, 1987, pp. 138–144. (Russian)
6. _____, *The vector shortest path problem and other vector optimization problems on graphs*, Proc. 17th Yugoslav Sympos. Oper. Res., Belgrade, 1990, pp. 169–172.

Lower Performance Bounds for On-line Algorithms in the Simple Two-dimensional Rectangle Packing Problems

A. D. VAĬNSHTEĬN

In the familiar two-dimensional rectangle packing problem first studied in [2], a finite list of rectangles must be packed without rotations or overlapping into a rectangular bin of given width and infinite height, so as to minimize the total height used. Since the problem of finding an optimal packing is obviously NP-hard, it becomes necessary to study rather fast (e.g. polynomial with a small degree) approximation algorithms.

Given such an algorithm A and a list L of rectangles, we can define $H_A(L)$, the height used by A to pack L, and $H_{\text{OPT}}(L)$, the height of the optimal packing of L. The supremum of the ratio $H_A(L)/H_{\text{OPT}}(L)$ over all possible lists L is said to be the *upper performance bound* U_A of the algorithm A. Similarly, the *asymptotic upper performance bound* V_A is defined as the maximal limit point of the set of these ratios.

Within this framework, the purpose is to obtain algorithms with the smallest values of U_A (or V_A). This was done for the two-dimensional rectangle packing problem in [1], [2], [5], [7], and [13], resulting in $U = 2.5$ (see [13]) and $V = 5/4$ (see [1]). Better bounds can be achieved for special cases of the problem, namely: (1) for each rectangle, its width equals its height (packing of squares); (2) all rectangles have the same height (bin packing); and (3) all rectangles have the same width (multiprocessor scheduling). For the first case, the bound $U = 2$ is obtained in [2] and [7]. For the bin packing problem, $V = 11/9$ was obtained in [9] and [10], and slightly improved in [14]. Polynomial approximation schemes for the problem are obtained in [6] and [11]. Finally, for the multiprocessor scheduling problem, the bound $U \leqslant 1.22$ is obtained in [4].

1991 *Mathematics Subject Classification*. Primary 05B40, 68R05.
Translation of Theory and Methods of Automating the Design of Complex Systems and the Automation of Scientific Research, Inst. Tekhn. Kibernet. Akad. Nauk BSSR, Minsk, 1985, pp. 22–25.

Another interesting problem arises when the class of possible algorithms is restricted—for instance, when we are interested only in on-line algorithms. An on-line algorithm works as follows: the ith rectangle can be specified only after the previous $i-1$ rectangles are already packed. In general, let $\mathscr{A}$ be the class of algorithms; then we define the *lower performance bound* $u_{\mathscr{A}}$ of the class as the infimum of U_A over all the algorithms $A \in \mathscr{A}$. The *asymptotic lower performance bound* $v_{\mathscr{A}}$ is defined similarly.

The lower bound $v \geqslant 3/2$ for on-line bin packing is obtained in [14] and improved to $v > 1.536$ in [12], while the best known on-line bin-packing algorithms provides $V = 5/3$ (see [14]). A thorough study of on-line lower bounds for the general two-dimensional rectangle packing problem and for packing of squares is carried out in [3]. In the present note we study on-line lower bounds for packing of rectangles with the same width 1 and of squares of two given sizes.

Let us consider the problem of packing an arbitrary list of rectangles having the same width 1 into a rectangular bin of width $m \geqslant 2$. (Without loss of generality, we may assume m to be an integer, since otherwise it would suffice to take a bin of width $\lfloor m \rfloor$.) Let u_1 be the lower performance bound for the class of on-line algorithms for this problem.

THEOREM 1. *The following relations hold:*

$$u_1 = 3/2 \quad \text{for } m = 2;$$
$$u_1 = 5/3 \quad \text{for } m = 3;$$
$$u_1 \geqslant 1 + \sqrt{2}/2 \quad \text{for } m \geqslant 4.$$

PROOF. Let $m = 2$. Let us consider the following two lists: $L = (1, 1)$, $L' = (1, 1, 2)$ (the numbers in parentheses represent the heights of rectangles). Let A be an arbitrary on-line algorithm. The following alternative is obvious: either $H_A(L) \geqslant 2$, or $H_A(L') \geqslant 3$. Now, it is easy to see that $H_{\text{OPT}}(L) = 1$ and $H_{\text{OPT}}(L') = 2$. Hence,

$$u_1 \geqslant \min\left\{\frac{H_A(L)}{H_{\text{OPT}}(L)}, \frac{H_A(L')}{H_{\text{OPT}}(L')}\right\} = \frac{3}{2}.$$

Next, the familiar "bottom–left" algorithm of [8] guarantees $U = 2 - 1/m = 3/2$ in the case $m = 2$, and it belongs to the class of on-line algorithms, hence $u_1 \leqslant 3/2$. Comparing the last two inequalities, we obtain the desired result.

Let $m = 3$. Let us consider the following four lists: $L_1 = (1, 1, 1)$, $L_2 = (1, 1, 1, 2, 2, 2)$, $L_3 = (1, 1, 1, 2, 2, 2, 7, 7, 7)$, $L_4 = (1, 1, 1, 2, 2, 2, 7, 7, 7, 15)$. Given an arbitrary on-line algorithm A, one can see easily that either $H_A(L_1) \geqslant 2$, or $H_A(L_2) \geqslant 5$, or $H_A(L_3) \geqslant 17$, or $H_A(L_4) \geqslant 25$. Taking into account that $H_{\text{OPT}}(L_1) = 1$, $H_{\text{OPT}}(L_2) = 3$, $H_{\text{OPT}}(L_3) = 10$, and $H_{OPT}(L_4) = 15$, we obtain

$$u_1 \geq \min\left\{\frac{H_A(L_i)}{H_{OPT}(L_i)}, i = 1, \ldots, 4\right\} = \frac{5}{3}.$$

The opposite inequality $u_1 \leq 5/3$ is derived as before from Graham's bound $U = 2 - 1/m$ for $m = 3$.

Now let m be greater than 3. Let us consider the following four lists:

$$L_1 = (1, \ldots, 1),$$
$$L_2 = (1, \ldots, 1, c, \ldots, c),$$
$$L_3 = (1, \ldots, 1, c, \ldots, c, c(c+1), \ldots, c(c+1)),$$
$$L_4 = (1, \ldots, 1, c, \ldots, c, c(c+1), \ldots, c(c+1), 2c(c+1)).$$

Here L_1 contains m rectangles, L_2 contains $2m$ rectangles, L_3 contains $3m$ and L_4 contains $3m + 1$; the constant $c > 1$ will be specified later. Similarly to the previous cases, we have the following possibilities: either $H_A(L_1) \geq 2$, or $H_A(L_2) \geq 1 + 2c$, or $H_A(L_3) \geq (1+c)(1+2c)$, or $H_A(L_4) \geq (1+c)(1+3c)$, whereas $H_{OPT}(L_1) = 1$, $H_{OPT}(L_2) = 1 + c$, $H_{OPT}(L_3) = (1+c)^2$, and $H_{OPT}(L_4) = 2c(1+c)$. Hence,

$$u_1 \geq \min\{2, (1+2c)/(1+c), (1+3c)/2c\}.$$

The best lower bound is obtained when $(1 + 2c)/(1 + c) = (1 + 3c)/2c$, thus producing $c = 1 + \sqrt{2}$ and $u_1 \geq 1 + \sqrt{2}/2$.

Now let v_1 be the asymptotic lower performance bound for the class of on-line algorithms solving the problem.

THEOREM 2. $v_1 \geq m^2/(m^2 - m + 1)$.

PROOF. For an arbitrary integer n, consider the following two lists: L_n, consisting of mn rectangles of height 1, and L'_n, obtained from L_n by adding (at the end of L_n) $m - 1$ rectangles, each of mn units height. Let A be an arbitrary on-line algorithm. Consider a packing of L_n produced by A. This packing consists of m columns of heights $x_1^n, \ldots, x_m^n$, and $\sum_{i=1}^{m} x_i^n \geq mn$ (because the columns may have gaps). Suppose without loss of generality that $x_1^n \geq x_2^n \geq \cdots \geq x_m^n$. Then $H_A(L_n) = x_1^n$. Evidently, $H_{OPT}(L_n) = n$, and so

$$v_1 \geq \limsup_{n \to \infty} \frac{x_1^n}{n}. \tag{1}$$

Consider now the packing of L'_n. Clearly, the first mn rectangles of L'_n must be packed in the same columns of heights $x_1^n, \ldots, x_m^n$ as the list L_n was. Therefore, $H_A(L'_n) \geq x_2^n + mn$. Evidently, $H_{OPT}(L'_n) = mn$; thus

$$v_1 \geq \limsup_{n \to \infty} \frac{x_2^n + mn}{mn}. \tag{2}$$

Combining (1) and (2), we obtain

$$v_1 \geq \limsup_{n \to \infty} f^n, \qquad (3)$$

where f^n is the optimal value of the problem

$$F(x_1^n, \ldots, x_m^n) = \max\{x_1^n/n, x_2^n/mn + 1\} \to \min$$

$$\sum_{i=1}^m x_i^n \geq mn, \qquad x_1^n \geq x_2^n \geq \cdots \geq x_m^n,$$

the x_i^n are integers, $i = 1, \ldots, m$.

Let us consider the linear relaxation of the above problem, and let $\bar{f}^n$ be the optimal value of the latter. It is easy to verify that

$$\bar{f}^n = m^2/(m^2 - m + 1),$$

and the optimal solution is given by

$$\bar{x}_1^n = m^2 n/(m^2 - m + 1),$$
$$\bar{x}_2^n = \bar{x}_3^n = \cdots = \bar{x}_m^n = m(m-1)n/(m^2 - m + 1).$$

Hence, for $n = (m^2 - m + 1)t$, $t = 1, 2, \ldots$, we have $f^n = \bar{f}^n$. Recall now that $f^n \geq \bar{f}^n$, and therefore

$$\limsup_{n \to \infty} f^n \geq m^2/(m^2 - m + 1).$$

Combining this with (3), we get the desired bound $v_1 \leq m^2/(m^2 - m + 1)$.

Let us consider now the problem of packing an arbitrary list of squares of sizes 1×1 (small squares) and 2×2 (large ones) into a rectangular bin of width m. The following simple on-line algorithm turns out to be rather efficient.

ALGORITHM A. Pack the current small square at the leftmost position of the lowest possible level, and the current large square at the rightmost position of the lowest possible level.

Let V_A be the asymptotic lower performance bound for this algorithm.

THEOREM 3. *The following relations hold*:
1) $V_A = 1$ *for* $m = 2k$;
2) $1 + 1/m \leq V_A \leq 1 - 1/m$ *for* $m = 2k + 1$.

PROOF. 1) One can easily prove an even stronger result: $H_A(L) \leq H_{\text{OPT}}(L) + 1$ for any list L. Details are left to the reader.

2) The second inequality is obvious. It means that, asymptotically, at most one column could be wasted. Again, details are left to the reader.

To prove the first inequality, consider for any integer n the list L_n consisting of $2n(2k+1)$ small rectangles following by $nk(2k+1)$ large ones.

Evidently, $H_A(L_n) = 4n(k+1)$ and $H_{OPT}(L_n) = 2n(2k+1)$; hence $V_A \geq (2k+2)/(2k+1) = 1 + 1/m$.

Finally, let v_2 be the asymptotic lower performance bound for the class of on-line algorithms solving this problem.

THEOREM 4. $v_2 \geq m^2/(m^2 - m + 1)$ for $m = 2k + 1$.

PROOF. For any integer $n > 0$, consider the following two lists: L_n defined as in the proof of Theorem 3, and the part L'_n of L_n containing only small squares. From here the proof is similar to that of Theorem 2.

REFERENCES

1. Brenda S. Baker, Donna J. Brown, and Howard P. Katseff, *A 5/4 algorithm for two-dimensional packing*, J. Algorithms **2** (1981), 348–368.
2. Brenda S. Baker, E. G. Coffman, Jr., and Ronald L. Rivest, *Orthogonal packings in two dimensions*, SIAM J. Comput. **9** (1980), 846–855.
3. Donna J. Brown, Brenda S. Baker, and Howard P. Katseff, *Lower bounds for on-line two-dimensional packing algorithms*, Acta Inform. **18** (1982/83), 207–225.
4. E. G. Coffman, Jr., M. R. Garey, and D. S. Johnson, *An application of bin-packing to multiprocessor scheduling*, SIAM J. Comput. **7** (1978), 1–17.
5. E. G. Coffman, Jr., et al., *Performance bounds for level-oriented two-dimensional packing algorithms*, SIAM J. Comput. **9** (1980), 808–826.
6. W. Fernandez de la Vega and G. S. Lueker, *Bin packing can be solved within $1 + \varepsilon$ in linear time*, Combinatorica **1** (1981), 349–356.
7. Igal Golan, *Performance bounds for orthogonal oriented two-dimensional packing algorithms*, SIAM J. Comput. **10** (1981), 571–585.
8. R. L. Graham, *Bounds for certain multiprocessing anomalies*, Bell System Tech. J. **45** (1966), 1563–1581.
9. David S. Johnson, *Fast algorithms for bin-packing*, J. Comput. System Sci. **8** (1974), 272–314.
10. D. S. Johnson et al., *Worst-case performance bounds for simple one-dimensional packing algorithms*, SIAM J. Comput. **3** (1974), 299–325.
11. Narendra Karmarkar and Richard J. Karp, *An efficient approximation scheme for the one-dimensional bin-packing problem*, 23rd Annual Sympos. Foundations of Computer Sci. (Chicago, IL, 1982), IEEE, New York, 1982, pp. 312–320.
12. Frank M. Liang, *A lower bound for on-line bin packing*, Inform. Processing Lett. **10** (1980), 76–79.
13. Daniel D. K. D. B. Sleator, *A 2.5 times optimal algorithm for packing in two dimensions*, Inform. Processing. Lett. **10** (1980), 37–40.
14. Andrew Chi Chih Yao, *New algorithms for bin packing*, J. Assoc. Comput. Mach. **27** (1980), 207–227.

Recent Titles in This Series

(*Continued from the front of this publication*)

- 119 **V. A. Dem′janenko, et al.,** Twelve Papers in Algebra
- 118 **Ju. V. Egorov, et al.,** Sixteen Papers on Differential Equations
- 117 **S. V. Bočkarev, et al.,** Eight Lectures Delivered at the International Congress of Mathematicians in Helsinki, 1978
- 116 **A. G. Kušnirenko, A. B. Katok, and V. M. Alekseev,** Three Papers on Dynamical Systems
- 115 **I. S. Belov, et al.,** Twelve Papers in Analysis
- 114 **M. Š. Birman and M. Z. Solomjak,** Quantitative Analysis in Sobolev Imbedding Theorems and Applications to Spectral Theory
- 113 **A. F. Lavrik,** Twelve Papers in Logic and Algebra
- 112 **D. A. Gudkov and G. A. Utkin,** Nine Papers on Hilbert's 16th Problem
- 111 **V. M. Adamjan, et al.,** Nine Papers on Analysis
- 110 **M. S. Budjanu, et al.,** Nine Papers on Analysis
- 109 **D. V. Anosov, et al.,** Twenty Lectures Delivered at the International Congress of Mathematicians in Vancouver, 1974
- 108 **Ja. L. Geronimus and Gábor Szegő,** Two Papers on Special Functions
- 107 **A. P. Mišina and L. A. Skornjakov,** Abelian Groups and Modules
- 106 **M. Ja. Antonovskiĭ, V. G. Boltjanskiĭ, and T. A. Sarymsakov,** Topological Semifields and Their Applications to General Topology
- 105 **R. A. Aleksandrjan, et al.,** Partial Differential Equations, Proceedings of a Symposium Dedicated to Academician S. L. Sobolev
- 104 **L. V. Ahlfors, et al.,** Some Problems on Mathematics and Mechanics, On the Occasion of the Seventieth Birthday of Academician M. A. Lavrent′ev
- 103 **M. S. Brodskiĭ, et al.,** Nine Papers in Analysis
- 102 **M. S. Budjanu, et al.,** Ten Papers in Analysis
- 101 **B. M. Levitan, V. A. Marčenko, and B. L. Roždestvenskiĭ,** Six Papers in Analysis
- 100 **G. S. Ceĭtin, et al.,** Fourteen Papers on Logic, Geometry, Topology and Algebra
- 99 **G. S. Ceĭtin, et al.,** Five Papers on Logic and Foundations
- 98 **G. S. Ceĭtin, et al.,** Five Papers on Logic and Foundations
- 97 **B. M. Budak, et al.,** Eleven Papers on Logic, Algebra, Analysis and Topology
- 96 **N. D. Filippov, et al.,** Ten Papers on Algebra and Functional Analysis
- 95 **V. M. Adamjan, et al.,** Eleven Papers in Analysis
- 94 **V. A. Baranskiĭ, et al.,** Sixteen Papers on Logic and Algebra
- 93 **Ju. M. Berezanskiĭ, et al.,** Nine Papers on Functional Analysis
- 92 **A. M. Ančikov, et al.,** Seventeen Papers on Topology and Differential Geometry
- 91 **L. I. Barklon, et al.,** Eighteen Papers on Analysis and Quantum Mechanics
- 90 **Z. S. Agranovič, et al.,** Thirteen Papers on Functional Analysis
- 89 **V. M. Alekseev, et al.,** Thirteen Papers on Differential Equations
- 88 **I. I. Eremin, et al.,** Twelve Papers on Real and Complex Function Theory
- 87 **M. A. Aĭzerman, et al.,** Sixteen Papers on Differential and Difference Equations, Functional Analysis, Games and Control
- 86 **N. I. Ahiezer, et al.,** Fifteen Papers on Real and Complex Functions, Series, Differential and Integral Equations
- 85 **V. T. Fomenko, et al.,** Twelve Papers on Functional Analysis and Geometry
- 84 **S. N. Černikov, et al.,** Twelve Papers on Algebra, Algebraic Geometry and Topology
- 83 **I. S. Aršon, et al.,** Eighteen Papers on Logic and Theory of Functions

(See the AMS catalog for earlier titles)